Gender, Intersectionality and Climate Institutions in Industrialised States

This book explores how climate institutions in industrialised countries work to further the recognition of social differences and integrate this understanding in climate policy-making.

With contributions from a range of expert scholars in the field, this volume investigates policy-making in climate institutions from the perspective of power as it relates to gender. It also considers other intersecting social factors at different levels of governance, from the global to the local level and extending into climate-relevant sectors. The authors argue that a focus on climate institutions is important since they not only develop strategies and policies, they also (re)produce power relations, promote specific norms and values, and distribute resources. The chapters throughout draw on examples from various institutions including national ministries, transport and waste management authorities, and local authorities, as well as the European Union and the UNFCCC regime. Overall, this book demonstrates how feminist institutionalist theory and intersectionality approaches can contribute to an increased understanding of power relations and social differences in climate policy-making and in climate-relevant sectors in industrialised states. In doing so, it highlights the challenges of path dependencies, but also reveals opportunities for advancing gender equality, equity, and social justice.

Gender, Intersectionality and Climate Institutions in Industrialised States will be of great interest to students and scholars of climate politics, international relations, gender studies and policy studies.

Gunnhildur Lily Magnusdottir is an Associate Professor in Political Science and is Deputy Head of The Department of Global Political Studies at Malmö University, Sweden.

Annica Kronsell is Professor and Chair of Environmental Social Science at the School of Global Studies at Gothenburg University, Sweden.

Routledge Studies in Gender and Environments

With the European Union, United Nations, UN Framework Convention on Climate Change, and national governments and businesses at least ostensibly paying more attention to gender, including as it relates to environments, there is more need than ever for existing and future scholars, policy-makers, and environmental professionals to understand and be able to apply these concepts to work towards greater gender equality in and for a sustainable world.

Comprising edited collections, monographs and textbooks, this new *Routledge Studies in Gender and Environments* series will incorporate sophisticated critiques and theorisations, including engaging with the full range of masculinities and femininities, intersectionality, and LGBTQI perspectives. The concept of 'environment' will also be drawn broadly to recognise how built, social, and natural environments intersect with and influence each other. Contributions will also be sought from global regions and contexts which are not yet well represented in gender and environments literature, in particular Russia, the Middle East, and China, as well as other East Asian countries such as Japan and Korea.

Series Editor: Professor **Susan Buckingham**, an independent researcher, consultant, and writer on gender and environment-related issues.

Gender and Religion in the City
Women, Urban Planning and Spirituality
Edited by Clara Greed

Transecology
Transgender Narratives on Environment and Nature
Edited by Douglas A. Vakoch

Negotiating Gender Expertise in Environment and Development
Voices from Feminist Political Ecology
Edited by Bernadette P. Resurrección and Rebecca Elmhirst

Gender, Intersectionality and Climate Institutions in Industrialised States
Edited by Gunnhildur Lily Magnusdottir and Annica Kronsell

For more information about this series, please visit: www.routledge.com/Routledge-Studies-in-Gender-and-Environments/book-series/RSGE

Gender, Intersectionality and Climate Institutions in Industrialised States

Edited by
Gunnhildur Lily Magnusdottir and
Annica Kronsell

Routledge
Taylor & Francis Group
LONDON AND NEW YORK

earthscan
from Routledge

First published 2021
by Routledge
2 Park Square, Milton Park, Abingdon, Oxon OX14 4RN

and by Routledge
605 Third Avenue, New York, NY 10158

Routledge is an imprint of the Taylor & Francis Group, an informa business

British Library Cataloguing-in-Publication Data
A catalogue record for this book is available from the British Library

Library of Congress Cataloging-in-Publication Data
Names: Gunnhildur Lily Magnúsdóttir, editor. | Kronsell, Annica, editor.
Title: Gender, intersectionality and climate institutions in industrialized states/edited by Gunnhildur Lily Magnusdottir and Annica Kronsell.
Description: Abingdon, Oxon; New York, NY: Routledge, 2021. | Includes bibliographical references and index.
Identifiers: LCCN 2020056142 (print) | LCCN 2020056143 (ebook) | ISBN 9780367512057 (hardback) | ISBN 9781003052821 (ebook)
Subjects: LCSH: Women and the environment–Developed countries. | Environmental policy–Social aspects–Developed countries. | Women's rights–Developed countries. | Climatic changes–Social aspects–Developed countries.
Classification: LCC GE195.9 .G454 2021 (print) | LCC GE195.9 (ebook) | DDC 363.738/746082091722–dc23
LC record available at https://lccn.loc.gov/2020056142
LC ebook record available at https://lccn.loc.gov/2020056143

ISBN: 978-0-367-51205-7 (hbk)
ISBN: 978-1-032-01370-1 (pbk)
ISBN: 978-1-003-05282-1 (ebk)

DOI: 10.4324/9781003052821

Typeset in Goudy
by Deanta Global Publishing Services, Chennai, India

Contents

List of figures   vii
List of tables   viii
About the contributors   ix
Acknowledgements   xvi

1 Gender, intersectionality and institutions   1
   ANNICA KRONSELL AND GUNNHILDUR LILY MAGNUSDOTTIR

PART 1
Intergovernmental and governmental climate institutions   15

2 Gender in the global climate governance regime: A day late and
   a dollar short?   17
   KAREN MORROW

3 EU external climate policy   36
   GILL ALLWOOD

4 How to make Germany's climate policy gender-responsive:
   Experiences from research and advocacy   52
   GOTELIND ALBER, DIANA HUMMEL, ULRIKE RÖHR, AND IMMANUEL STIEß

5 Promoting a gender agenda in climate and sustainable
   development: A civil servant's narrative   69
   GERD JOHNSSON-LATHAM AND ANNICA KRONSELL

6 Take a ride into the danger zone?: Assessing path dependency
   and the possibilities for instituting change at two Swedish
   government agencies   86
   BENEDICT E. SINGLETON AND GUNNHILDUR LILY MAGNUSDOTTIR

PART 2
Sectoral climate institutions 103

7 Towards a climate-friendly turn?: Gender, culture and
performativity in Danish transport policy 105
HILDA RØMER CHRISTENSEN AND MICHALA HVIDT BREENGAARD

8 Wasting resources: Challenges to implementing existing policies
and tools for gender equality and sensitivity in climate
change–related policy 124
SUSAN BUCKINGHAM

9 Gender analysis of policy-making in construction and
transportation: Denial and disruption in the Canadian green
economy 143
BIPASHA BARUAH AND SANDRA BISKUPSKI-MUJANOVIC

10 Why radical transformation is necessary for gender equality and
a zero carbon European construction sector 164
LINDA CLARKE AND MELAHAT SAHIN-DIKMEN

PART 3
Local, community institutions, and climate practices 181

11 Addressing climate policy-making and gender in transport plans
and strategies: The case of Oslo, Norway 183
TANU PRIYA UTENG, MARIANNE KNAPSKOG, ANDRÉ UTENG,
AND JOMAR SÆTERØY MARIDAL

12 When gender equality and Earth care meet: Ecological
masculinities in practice 207
ROBIN HEDENQVIST, PAUL M. PULÉ, VIDAR VETTERFALK, AND MARTIN HULTMAN

13 Pathways for inclusive wildfire response and adaptation in
northern Saskatchewan 226
HEIDI WALKER, MAUREEN G. REED, AND AMBER J. FLETCHER

14 Moving forward: Making equality, equity, and social justice central 245
GUNNHILDUR LILY MAGNUSDOTTIR AND ANNICA KRONSELL

Index 249

Figures

2.1 Average percentage female and male membership of UNFCCC
constituted bodies drawn from gender composition reports 21
2.2 UNFCCC highest and lowest percentages of female
representation on constituted bodies 2013–2019 21
2.3 Percentage of women members of UNFCCC constituted bodies 2019 22
2.4 Numbers of UNFCCC documents pertaining per gender 26
7.1 Number of articles about Westh and Mikkelsen during their
time in office 109
8.1 Gendered use of 'wasteapp' (More details of the evaluation of all
innovations can be found in Buckingham et al., 2020.) 135
11.1 Oslo population by gender and age groups, 2020 versus 2040 194
11.2 Oslo mean number of daily long trips per capita, BAU scenario 195
11.3 Oslo mean number of daily medium trips per capita, BAU scenario 195
11.4 Oslo mean number of daily short trips per capita, BAU scenario 196
11.5 ZGO Scenario number of trips on select modes by age group and
gender, long trips 199
11.6 ZGO Scenario number of trips on select modes by age group and
gender, medium trips 200
11.7 ZGO Scenario number of trips on select modes by age group and
gender, short trips 201

Tables

8.1 Gender considerations at each stage of the project, following
EIGE methodology 125

8.2 Employment by gender in water supply, sewerage, waste
management and remediation activities, 2017 131

8.3 Gendered attitudes and behaviour towards avoiding waste 132

8.4 EIGE Gender Equality Index ranking of participating countries
(2019) 133

8.5 Case studies' environmental performance relative to gender
equality and sensitivity 136

10.1 Graduates in engineering, manufacturing and construction 2018 165

11.1 Strategic climate policy framework for Oslo 2020 189

11.2 The public transport company Ruter's estimate of reaching the
zero-growth goal by 2030 192

11.3 ZGO Scenario growth rates and aggregate mode shares, long trips 199

11.4 ZGO Scenario growth rates and aggregate mode shares, medium trips 200

11.5 ZGO Scenario growth rates and aggregate mode shares, short trips 201

13.1 Characteristics of adaptation pathways (based on Nalau and
Handmer, 2015; Pelling et al., 2015; Heikkinen et al., 2018) 229

Contributors

Gotelind Alber is an independent researcher and advisor on sustainable energy and climate change policy with a special focus on gender and climate justice. She is co-founder and board member of the global network GenderCC. She holds an advanced university degree in physics and has many years of working experience in research, policy, and management. Her current work focuses on the integration of gender dimensions into national and urban climate policy, as well as the UNFCCC process. Her current activities include research projects and GenderCC's 'Gender into Urban Climate Change Initiative'.

Gill Allwood is Professor of Gender Politics at Nottingham Trent University, UK. She has published widely on gender and EU policy, in particular climate change, migration, and development policy. Her publications include 'Gender Equality in European Union Development Policy in Times of Crisis', *Political Studies Review*, 2019; *Transforming Lives? CONCORD Report EU Gender Action Plan II: From Implementation to Impact*, Brussels: CONCORD Europe, 2018; 'Gender Mainstreaming and EU Climate Change Policy' *European Integration Online Papers*, 18 (1), 2014; and 'Gender Mainstreaming and Policy Coherence for Development: Unintended Gender Consequences and EU Policy', *Women's Studies International Forum*, 39, 2013, pp. 42–52.

Bipasha Baruah is Professor and Canada Research Chair on Global Women's Issues at Western University. She earned her PhD in Environmental Studies from York University, Toronto in 2005. Her current research aims to understand how a global low-carbon economy can be more gender-equitable and socially just than its fossil fuel–based predecessor and how that can be ensured. Bipasha Baruah frequently serves as an expert reviewer and advisor to Canadian and intergovernmental environmental protection and international development organisations. She was named to the Royal Society of Canada's College of New Scholars, Artists and Scientists in 2015.

Michala Hvidt Breengaard is a Post-Doctoral Researcher at the Department of Sociology, University of Copenhagen. Her research focuses on the intersection of gender and diversity studies and climate and sustainability. She has a longstanding knowledge of gender issues in relation to transport and mobility patterns and needs and has recently been involved in research on biking and diversity. Breengaard has worked within sociological fields of family and parenting and has a PhD on practices of motherhood in the frame of China's one-child policy. Moreover, she has worked on several EU projects related to sustainability and climate.

Susan Buckingham is an independent researcher, consultant, activist, and writer working on achieving gender equality in environmental projects (from waste management to ocean sustainability, with groups as diverse as the Women's Environmental Network and the United Nations, and in locations across Europe and Pakistan). She has recently completely revised her seminal book 'Gender and Environment', which argues that we cannot achieve a sustainable environment without addressing gender – and other – inequalities. Susan is also the series editor for the Routledge series 'Studies in Gender and Environments'.

Hilda Rømer Christensen is Associate Professor and Head of the Coordination for Gender Research at the University of Copenhagen. She has written extensively on gender, culture, and – more lately – on gender in transport in comparative perspectives. She was Head of the EU Transgen Project (2007), co-ordinator of the research network Gendering Smart Mobilities in the Nordic Region (2017–2018), and is currently the scientific coordinator for the Horizon 2020 project: TInnGO – focused on gender, transport, and smart mobilities (2019–2021). Among her recent publications are: Hilda Rømer Christensen (with Tanu Uteng and Lena Levin) (eds.) (2020), *Gendering Smart Mobilities*. Routledge. Rømer Christensen. (2019), and 'Gendering Mobilities and (in) equalities in Post-socialist China' in Scholten, Joelsson (eds.) *Integrating Gender into Transport Planning*. Palgrave.

Linda Clarke is Professor of European Industrial Relations and Co-Director of the Centre for the Study of the Production of the Built Environment (ProBE) at the University of Westminster. She is a board member of the European Institute for Construction Labour Research (CLR) and Associate Director of Canada's York University *Climate Change and Work* programme. She has extensive experience of comparative research on labour, equality, vocational education and training (VET), employment and wage relations, with particular emphasis on the European construction sector and on climate change.

Melahat Sahin-Dikmen is a Research Fellow at the Centre for the Study of the Production of the Built Environment (ProBE), University of Westminster. Her current work sits at the intersection of the world of work and climate change. She is engaged in research examining the social, employment, and educational implications of green transition policies in the construction sector and the role of worker agency in climate action at local, national, and global levels. Her broader expertise lies in the sociology of work and employment particularly in relation to gender, ethnicity, and the professions.

Amber J. Fletcher is Associate Professor of Sociology and Social Studies at the University of Regina, Canada. Her research examines the social and gender dimensions of climate disaster in agricultural and Indigenous communities in the Prairie region of Canada. She is interested in how gender and intersectional forms of inequality shape the experience of environmental crises.

Robin Hedenqvist is an environmental justice scholar and activist. He earned a Master's Degree from the *Globalisation, Environment and Social Change* programme at Stockholm University. Robin specialises in social science approaches to the ecological crisis. As a former research assistant at the Centre for Climate Science & Policy Research, Linköping University, he contributed to urban climate adaptation projects. Currently, he considers the nexus between gender equality, masculinities, and the environment. He is a core team member of the Gender and Environment Network for Inspiration, which provides a meeting place to explore, educate, and learn about these interlinked issues.

Martin Hultman, Associate Professor at Chalmers University, is a scholar of energy, climate and environmental issues. His publications include journal articles in Environmental Humanities, NORMA: International Journal for Masculinity Studies, and History and Technology along with numerous books such as *Discourses of Global Climate Change*; *Ecological Masculinities*; *Men, Masculinities and Earth*. Hultman is a public intellectual and a frequent contributor to newspapers, public discussions, and political policy in Sweden and beyond.

Diana is a research scientist at ISOE and has been a member of the executive board since April 2014. Prior to that, she was head of the Biodiversity and People research unit and coordinator for the cross-divisional area 'Academic Cooperation and Qualification of Young Scientists'. Her research revolves around demography, gender, and environment. She is an assistant professor in social science at the Goethe University, Frankfurt am Main, with a focus on international development. She received her doctorate from Goethe University in 1999 with a thesis on 'Population Discourse: Demographic Knowledge and Political Power'. In 2009, she did her habilitation at the Goethe University.

Marianne Knapskog is a research planner at the Norwegian Institute of Transport Economics and part of the research group working with Sustainable Urban Development and Mobility. Before becoming a researcher, she worked in planning practice for more than ten years. Her research interests include walkability, location policies, accessibility, and how planning is materialised through agreements and packaged and the resulting regulations and practices in cities and towns. She is part of a Scandinavian case study in cemeteries and crematoria as public spaces of belonging in Europe; a study of migrant and minority cultural inclusion, exclusion, and integration (CeMi).

Annica Kronsell holds the Chair of Environmental Social Science at the School of Global Studies at Gothenburg University and has an interest in how public institutions can govern climate and sustainability issues in Scandinavia and Europe. Kronsell uses (eco)feminist theorising and intersectionality to address governance issues in relation to power and inclusiveness. Her publications on climate governance in the public sector include: *Rethinking the Green State. Environmental Governance Towards Environmental and Sustainability Transitions*, with Bäckstrand, Routledge (2015) and on how gender and intersectionality are implicated in climate governance in 'Climate Change Through the Lens of Intersectionality', with Kaijser, *Environmental Politics* (2014).

Gerd Johnsson-Latham has 40 years of experience as a civil servant within Swedish Government Offices. Her work revolves around gender, human rights, and sustainable development. She has also worked for the UN and the World Bank, with postings in Hanoi and Washington DC. She has been Chair of the Swedish NGO Kvinna till Kvinna and is currently a board member of *Klimatriksdagen* and, furthermore, a chief-editor for the net-magazine *Human Security*. Johnsson-Latham has written several studies on gender for the government offices, such as 'Power and Privileges – on Gender Discrimination and Poverty', 'Patriarchal Violence as a Threat to Human Security', and 'Gender Equality as a Prerequisite for Sustainable Development'.

Gunnhildur Lily Magnusdottir is an Associate Professor in Political Science and Deputy Head of the Department of Global Political Studies at Malmö University. Her current research focuses on climate and energy policy-making in Scandinavia, with a specific focus on the role of governmental institutions and how they work with issues of environmental justice and social differences. Her other research interests are small states in the European Union and their environmental considerations, and public administrations in Scandinavia. Relevant publications include, with A. Kronsell 'The In(visibility) of Gender in Scandinavian Climate Policy-making' in *The International Feminist Journal of Politics,* 'The Double Democratic Deficit in Climate Policy-making by the EU Commission' in *Femina Politica*, and with J. Bendel, 'Opportunities and Challenges of a Small State Presidency: The Estonian Council Presidency' in *Administrative Culture.*

Jomar Sæterøy Maridal is the manager of Panda Analyse, a foundation that develops tools for analysing, projecting, and impact-assessing regional development and demographic shifts. He graduated from the Norwegian University of Science and Technology in 2018 with a Master's Degree in Entrepreneurship, Innovation, and Society. He wrote his master's thesis in cooperation with the Institute of Transport Economics on the potential of green transitions in Oslo's mobility. After graduating, he worked at Trøndelag County Authority. There, he participated in various projects on urban development and green mobility in Northern Trøndelag and ran travel surveys in the region.

Karen Morrow has been a Professor of Environmental Law at Swansea University since 2007. Her research interests focus on theoretical and practical aspects of public participation in environmental law and policy and gender and the environment. She has published extensively in these areas. She is on the editorial boards of the Journal of Human Rights and the Environment, the Environmental Law Review, and the University of Western Australia Law Review. She is a series editor for Critical Reflections on Human Rights and the Environment (Edward Elgar) and sits on the International Advisory Board for Gender and Environment book series (Routledge).

Sandra Biskupski-Mujanovic is a PhD candidate in the Department of Women's Studies and Feminist Research at Western University. Her areas of specialisation include gender and human security; militarisation and peacekeeping; and gender and work.

Paul M. Pulé is an Australian social and environmental justice activist, scholar, and ecopreneur. He currently serves as a post-doctoral researcher at Chalmers University in Göteborg, Sweden, where he works on the intersections between men, masculinities, and Earth. His research work appears in numerous academic texts and popular science reports on the theory and practices of ecological masculinities as they apply to government, corporations, NGOs, with applications in Transitions activism in civil society as well. His work addresses the social constructions of masculinities that dwell in us all, with a particular focus on boys and men in groups and as individuals.

Maureen G. Reed is a Professor and UNESCO Chair in Biocultural Diversity, Sustainability, Reconciliation and Renewal at the School of Environment and Sustainability at the University of Saskatchewan. She has a longstanding research interest in the gendering of forestry work and forest management, with some comparative analysis of Canada and Sweden. Her research in environmental governance and social dimensions of sustainability has driven her to learn more about how climate hazards have been experienced by women and men in Indigenous and settler communities of rural Saskatchewan, Canada.

Ulrike Röhr has a background as a civil engineer and sociologist and has been working on gender issues in planning, Local Agenda 21, environment, and especially in energy and climate policy for about 30 years. She is committed to mainstreaming gender into environment and climate policy on local and national levels and has been involved in gendering the UNFCCC process since the very beginning. She heads the German focal point gender, environment, sustainability (genanet) which aims to support gender mainstreaming in environmental policy. Most recently she was involved in two German research projects on gender and climate change respectively and gender and consumption.

Benedict E. Singleton is a post-doctoral researcher at the School of Global Studies, University of Gothenburg, where he conducts research on the project

Intersectionality and Climate Policy Making: Ways Forward to a Socially Inclusive and Sustainable Welfare State. His previous research publications include explorations of 'nature-based integration' projects aimed at migrants to central Sweden; investigations of Faroese whaling (*grindadráp*); research on carers of sick and disabled relatives in the UK; and HIV stigma and treatment in Jamaica.

Immanuel Stieß has been working at ISOE for many years as a research scientist. He is head of the research unit Energy and Climate Protection in Everyday Life. He is a sociologist and planning expert and received his PhD in Architecture, Urban Planning and Landscape Planning from the University of Kassel, with a thesis on communication with tenants during modernisation processes. His focus is on social-ecological lifestyle research in the areas of housing and nutrition, sustainability and social inclusion, and gender impact assessment. Stieß has wide experience in the design and evaluation of target group-oriented sustainability communication.

André Uteng is an expert on transportation, land-use, and business dynamics at Rambøll Norway. He has a PhD degree in transport engineering from the Norwegian University of Science and Technology (NTNU). Uteng has worked with mathematical, statistical, and economic analysis, transport modelling, and model development. Among his main accomplishments has been the development of INMAP, a model which is currently in use in Norway for estimating the mutual relationships between land-use and transportation, as well as developing accessibility-based models for assessing the potential effects of E-bike and E-scooter-integration into the public transportations systems in Norway's larger cities.

Tanu Priya Uteng is a Senior Researcher at the Institute of Transport Economics in Oslo, Norway. She has worked across a host of cross-cutting issues in the field of urban-transport planning in the past 16 years. A few of her areas of expertise include: mobilities, travel behaviour studies, inclusive cities, and gender studies. She is currently leading several research projects looking at topics like first–last mile, climate and travel behaviour, green-shift, and sharing. In her research undertakings, she explores the 'policy-making' dimension and plans to build a structured understanding of the topic in creating inclusive cities. She has edited *Gendered Mobilities*, *Urban Mobilities in the Global South* and *Gendering Smart Mobilities*.

Vidar Vetterfalk is a certified psychologist and international project manager at MÄN, Sweden. He has been active in the feminist organisation MÄN since its inception in 1993, working with and engaging boys and men for gender equality and ending men's violence. For Vetterfalk, as for MÄN, holding men accountable through close cooperation with women's movements is central to the holistic work with primary violence prevention, support to women survivors and work with perpetrators. Vetterfalk grew up on a biodynamic farm and is passionate about the intersections between gender equality, the environment, the climate crisis, and how to engage more men to care for others.

Heidi Walker is a PhD candidate in the School of Environment and Sustainability at the University of Saskatchewan. Her primary research interests relate to the social dimensions of climate hazards, climate change adaptation, and environmental assessment processes. Applying an intersectional lens, her current research explores how people experience and respond to wildfire in a socially diverse and jurisdictionally complex region of northern Saskatchewan.

Acknowledgements

This edited book is the result of several years of research on gender, intersectionality, and climate change – a new but growing research field in which the authors of this volume have done important groundwork. We are grateful to the authors for their work and valuable contribution to this emerging research field. We would also like to thank the editorial team at Routledge for their interest in the topic as well as encouragement and support.

The book was finalised during the COVID-19 pandemic, which has entailed many challenges for all of us. The contributors submitted their drafts promptly, discussed them and received feedback in online workshops, and prepared their final manuscripts accordingly. Everyone helped by sticking to our many deadlines, and we are grateful for everyone's effort and dedication to this book. In light of this, we want to highlight the unequal effects of the pandemic on academics and we are especially thankful for the authors' commitment during these challenging times. In addition, the pandemic has clearly exposed how different social groups are affected, which relates to their differential status and power, and it stresses further, not only the urgency of transformative climate actions, but also the general need to include gender and intersectional approaches in policy-making.

We are also indebted to the Swedish Research Council for Sustainable Development, FORMAS and research grant 2018-01704, which has financed the time we have invested in the book language editing and a part of the open access fees.

Gunnhildur Lily Magnusdottir and Annica Kronsell

1 Gender, intersectionality and institutions

Annica Kronsell and
Gunnhildur Lily Magnusdottir

Introduction

Climate change is one of the main security challenges of the 21st century and a growing concern among the public, politicians and civil societies around the globe. Climate objectives that have been agreed upon globally and nationally require substantial changes in most societal sectors and call for broad societal engagement. Proposed strategies tend to focus on technical and economic solutions, while research stresses the importance of social inclusion to achieve a climate transition in line with current sustainability goals. Political and administrative institutions at different levels, i.e. international organisations, the EU Commission and its directorates, national, regional, as well as municipal governments and administrations, are authoritative key actors in climate policy-making. However, there are often shortcomings in their climate policy-making, most prominently their prevalent focus on technical innovations and economic incentives, and the lack of attention to social dimensions. There are important social differences that deserve recognition and this book will address this lack by providing gender and intersectional perspectives on climate policy and institutions.

Research shows how greenhouse gas emissions, vulnerability to impacts and political participation vary considerably across the population, according to gender, race, class, age, geography and other intersectional factors (IPCC, 2014; Kaijser and Kronsell, 2014; Djoudi et al., 2016; Alber et al., 2017; Buckingham and Le Masson, 2017, pp. 3–5). Recognition of social differences needs to inform climate policy and the incorporation of the UN's Sustainable Development Goals provides an impetus. If social differences are left unattended to, climate policy will likely reinforce existing inequalities, but it will also risk overlooking differential effects and ending up becoming ineffective and give rise to protests among groups who feel unjustifiably challenged by climate policies and decisions.

Research on climate institutions in industrialised states – in the Global North – indicates a lack of knowledge among policy-makers on social differences and how to include climate-relevant social factors into climate objectives, such as in governmental agencies in Scandinavian countries and in the EU Commission's Climate Action Directorate (Allwood, 2014; Magnusdottir and Kronsell, 2015, 2016; Buckingham and Le Masson, 2017). There is clearly a need for more

DOI: 10.4324/9781003052821-1

knowledge on how social equity, equality and justice issues are and can be included in climate policy-making. In turn, this requires more knowledge about how social differences are relevant to the policy fields that are most significant for realising climate objectives, going beyond the environmental sector to include transport, energy, industry, building, land use and waste, sectors where activities that lead to carbon emissions are carried out. Thus, climate institutions in industrialised states are the primary focus of the book. Climate institutions can be narrowly defined as governing bodies at different levels, such as global and regional intergovernmental organisations as well as state and local authorities. Inspired by institutionalist scholarship, the concept institution does not merely refer to formal aspects but includes the people, i.e. climate policy-makers and, most importantly, we include the norms, rules and practices of the governing climate bodies in our definition of climate institutions (March and Olsen, 1989; Mackay et al., 2010).

Gender analysis, developed from feminist theory, is the theoretical starting point of the book but an intersectional approach increases our understanding 'that there are different ways in which different women and men are affected by and relate to climate change, mediated by power structures which they experience differently' (Buckingham and LeMasson, 2017, p. 5). Intersectionality has been recognised in the broader feminist environmental literature as a useful tool to analyse 'capitalism, rationalist science, colonialism, racism, (hetero) sexism and speciesism' (MacGregor, 2017, pp. 1–2). Feminist intersectional research on climate change has to a large extent focused on developing states, and on the violence of climate events, storms, hurricanes, fires and on gendered vulnerabilities of such climate change events (Kinnvall and Rydstrom, 2019). While such research is relevant and highlights the serious effects of climate change, our work zooms in on industrialised states and on how their formal and informal climate institutions shape and develop their climate strategies. For one, this is because the industrialised countries need to reduce their carbon emissions drastically and do so with increasing urgency (UNFCCC, 2015). Second, climate policies developed in rich industrialised countries tend to become normative for other countries that then follow suit (Sommerer and Lim, 2016; Griffin Cohen, 2017; Tobin, 2017). Third, rich industrialised countries, e.g., in Europe and North America, are interesting since they have developed welfare systems and the resources needed to become green decarbonised states (Duit et al., 2016). Moreover, as shown in previous studies, equality and social issues, the diversity of the public and their various needs and behavioural patterns, such as transport behaviour of different groups, have not been sufficiently recognised by industrialised states (Allwood, 2014; Magnusdottir and Kronsell, 2015, 2016; Alber et al., 2017; Buckingham and Kulcur, 2017).

Accordingly, the overall aim of the book is to explore if and how climate institutions in industrialised states work with recognition and understanding of social differences in climate policy-making. We argue that the focus on climate institutions is important since they not only develop strategies and policies, but they also produce power relations, for example, by distributing resources such as through strategies for a green economy. Climate institutions also affect power

relations by promoting specific norms and values such as ecological modernisation and by including specific types of knowledge only, for example by privileging technical knowledge. The different chapters of this book investigate
policy-making in climate institutions and the role of informal institutions and
norms, from the perspective of power related to gender and other intersecting
social factors. We start by discussing the international institution of the United
Nations Framework Convention on Climate Change (UNFCCC), because its
norms established through international negotiations are significant as part of
the norms that guide policies. All the institutions discussed in this book adhere
and relate to the UNFCCC regime (see further in chapter two, this volume). The
regime has become normative at national, as well as local levels, and in different
sectors in many industrialised states. The existence of climate norms at multiple levels and within different governmental or sector institutions, is evident in
chapters both focusing on intergovernmental organisations, such as the UN and
the EU as well as state and local authorities. As we are highly concerned with
what the enactment of relevant norms like climate emission targets imply for
gender equality, equity and justice, we investigate several different institutions
where the actions required from climate agreements take place. In studying these
and other relevant institutions the book explores and addresses the following two
important questions:

> What are the main institutional challenges and barriers to the inclu-
> sion of climate relevant social factors and in particular gender in climate
> policy-making?
> What and where are the main opportunities to advance gender equality,
> equity and social justice within climate institutions?

Our main ambition is to study climate institutions, which makes feminist institutional theory a particularly relevant frame. Below we outline how institutional
approaches combined with gender and intersectional perspectives can contribute to further an understanding of power relations within climate institutions in
industrialised states and thereby highlight the challenges and opportunities for
advancing gender equality, equity and social justice.

Representation in climate policy-making

How gender and other social categories are represented in climate institutions is
important for discussions of inclusive climate strategies. If gender representation is
imbalanced in climate authorities it can be a sign of an ill-functioning democracy
or a democratic deficit, preventing the development of inclusive and gendered
climate policies. Drawing on Anne Philip's (1995) politics of presence, two main
categories of representation have been developed, descriptive and substantive
redundant (Lovenduski, 2005; Chaney, 2012). Descriptive representation relates
to the number of women and men who can influence policy-making, that is,
the relationship and balance between males and females in political institutions

(Wängnerud, 2009, p. 53). When women are represented in greater numbers it is often the result of formal or informal quotas (Dahlerup and Leyenaar, 2013). Gender equity guidelines aim for a balance between male and female bodies, falling within the 40–60% range and qualify descriptive representation with the argument that the presence of only a few women in politics will not make a real difference (Lovenduski, 2005). Instead, it is argued that a certain number of women, a critical mass, must be present in order for their actions to produce substantial effects (Dahlerup, 1988, 2006). Concepts like critical mass and mechanisms involving gender quotas have helped to establish a growing female presence in politics and policy-making but can be problematic when reduced to mere numbers. Based on a democratic ethos as the baseline, representation should be inclusive and reflect the citizenry. Here, we broaden the perspective to include other climate-relevant social categories that intersect with gender, when analysing what kind of knowledge is included or excluded, and which voices are represented in the policy-making process.

Feminists have argued that rather than looking at numbers, it is critical acts – action that makes a difference – that matters. This is tricky as it is linked to the question of whether women will make a difference or not, once they have been included in the polity and often there is an idea that they will, although this is seldom explicit (Dahlerup, 2006; Mushaben, 2012). Normally, it is based on some vague idea of a certain difference between men and women acquired through diverse life experiences and conditions, hence, leading to expectations that they bring into policy, diverse interests, knowledge and perspectives. When we move beyond the gender binary and include other social categories in the discussion about substantive representation, the idea of what additional qualities the experiences of these groups will bring to institutions becomes even more complicated. Hence, we suggest that it suffices to argue for social inclusion on the basis of democratic rights and social justice. The degree of representation for women and other social groups will remain a pertinent issue also for the chapters in this book, not least due to the rather extreme conditions of male dominance exemplified in the different sectors, i.e. transport, energy, building or the 'green economy', which are taking on climate mitigation challenges.

Beyond this, we are mostly interested in critical climate acts. What are some of those critical acts that policy-makers and civil servants can perform in order to elicit change? Gender mainstreaming is a strategy endorsed in the EU (since 1996) and by many institutions of its member states, which pushes the work with gender issues beyond adding women. It rests on the idea that most activities have a gender dimension found in the underlying norms of institutions (Woodward, 2003; Kantola, 2010; Lowndes, 2020). The strategy is to 'mainstream' women's experiences and needs by incorporating gender perspectives into all policy areas, at all stages, at all levels. Gender experts are often assigned the task of gender mainstreaming (Ferguson, 2015; Bustelo et al., 2016). Gender experts and advisors have a particular role in the institutions and are assumed to have specific knowledge of gender relations. They may perform critical acts. The concept femocrats, captures the more transformational aspect of such critical acts as it names feminist civil servants, inspired by feminist movements, who work to further gender and feminist concerns within government and administrative institutions (Eisenstein, 1996; Yeatman, 2020). Similarly, green activists within administrations perform in the same way on the basis of their dedication to environmental concerns (Hysing and Olsson, 2017). A related concept is that of 'outsiders within' originally from black feminist theory. Patricia Hill Collins (1986) points to the importance of individual actors in policy development and how they can bring in new and alternative perspective and knowledge on, for example, gender and race to climate policy-making from their positions as 'others'. These questions about if and how an increased number of female policy-makers in climate institutions and in climate-relevant sectors matters are addressed in various chapters.

Path dependence in climate institutions

Feminist institutionalism argues that it is important for feminist scholars to study institutions because they organise power inequalities through formal as well as informal rules and practices (Mackay et al., 2010; Krook and Mackay, 2011; Ljungholm, 2017; Miller, 2020). Through historical processes, power inequalities become deeply embedded in organisations and are reinforced over time and the resulting institutional arrangements then steer or guide political actions (Pierson, 2004, p. 11). Institutions tend to have particular 'pattern-bound' effects over time, caused by locking into place certain rules and norms of behaviour which also give institutions resilience (Krook and Mackay, 2011; Waylen, 2014; Lowndes, 2020). This explains why historically derived notions of gender tend to persist. Path dependence makes institutions 'sticky' and opportunities for innovation and change are thereby constrained by previous choices (Kenny, 2007, p. 93; Miller, 2020). In line with this historical institutional perspective, climate institutions can be viewed as path-dependent when it comes to both gender recognition and the acknowledgement of other intersecting social factors such as class, ethnicity, age, location and education. The stickiness or path dependence of climate institutions that have significance here is related to how climate change was originally put on the global agenda as a highly scientific, elitist, technical and masculinised issue, privileging mainly the natural science community to define the problem, thus downplaying and making irrelevant social interpretations (Skutsch, 2000; Hemmati and Röhr, 2009). In their study of the Commission Directorate-General for Climate Action, Magnusdottir and Kronsell (2016) concluded that the existing power order was reproduced within a new institutional environment –in the then newly established DG Climate Action of the Commission – where path dependence meant that policy-makers in DG Climate Action largely accepted and adapted to the established masculinised institutional environment in which EU climate policies were formulated.

Path dependence often has to do with normalisation, for example of masculinity and of gender binaries. Normalisation reproduces power through simple everyday acts that are perceived as normal and it does not require overt struggles (Kronsell, 2016). From the lens of feminist institutionalism, institutional practices are based on a logic that may make the inclusion of gender and other intersecting social factors appear less appropriate and less desirable. It builds on March and Olsen's (1989, pp. 21–38, p. 161) ideas about how institutions are reproduced through patterns of action in a 'logic of appropriateness'. The logic is that individual policy-makers follow embedded rules and routines, according to what is appropriate for their social and professional role and individual identity. Feminist institutionalists coined the more specific '*gendered* logic of appropriateness', which: 'prescribes (as well as proscribes) acceptable masculine and feminine forms of behaviour, rules and values for men and women within institutions' (Chappell and Waylen, 2013, p. 601). Accordingly, it is not the individual interest or personality of political actors in climate institutions that is important but how individuals give their actions meaning. What will be considered appropriate in any given situation is not trivial. 'Actions are fitted to situations by their appropriateness within a conception of identity' (March and Olsen, 1989, p. 38). The outcomes of these institutional practices, thus the actual climate actions, policies and the included parameters, aspects, issues and knowledge, depend on the role expectations of individual policy-makers and how they make their role meaningful. If climate policy-making is taking place in the energy, environment or transport agencies, as is the case in most industrial states, then the rules and culture of those institutions as well as the professional identity of the policy-makers, perhaps educated as engineers or economists, are likely to affect whether and how they view social power relations in policy-making. Within engineering, technical solutions may seem most appropriate while economists may focus on societal costs. Several chapters in this book critically assess climate policy-making from the feminist institutional perspective. They reveal path dependencies, while at the same time remaining open to the possibility of change, in search of transformations that challenge the 'stickiness' of climate institutions. Thus, the book raises questions about how climate policy-makers are affected by this stickiness and bound by a gendered logic of appropriateness and if or how institutional path dependence can be challenged.

Intersectionality as a tool in institutional analysis

Intersectionality was developed in feminist theory as a way to complicate the analysis of gender, arguing that gender is hardly ever a power relation that stands alone, but is related to other power differences, such as class, ethnicity and age (Davis, 2008). Intersectionality is an important concept and an analytical tool that can shed light on how power structures around social differences emerge and interact (see e.g., Lykke, 2010; Cho et al., 2013; Buckingham and Le Masson, 2017). Surely, it is applicable to climate issues. While gender is highly relevant (Alston, 2013; Resurrección, 2013), it is also nested into other power categories, for example, depending on context and place (Nagel, 2012; Kaiser and Kronsell, 2014) and related to economic status, e.g., in explaining carbon emissions (Ergas and York, 2012; MacGregor, 2017, p. 22) and ecological footprints. Although our starting point is in exploring gender difference, many of the chapters in this book reflect on which social categories are recognised and conceptualised in policy-making and ask questions about whether intersectional aspects are visible and/or included.

The contributions to this book are placed in three interconnected thematic parts, which focus on climate institutions, both formal authorities and informal, normative institutions, in industrialised states.

Part 1: Intergovernmental and governmental climate institutions

The first part of the book outlines the broader scene of climate policy-making with its focus on formal intergovernmental and governmental climate institutions and how they work – e.g. with recognition of gender and other intersectional factors, gender representation, and gender equality in climate policy-making. The chapters in this part discuss the importance of formal institutions as well as the role of individual civil servants who might be dictated by a gendered logic of appropriateness. The chapters in Part 1 highlight the importance of representation in policy-making and the effects of path dependence as well as how formal institutions shape power relations and produce knowledge and norms in climate policy-making.

In Chapter 2, 'Gender in the global climate governance regime: a day late and a dollar short?', Karen Morrow outlines the evolution of gender representation in the global climate governance regime (the UNFCCC), which is important for furthering international cooperation and for its prominent influence on climate agendas across the world. The UNFCCC's tardiness to act on gender is puzzling, not the least in comparison to activities within other Rio conventions. Although a laggard on gender, Morrow reflects on the efforts made since its belatedly official recognition to the Gender Constituency in 2011. She discusses impediments to this process of improvements, which is conditioned by systemic features of the UNFCCC regime in international law and politics and an economics-driven, technocratic stance on climate issues. Such path dependencies influence both the treatment of gender and broader social inclusion and thus, also the efficacy of initiatives taken so far.

In Chapter 3 on the EU external climate policy, Gill Allwood notes that EU is committed to gender equality – a fundamental value in most policy areas, including EU's external relations – but EU climate policy is still largely gender-blind. It is the strong impact and vulnerability to climate change in countries outside Europe, that has provided a powerful framing of climate change as a development issue and where it has related to gender concerns. In turn, this framing has served the interest of a security-focused foreign policy which downplays any efforts to mainstream gender or prioritise social inclusion. The chapter stresses the importance of issue framing resulting in an institutional stickiness that opens for certain actors to engage and excludes others. This trajectory means that the gender agenda and the climate agenda remain separate tracks although they are both embedded in the EU's legal framework and institutions.

In Chapter 4, 'Making Germany's climate change policy gender-responsive', Gotelind Alber, Diana Hummel, Ulrike Röhr and Immanuel Stieß, highlight how German climate institutions have worked with gender, intersectionality and civil society. German climate policy-makers have only recently begun addressing gender and have primarily been inspired by international agreements such as the UNFCCC Gender Action Plan. Recognising that research on these issues was lacking, the authors engaged in a research project which developed a Gender Impact Assessment tool as an instrument to analyse gender equality which was made available to policy-makers. The chapter accounts for these experiences and also sheds light on critical acts or the efforts made by the (female) minister for the environment to motivate and educate senior staff and climate change experts on gender and climate policy and how cooperation with female leaders in environmental organisations was key for gender to gain traction in German climate policy-making.

In Chapter 5, 'Promoting a Gender Agenda in Climate and Sustainable Development, Gerd Johnsson-Latham and Annica Kronsell offer valuable insights into the institutional challenges government officials face when initiating an inclusive gender-equal approach. The view is from 40 years of experience as a 'femocrat' working within the Swedish government, the Ministry of Foreign Affairs and in international institutions. The chapter emerged as a dialogue between a feminist academic and a feminist bureaucrat and the resulting narrative is analysed against reflections from feminist institutional theory. It offers an insider's perspective of institutional practices by highlighting the institutional challenges that government officials face when promoting gender equality in sustainability and climate issues and how they navigate that space in order to gender mainstream the agenda, e.g. through several different types of critical acts elaborated in the text.

In Chapter 6, 'Take a ride into the danger zone? Assessing path dependency and the possibilities for instituting change at two Swedish government agencies', Ben Singleton and Gunnhildur Lily Magnusdottir argue that government agencies are important sites for climate change action. Yet, the action space available to civil servants reflects historically embedded norms in those institutions, i.e., path dependencies that relate to the prevalence of ecological modernisation and the preference for technical solutions. This frames subsequent policies with direct consequences for what type of climate action is possible. Using interview data, it explores how Swedish civil servants frame possibilities for institutional action and change on social justice in climate action. The civil servant is positioned in an ambiguous space, acting according to the logic of appropriateness in the context of the general ethos of bureaucracy within the specific logic of the agency's remit. On the other hand, in line with another task of public officers to act and serve the will of the democratically elected government transmitted from the political level.

Part 2: Sectoral climate institutions

The second thematic part of the book discusses how different climate-relevant sectoral institutions in transport, energy, construction and the green economy work with the inclusion of gender and other intersecting social factors. These chapters discuss the challenges and barriers related to path-dependent norms and how they can be challenged, and address questions on how an increased number of female policy-makers and/or professionals have a bearing on climate institutions and in climate-relevant sectors.

In Chapter 7, 'Towards a climate-friendly turn? Gender, culture and performativity in Danish transport policy', Hilda Rømer Christensen and Michala Hvidt Breengaard stress that configurations of environmentally friendly transport and gender are vital in policy-making. The authors demonstrate how such alignments have unfolded in Denmark by examining a range of communicative events related to gender and the enhancement of more climate-friendly transport practices. Using digital media archives and drawing on recent network analysis of the Danish political elite, they disclose how the car-centred society has been constantly re-constituted and maintained in both societies and the particular culture of transport policy. Further, they expose an institutional path dependence around car-centrism and masculine dominance that reconstitutes existing norms in transport policy-making and culture. Through the analysis, both the potentials and limitations of change are considered. It demonstrates how the longstanding alliances of car culture and hegemonic masculine norms were performed at a critical moment in the development of Danish transport policy. It locates the gendered and cross political character of the alliances and hegemonies implicated in policy-making and shows how various femininities and masculinities have been shaped and nurtured within these institutional and hegemonic structures and cultures.

In Chapter 8, 'Wasting resources: challenges to implementing existing policies and tools for gender equality and sensitivity in climate change-related policy', Susan Buckingham explores how waste management can make a significant contribution to curbing greenhouse gas emissions across Europe. However, given waste management's record as being historically male-dominated, engineering-focused and having deeply embedded masculinist structures and path dependencies, it is not clear that it is yet up to the task. The author's involvement in two research projects exploring the relationship between gender equality and waste minimisation provides a unique experience from which she reflects on the capacity for, and challenges to, gender mainstreaming in waste management, and its associated environmental benefits. The chapter sets this exploration within the broader context of gender mainstreaming in the EU and explains how commitment to and awareness of gender equality and sensitivity vary widely depending on prior engagement with gender equality.

In Chapter 9, 'Gender analysis of policy-making in construction and transportation', Bipasha Baruah and Sandra Biskupski-Mujanovic focus on two sectors (construction and transportation) deemed critical to Canada's green economy, though sectors in which women are severely underrepresented. Using a gender equality perspective, they analyse existing government policies and programmes as well as corporate and civil society initiatives. Taking a starting point in Anne Phillip's work (1995), they emphasise the importance of promoting employment equity measures in these sectors where women are marginalised. They discuss institutionalised challenges that women face in these 'masculine sectors' and explore the role of different actors, including the limited role of the federal government, in framing and implementing effective policies to enable the transition to a green economy.

Chapter 10, 'Why radical transformation is necessary for gender equality and a zero carbon European construction sector', reveals how gendered analysis can help identify structural problems, in this case for the construction sector. Linda Clarke and Melahat Sahin-Dikmen argue that these structural problems need to be solved for a climate transition and gender equality to be accomplished, and in tandem, because they are interdependent. The context of their study is the European Union's (EU) strategy to reduce environmental carbon emissions in building construction – which of considerable importance as the construction industry is responsible for 30–40% of end-use emissions in EU. As a heavily white, male-dominated industry, the representation and participation of women across the construction sector are very low and the industry remains (gender) blind to this fact. Also, this sector works under the path dependence of ecological modernisation and is technology-driven.

Part 3: Local, community institutions and climate practices

The third theme explores normative climate institutions and community authorities in local policy-making. The chapters reflect on which social categories are conceptualised and recognised in policy-making and ask questions about whether and how intersectional aspects are visible and/or included. They also raise questions about how climate policy-makers are affected by path dependence and gendered logic of appropriateness and if or how institutional path dependence – e.g., existing norms on ecomodern masculinity – can be challenged.

In Chapter 11, 'Addressing climate policy-making and gender in transport plans and strategies', Tanu Priya Uteng, Marianne Knapskog, André Uteng and Jomar Sæterøy Maridal argue for the need to recognise gendered daily mobilities in developing climate policies due to the way gender/intersectionality is linked to transport patterns. Questions such as 'Which transport modes are women using?' and 'What are its systematic benefits for climate policy-making?' demand further consideration in current and future scenarios, in light of demographic trends like the increase of elderly in the population. Climate policy-making in urban Norway has a zero growth objective (ZGO) and stipulates that future traffic growth should be absorbed by public transport, walking and cycling. Social differences in travel behaviour along the lines of e.g. gender and age, in the willingness to substitute car and an uneven population growth rate can all pose challenges in achieving climate goals if not addressed through stratified interventions. Considering Oslo's population growth, shifts in age distribution and combining it with empirical evidence on different demographic groups' travel habits, this chapter investigates the overlaps between climate policy-making, gender and ZGO.

In Chapter 12, 'When gender equality and Earth care meet: ecological masculinities in practice', Robin Hedenqvist, Paul M. Pulé, Vidar Vetterfalk and

Martin Hultman stress that environmental considerations, such as global warming, have traditionally been marginal issues in masculinities research although men are the main perpetrators of the slow violences of social inequalities and ecological destruction. These violences are instigated and maintained by industrial/breadwinner and ecomodern masculinities, which present considerable barriers for men to engage in social and environmental care. The authors suggest ecological masculinities as an alternative for the necessary reconfiguration of masculinities (particularly in the Global North). Interviews from a selected group of men in a progressive community in rural Sweden are used to identify several themes deemed important for a transition from passive awareness to deeper social and ecological actions.

In Chapter 13, 'Diverging priorities for adaptation: towards inclusive wildfire planning in northern Saskatchewan', Heidi Walker, Maureen G. Reed and Amber J. Fletcher explore wildfire management in the province of Saskatchewan, Canada. The region, with one First Nation and two municipal jurisdictions, was among the areas most significantly affected by wildfires in 2015. The chapter examines how institutions for emergency and wildfire management shaped pathways for adaptation, as well as how gender and other intersecting social structures and power influenced these pathways. Drawing from semi-structured interviews with local government representatives and community residents, the authors argue that there is a need for deeper transformative change towards an inclusive wildfire adaptation as the current institutional approaches focus primarily on technical, physical and economic impacts and fail to address many of the more intangible impacts experienced by community residents, many of which were experienced differently across intersections of gender, race, socioeconomic status, and age.

In Chapter 14, which is the final chapter of the book, we reflect on the main findings of the different chapters and the main research questions presented in the introduction.

References

Alber, G., Cahoon, K. and Röhr, U. (2017) 'Gender and urban climate change policy: Tackling cross-cutting issues towards equitable, sustainable cities', in Buckingham, Susan and Le Masson, Virginie (eds.) *Understanding Climate Change through Gender Relations*. London and New York: Routledge, pp 64–86.

Allwood, G. (2014) 'Gender mainstreaming and EU climate change policy', *The Persistent Invisibility of Gender in EU Policy. European Integration Online Papers*, 1(18), pp. 1–26.

Alston, M. (2013) 'Women and adaptation', *Wiley Interdisciplinary Reviews: Climate Change*, 4(5), pp. 351–358.

Buckingham, S. and Kulcur, R. (2017) 'It´s not just the numbers: Challenging masculinist working practices in climate-change decision-making in UK government and environmental non-governmental organizations', in Griffin Cohen, M. (ed.) *Climate Change and Gender in Rich Countries*. London and New York: Routledge, pp 35–51.

Buckingham, S. and Le Masson, V. (eds.) (2017) *Understanding Climate Change through Gender Relations*. London and New York: Routledge.

Bustelo, M., Ferguson, L. and Forest, M. (eds.) (2016) *The Politics of Feminist Knowledge Transfer: Gender Training and Gender Expertise*. London: Palgrave.

Chaney, P. (2012) 'Critical actors vs. critical mass: The substantive representation of women in the Scottish Parliament', *British Journal of Politics and International Relations*, 14(3), pp. 441–457.

Chappell, L. and Waylen, G. (2013) 'Gender and the hidden life of institutions', *Public Administration*, 91(3), pp. 599–615.

Cho, S., Crenshaw, K. and McCall, L. (2013) 'Toward a field of intersectionality studies: Theory, applications, and praxis', *Journal of Women in Culture and Society*, 38(4), pp. 785–810.

Collins, P.H. (1986) 'Learning from the outsider within: The sociological significance of Black feminist thought', *Social Problems*, 33(6), pp. 14–32.

Dahlerup, D. (1988) 'From a small to a large minority: Women in Scandinavian Politics', *Scandinavian Political Studies*, 11(4), pp. 275–97.

Dahlerup, D. (2006) 'The story of the theory of a critical mass', *Politics & Gender*, 2(4), pp. 511–522.

Dahelrup, D. and Leyenaar, M. (eds.) (2013) *Breaking Male Dominance in Old Democracies*. Oxford: Oxford University Press.

Davis, K. (2008) 'Intersectionality as buzzword: A sociology of science perspective on what makes a feminist theory successful', *Feminist Theory*, 9(1), pp. 67–85.

Djoudi, H., Locatelli, B., Vaast, C., Asher, K., Brockhaus, M. and Sijapati, B.B. (2016) 'Beyond dichotomies: Gender and intersecting inequalities in climate change studies', *Ambio*, 45(3), pp. 248–262.

Duit, A., Feindt, P.H., Meadowcroft, J. (2016) 'Greening Leviathan: The rise of the environmental state?', *Environmental Politics*, 25(1), pp. 1–23.

Eisenstein, H. (1996) *Inside Agitators: Australian Femocrats and the State*. Philadelphia: Temple University Press.

Ergas, C. and York, R. (2012) 'Women's status and carbon dioxide emissions: A quantitative cross-national analysis', *Social Science Research*, 41(4), pp. 965–976.

Ferguson, L. (2015) 'This Is Our Gender Person', *International Feminist Journal of Politics*, 17(3), pp. 380–397.

Griffin Cohen, Marjorie. (ed.) (2017) *Climate Change and Gender in Rich Countries*. London and New York: Routledge.

Hemmati, M. and Röhr, U. (2009) 'Engendering the climate-change negotiations: Experiences, challenges, and steps forward', *Gender & Development*, 17(1), pp. 19–32.

Hysing, E. and Olsson, J. (2017) *Green Inside Activism for Sustainable Development: Political Agency and Institutional Change*. New York: Springer.

IPCC (2014) Climate change: Impacts, adaptation and vulnerability. https://www.ipcc.ch /report/ar5/wg2/ (Accessed 15 November 2020).

Kaijser, A. and Kronsell, A. (2014) 'Climate change through the lens of intersectionality', *Environmental Politics*, 23(3), pp. 417–433.

Kantola, J. (2010) *Gender and the European Union*. New York: Palgrave McMillan.

Kenny, M. (2007) 'Gender, institutions and power. A critical review', *Politics*, 27(2), pp. 91–100.

Kenny, M. (2013) *Gender and Political Recruitment: Theorizing Institutional Change*. Basingstoke: Palgrave Macmillan.

Kinnvall, C. and Rydstrom, H. (eds.) (2019) *Climate Hazards, Disasters, and Gender Ramifications*. London and New York: Routledge.

Kronsell, A. (2016) 'Sexed bodies and military institutions: Gender path dependency in EU's common security and defense policy', *Men and Masculinities*, 19(3), pp. 311–336.

Krook, M.L. and Mackay, F. (eds.) (2011) *Gender, Politics and Institutions. Towards a Feminist Institutionalism*. New York: Springer.

Ljungholm, D.P. (2017) 'Feminist institutionalism revisited: The gendered features of the norms, rules and routines operating within institutions', *Journal of Research in Gender Studies*, 1(7), pp.248–254.

Lovenduski, J. (2005) *State Feminism and Political Representation*. Cambridge: Cambridge University Press.

Lowndes, V. (2020) 'How are political institutions gendered?', *Political Studies*, 68(3), pp. 543–564.

Lykke, N. (2010) *Feminist Studies: A Guide to Intersectional Theory, Methodology and Writing*. London and New York: Routledge.

MacGregor, S. (ed.) (2017) *Routledge Handbook of Gender and Environment*. London and New York: Routledge.

Mackay, F., Kenny, M., and Chappell, L. (2010) 'New institutionalism through a gender lens. Towards a feminist institutionalism?', *International Political Science Review*, 31(5), pp. 573–588.

Magnusdottir, G.L. and Kronsell, A. (2015) 'The In(visibility) of gender in Scandinavian climate policy-making', *International Feminist Journal of Politics*, 17(2), pp. 308–326.

Magnusdottir, G.L. and Kronsell, A. (2016) 'The double democratic deficit in climate policy-making by the EU Commission', *Femina Politica: Zeitschrift für feministische Politikwissenschaft*, 25(2), pp. 64–77.

March, J.G. and Olsen, J.P. (1989) *Rediscovering Institutions: The Organizational Basis of Politics*. New York: Free Press.

Miller, C. (2020) 'Parliamentary ethnography and feminist institutionalism: Gendering institutions—but how?', *European Journal of Politics and Gender*.

Mushaben, J.M. (2012) 'Women on the move. EU migration and citizenship policy', in Abels, G. and Mushaben, J.M. (eds.) *Gendering the European Union: New Approaches to Old Democratic Deficits*. Houndsmill, Basingstoke: Palgrave MacMillan, pp. 208–227.

Nagel, J. (2012) 'Intersecting identities and global climate change', *Identities—Global Studies in Culture and Power*, 19(4), pp. 467–476.

Pierson, P. (2004) *Politics in Time. History, Institutions, and Social Analysis*. Princeton, NJ: Princeton University Press.

Phillips, A. (1995) *Politics of Presence*. Oxford: Oxford University Press.

Resurrección, B. (2013) 'Persistent women and environment linkages in climate change and sustainable development agendas', *Women's Studies International Forum*, 40, pp. 33–43.

Skutsch, M.M. (2000) 'Protocols, treaties, and action: The 'climate change process' viewed through gender spectacles', *Gender & Development*, 10(2), pp. 30–39.

Sommerer, T. and Lim, S. (2016) 'The environmental state as a model for the world? An analysis of policy repertoires in 37 countries', *Environmental Politics*, 25(1), pp. 92–115.

Tobin, P. (2017) 'Leaders and laggards: Climate policy ambition in developed states', *Global Environmental Politics*, 17(4), pp. 28–47.

UNFCCC (2015) United Nations framework convention on climate change, The Paris Agreement. https://unfccc.int/process-and-meetings/the-paris-agreement/the-paris-agreement (Accessed: 22 November 2019).

Yeatman, A. (2020) *Bureaucrats, Technocrats, Femocrats: Essays on the Contemporary Australian State*. London and New York: Routledge.

14   *Kronsell and Magnusdottir*

Wängnerud, L. (2009) 'Women in parliaments: Descriptive and substantive representation', *Annual Review of Political Science*, 12, pp. 51–69.
Waylen, G. (2014) 'Informal institutions, institutional change, and gender equality', *Political Research Quarterly*, 67(1), pp. 212–223.
Woodward, A. (2003) 'European gender mainstreaming. Promises and pitfalls of transformative policy', *Review of Policy Research*, 20, pp. 65–88.

Part 1

Intergovernmental and governmental climate institutions

2 Gender in the global climate governance regime

A day late and a dollar short?

Karen Morrow

Introduction: Getting gender on the agenda

The constitution, form, practices and culture of institutions generally (international institutions among them) are increasingly understood as important not only in mandating and structuring activities within their particular remits but also in exercising a gate-keeping role. The latter is exhibited in the voices and interests that they include/exclude from participation in their processes. Institutions and their operation ultimately embody and emphasise the broader contours of our societies and thus exhibit strongly gendered structural characteristics, which have long been the subject of feminist inquiry and analysis (Prugl and Meyer, 1999). In this context, Meryl Kenny's application of a feminist lens to path dependency (which suggests that early institutional choices shape and limit subsequent regime developments) and functionalist views of institutional change provides particularly fruitful insights for considering the shaping of international climate governance (Kenny, 2007). While often presented as alternative explanatory models, it can be said that the United Nations Framework Convention on Climate Change (UNFCCC, 1992) regime exhibits features of both approaches. In terms of path dependency, as a novel manifestation of a historically gendered international legal system, the UNFCCC regime's roots shape its institutions and institutional culture. Put briefly, it has long been argued by feminist scholars that, while ostensibly presenting itself as gender-neutral, 'international law has a gender … that gender is a male one, and … this skews the discipline' (Charlesworth, 2002, p. 94). Furthermore, it has been observed that it remains the case that '[t]here is, by and large, a disproportionate representation of men in the institutions of international law' (ibid.). The usually uninterrogated privileged, male-dominated and largely masculinist nature of international legal institutions signally hampers their ability to evolve and to develop creative approaches to new areas, the approach on offer being to all intents and purposes 'more of the same'.

Continuing gender disparity within UN institutions raises serious questions as to the profound and seemingly intractable nature of this aspect of structural inequality, particularly in the face of much-vaunted systemic recognition and serial attempts to engage with it (Morrow, 2006). The fact that the UN has, since its inception, raised the need to address gender equality, not least as a foundational and supposedly integral part of its core human rights agenda in both the 1945 UN

DOI: 10.4324/9781003052821-2

Charter and the 1948 Universal Declaration of Human Rights, seemingly augurs well (UN OHCHR, 2014). On one level this positive impression is augmented by specific and ongoing coverage for rights-based approaches to women's participation in and across the UN and states. Rights-based coverage, for example, features strongly in the 1979 Convention on the Elimination of All Forms of Discrimination Against Women (CEDAW), Article 8 of which is germane in the current context, providing that:

> States Parties shall take all appropriate measures to ensure to women, on equal terms with men and without any discrimination, the opportunity to represent their Governments at the international level and to participate in the work of international organizations.
>
> (UN CEDAW, 1979)

Women's participation rights were further amplified by the ambitious 1995 Beijing Declaration and Platform for Action (UN OHCHR, 2014). The document features dedicated coverage for women in power and decision-making, developing institutional mechanisms for the advancement of women, and addressing women's human rights (UN BDPA, 1995, Chapter IV, sections G, H and I, respectively). The rationale for including coverage of this nature was succinctly stated:

> Equality in political decision-making performs a leverage function without which it is highly unlikely that a real integration of the equality dimension in government policy-making is feasible.
>
> (UN BDPA, 1995, para. 181)

The observation remains as cogent today as when it was made. The same holds for international policy-making contexts as they are populated by representatives selected by states. The strategic objectives for governments under the BDPA included what have become familiar themes: data collection and monitoring in the pursuit of gender balance, both domestically and in regard to UN bodies (UN BDPA, 1995, para. 190). The UN itself was also charged with putting its own house in order by pursuing gender equality in its staffing, particularly at senior level, and with collecting and disseminating data on its progress (UN BDPA, 1995, para. 193). However, while the UN's engagement with gender has matured, developing in range and sophistication over time (Morrow, 2006), it is also the case that there has been a constant need for periodic high-profile re-engagement with gender issues. The latter throws a less positive light on matters, being prompted by a continuing paucity of progress on the ground. It remains the case that progress on realising rights is uneven and in consequence:

> Women around the world ... regularly suffer violations of their human rights throughout their lives, and realising women's human rights has not always been a priority.
>
> (UN OHCHR, 2014, p. 1)

The UNFCCC is firmly located in this long-established milieu and on this ground alone would have been ripe for feminist inquiry in the cause of the imperative need to address inclusivity and equality.

At the same time, there is also a strong functionalist dimension to the institutional character of the UNFCCC system, rooted in the manner in which climate change was characterised in the fledgling regime, which has also profoundly shaped it and its approach to its role. The elements of the regime are many and their inter-relationship complex. For present purposes, the main regime actors and roles are state signatories who make up the supreme governing body of the convention in question; in the UNFCCC this is known as the Conference of Parties (CoP). Signatory states also provide the members of the regime's specialist, limited membership constituted bodies, which are charged with carrying out a range of subject-specific technical roles. The activities of the CoP and the constituted bodies are supported by a number of enabling bodies, notably, the regime's Secretariat, which provides general technical advice and support and organisational services and the more focused Subsidiary Body for Implementation (SBI). Additionally, a number of recognised non-state actors now play a variety of roles in regime activities. In initially characterising climate change as a technical issue, suitable for traditional state-centric international law coverage and dominated by the search for scientific and economic 'fixes' (themselves reliant on male-dominated disciplines) its pervasive, cross-cutting complex nature was not fully addressed (UN FCCC, 1992; Morrow, 2017a, p. 31). The social dimensions of climate change were initially largely absent from consideration in the regime, though, in fairness to the UNFCCC Secretariat, it recognised this lack at a fairly early stage and expanded its stakeholder engagement beyond states to embrace (some) major societal groups in its activities. However, the approach adopted tended to augment the initial masculinist dominance of technical and economic concerns within the regime (Morrow, 2017b, p. 39). The various voids that this approach propagated and perpetuated now seem obvious, yet took many years for the UNFCCC regime to grasp, in part because:

> Essentially, people tend to assume that our own way of thinking about or doing things is typical. ... If the majority of people in power are men – and they are – the majority of people in power just don't see it. Male bias just looks like 'common sense' to them.
>
> (Perez, 2019, p. 270)

Tellingly, the newly expansive approach towards stakeholder groups did not, for many years and despite vigorous and sustained campaigning, extend to women. The regime's failure to act decisively on gender inclusion was mystifying on a number of levels, not least as the UNFCCC had long been on the record as recognising the fact that women were underrepresented in its activities and on its constituted bodies and that this needed to be addressed (UN FCCC, 2001). This apperception did not, however, prompt an effective response, with calls for data collection and dissemination with reference to the gender composition of the regime's constituted bodies failing to generate much impact. Alongside this, it took the best part of two decades for the constituency status enjoyed by other major groups under the regime to finally be accorded to the Women and Gender Constituency (WGC) in 2011 (Morrow, 2017a, pp. 33, 37–38). This development coincided with a higher profile for gender in the context of discussions about climate change among UN institutions more generally, notably in bodies with remits in gender and in environmental matters (Morrow, 2017a, pp. 34–35). It was more particularly prompted by activity in the cognate 'Rio Conventions', comprising the UNFCCC itself, the Convention on Biodiversity (CBD) and the subsequently adopted United Nations Framework Convention to Combat Desertification (UNCCD) in which gender featured prominently, in their preparations for the Rio+20 Conference in 2012. The activities of the CBD and the UNCCD had seen gender not only raised but also engaged with (albeit to varying degrees) at a regime level in a way that, until this point, those of the UNFCCC had not and the latter was, in comparison a laggard (CBD, UNCCD, UNFCCC, 2012, p. 5; Morrow, 2017a, p. 35).

Gender landmarks in the UNFCCC regime

Monitoring and reporting on gender representation

Decision 23/CP.18, adopted at the UNFCCC CoP in Doha in 2012, arguably signals the start of an attempt to promote serious engagement with gender in global climate change governance (UN FCCC, 2012). The Decision's title set out its stall very clearly: 'Promoting gender balance and improving the participation of Women in UNFCCC negotiations and in the representation of Parties in bodies established pursuant to the Convention or the Kyoto Protocol'. Decision 23/CP.18 was hugely important in both principle and evolving regime practice. It not only served to reiterate concerns previously (though sporadically) raised about the gender composition of the UNFCCC's constituted bodies; it also looked explicitly to the gender make-up of signatory state delegations. More importantly still, it set the scene for gender to become a regular (rather than merely occasional) regime agenda item. The crucial element of Decision 23/CP.18 lay in beginning to construct a factual foundation upon which to ground regime actions on gender, by regularising and enhancing monitoring, and reporting on gender representation in state delegations and constituted bodies. In consequence, annual reports on the composition have featured in the UNFCCC's regular diet of business since 2013 and this remains a central component of how the regime presents its coverage of gender issues. While a more granular analysis is not possible within the confines of the current chapter, even a basic exploration of material drawn from the regime's annual gender composition reports from 2013 to date is revealing (UN FCCC Gender and Climate Change Documents, n.d.). Figure 2.1 takes the average representation of women in the UNFCCC's constituted bodies[1] as an example and shows that taken overall, despite the UNFCCC's regime machinery shifting to and gradually ratcheting up the active

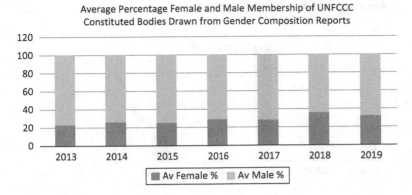

Figure 2.1 Average percentage female and male membership of UNFCCC constituted bodies drawn from gender composition reports

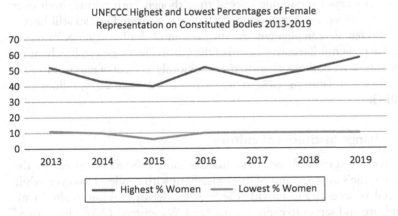

Figure 2.2 UNFCCC highest and lowest percentages of female representation on constituted bodies 2013–2019

promotion of gender equality, progress has been scant. This remains the case when the expansion over this period of the number of constituted bodies from 11 to 15 is accounted for.

Figure 2.2, showing the top and bottom of the range of percentages of women sitting on constituted bodies, is also disappointing. While, as Figure 2.2 shows, particularly from 2017 onwards, upper averages are improving, the lower register remains stubbornly on or below 10%. Figure 2.3 likewise shows a bifurcated picture: while in 2019 female representation was 30% and above on nine of the regime's 15 constituted bodies, closer inspection of the source material reveals that there is still a long way to go. Of these nine bodies, women's representation in eight sits between 30–39%, and of these, only three register above 35%.

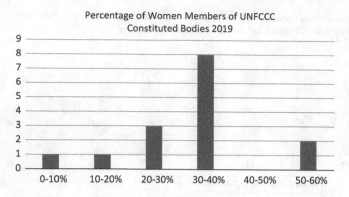

Figure 2.3 Percentage of women members of UNFCCC constituted bodies 2019

Only the Adaptation Committee (which has consistently been among the better performers on women's representation, and more so since 2016) and the Paris Committee on Capacity Building exceed this, though both include well over 50% female members. Furthermore, in 2019, five constituted bodies still have a less than 30% female complement. At the bottom of the lower range of representation, the powerful Executive Board of the Clean Development Mechanism, which, while now an outlier rather than, as previously, typical of many of its peer bodies, persists at a damning 10% (where it has stood since 2015, falling from 20% in 2014).

Acting to change institutional culture

The provision of statistical data remains the most prominent strand in the UNFCCC regime's own account of its engagement with gender. However, while such material is necessary to provide the basis for debate seeking to alter institutional culture and serves to capture what Lena Wängnerud (2009) has termed descriptive representation, it cannot suffice to promote effective engagement with gender issues. This requires a deeper, conscious and active engagement with equality, which, as revealed by work on domestic parliaments and gender, inevitably collides directly with gendered systems, structures and practices (Wängnerud, 2009, p. 52; Criado Perez, 2019, pp. 271–286). The figures above strongly suggest that the same is true for the regime machinery of international agreements such as the UNFCCC, where data production and dissemination have not prompted appreciable progress. While the UNFCCC constituted bodies have a shorter history than domestic institutions, they are, as discussed above, grafted on to a system hallmarked by entrenched gender inequality. In any event, even presence in numbers, while necessary to progress gender equality in a system, is not necessarily sufficient (Wängnerud, 2009, p. 59).

Thus, in addition to considering women's presence in regime bodies, it is also crucial to examine what Wängnerud (2009) terms substantive representation, that is, the effects of women's presence on the regime. In the context of the UNFCCC, these will be examined by considering indicative developments in the regime's gender culture through the changing nature of the coverage it offers to gender issues.

The UNFCCC has not confined its activities to recording the numbers of women present in its constituent bodies and national delegations. The search for increased efficacy saw the gender agenda coalesce in the activities of the Conference of Parties itself, with the adoption of a suite of important decisions shaping the regime's new approach. These documents are significant in that, while they continue to be led by quantitative elements, focusing on participation metrics, this is now coupled with commitments to substantive action with more qualitative dimensions. The latter demonstrates a shift, seeking to improve not just the numbers of women active in regime processes but also their ability to participate meaningfully in them. Decision 18/CP20 on the Lima Work Programme on Gender (LWPG) was a landmark in this process (UN FCCC, 2014). The document, weighing in at only two pages, was brief but foundational, focusing on improving the coherence of the regime's work to address gender through mainstreaming in the UNFCCC regime. Significantly, the contextual markers identified by the LWPG included CEDAW and the UN BDPA. The work programme was set up to pursue gender balance and develop gender-responsive climate policy (the meaning of which it promises to clarify) to improve women's participation in the regime's constituted bodies. The programme included strengthened reporting requirements, but also looked to substantive matters centred on training and capacity building and, perhaps most importantly of all, committed to a review in 2016, keeping progress under scrutiny (UN FCCC, 2014, paras. 3–6 and 16, respectively).

Incremental progress continued with the adoption of Decision 3/CP.23 (UNFCCC, 2017), which added a gender action plan (GAP) to the LWPG. The CoP had requested that the SBI develop the GAP in Decision 21/CP.22, with a view to supporting the implementation of gender-related decisions and mandates within the UNFCCC. A fairly broad range of potential courses of action was outlined therein, including 'identifying priority areas, key activities and indicators, timelines for implementation … responsible and key actors and indicative resource requirements for each activity' and enhanced monitoring and review processes (UN FCCC, 2016, para. 27). If the LWPG provided the skeleton of the regime's new, ostensibly more coherent approach to gender, the GAP put flesh on these bones. The GAP ran until 2019 and identified five priority areas, deliverables for each and the actors responsible for them. The five priority areas comprised knowledge-sharing, gender balance, coherence, implementation and monitoring and reporting. While some of these areas share a strong quantitative bent with earlier initiatives, the package taken as a whole now demonstrates a mixed approach, though with qualitative elements continuing to play a part across the board. The hybrid approach espoused is apparent, for example, in priority area B, on gender balance – which on the face of it would seem to be largely quantitative. However, while the deliverables identified for it included quantitative elements, such as the promotion of travel funds for female participants in regime process, and information about gender balance to accompany recruitment to constituted bodies, it also featured two activities centred on training, which also looked to qualitative considerations concerning capacity-building. The responsible actors identified included state parties, the UNFCCC Secretariat, other UN bodies and relevant external organisations (UNFCCC, 2017, Annex). The GAP is presented as a four-page annex to Decision 3/CP.23, the first of which outlined the five priority areas and the remaining three of which comprised closely written tables for each of them, detailing activities, the attribution of responsibilities, timelines, deliverables/outputs and the level at which implementation is required (local, national, regional and/or international). The task and action approach adopted in the GAP has a number of advantages in terms of expectation management and transparency for all actors within the UNFCCC system. While a rational and coherent approach does not of itself guarantee progress, it does at least serve to offer desirable consistency.

Pursuant to the LWPG and the GAP's stated priority to ensure that gender was integrated into the work of 11 of the UNFCCC's constituted bodies,[2] in 2019 the UNFCCC Secretariat produced a desk-based synthesis report on progress (or the lack of it) to date on reporting of gender coverage within the various mandates involved (UN FCCC, 2019a). The report identified basic progress through time, in that while in 2017 only six constituted bodies mentioned gender as part of their regular reporting processes, in 2018 this had increased to 11. However, most were acknowledged as offering at best cursory coverage, indicating only marginal improvement. Of the seven that reported in some depth on their progress in integrating gender into their processes and work, only three reported that they had instituted practical action such as setting up working groups and/or gender focal points or instituting their own gender action plans (UN FCCCC, 2019a, para. 8–10). Slight improvement on previous practice must, however, be set against the fact that, the UNFCCC's multiple efforts in the last few years notwithstanding, coverage was at best uneven. It ranged from comprehensive approaches by some through to two constituted bodies that provided such limited information that it was not possible to determine their progress. Furthermore, where gender was reported upon, the approach adopted was idiosyncratic, making comparison and the sharing of best practice challenging. The provision of UNFCCC guidance on the form and substance required for gender progress reports was suggested to improve the utility of the reporting process (UN FCCC, 2019a, para. 121).

The LWPG and GAP themselves fell to be reviewed by the SBI (UN FCCC, 2019b). The SBI's evaluation of progress on the five priorities, based on submissions from a range of state parties and observer organisations, also pointed to the importance of the LWPG to the regime's institutional culture as a framework for gender action and the GAP as the practical focus for action. In the latter context, the LMPG and GAP were identified as providing a platform for stakeholders to exchange information on and to advance the five priorities. Review participants also took the opportunity to suggest improvements, showing that the LWPG and GAP are not only important in themselves but also part of an ongoing and iterative process. This is confirmed by the adoption of Decision 3/CP.25 of the Enhanced Lima Work Programme on Gender (ELWPG) and its gender action plan (UN FCCC, 2019c), which are to run until 2024. The ELWPG rehearses the usual background mantra of the climate change regime: reiterating links with other UN remits (notably the SDGs), exhorting states to act and requesting that the Secretariat continue to provide support to enable them to do so. However, as anyone familiar with the UNFCCC's attempts to improve its practice on gender equality fully expects, it also clearly acknowledges that, previous efforts notwithstanding, there is a

> persistent lack of progress in and the urgent need for improving the representation of women in Party delegations and constituted bodies.
> (UN FCCC, 2019c, para. 2)

The augmented GAP is similar in length, content and organisation to its predecessor – and in many ways also indicates a 'more of the same' approach. Once again taking priority area B, gender balance, participation and women's leadership, as an example, travel funding and training remain as identified activities. An additional strand of activity concerns working with the Local Communities and Indigenous Peoples Platform Facilitative Working Group (LCIPPFWG) 'on women's leadership and enhancing the participation of local communities and indigenous women in climate policy and action … to the extent that it is consistent with the workplan of the … [LCIPPFWG] … and within existing resources' (UN FCCC, 2019c, B3). Given that women's participation is often contentious in these contexts (UN Secretariat of the Permanent Forum on Indigenous Issues [SPFII] et al., 2010), it is difficult to gauge how effective this activity will be, beyond initiating a dialogue on the issues. However, in a crucial exception to the prevalent 'more of the same' approach, the most significant development in the new GAP involves priority area D, gender-responsive implementation and means of implementation. This sees a shift in emphasis in the action plan, with seven activity areas identified, a significant increase on the three covered in the 2017 version. Priorities continue developing work with various women's organisations to better inform approaches to gender issues but extend to: improving the provision of gender-disaggregated data, sharing good practice on mainstreaming gender in climate policy, and support for integrating gender into the central funding and technical concerns of the regime. The fact that the implementation strand is now providing for the main thrust of activity in the GAP is significant in drawing attention to gender implications in some of the core practical elements of the UNFCCC.

In addition to the work programmes and action plans, the UNFCCC charged its Secretariat with securing practical improvements in the regime's operation and outputs by developing a series of technical papers to inform the regime's approach to gender. Such synchronous developments were geared to promote deeper integration of gender into the regime's architecture. One important example lay in identifying possible actions in the workstreams of the regime's constituted bodies that were, or could be, integrated to deliver informed reporting on progress towards the parties' goals on gender balance and gender-responsive climate policy (UN FCCC TP, 2018). The technical papers, in turn, form the basis for further work – in this example, the CoP in Decision 3/CP.23 (UNFCCC, 2017) requested that its outcomes and recommendations form the basis of a dialogue between the chairs of the constituted bodies, to share experience and expertise and build capacity for effective engagement with gender.

Increasing the profile and visibility of gender in the UNFCCC regime

If progress on gender representation and integration leaves much to be desired, it is nevertheless the case that gender has become an established (if unevenly realised) agenda item across UNFCCC processes. Gender coverage is now also a much more visible and accessible presence on the organisation's website. In addition to prominent positioning as one of the regime's headline topics, collated coverage of relevant regime decisions and documents pertaining to gender is now provided (UN FCCC Gender and Climate Change Documents, n.d.). The gender topic is also furnished with cross-cutting links to other core areas of the UNFCCC regime, namely adaptation, mitigation, capacity-building, technology, climate finance and cross-cutting topics (UN FCCC Topics, n.d.).

As illustrated in Figure 2.4, the coverage of gender in regime documents has broadly (if initially sporadically) increased over time, admittedly from paltry beginnings in 2001. More consistent development, prompted by shifts in the priority that the UNFCCC has accorded to gender in its institutional activities (discussed above), began in 2012 and relevant regime activity since has been considerably more frequent and, while not constant, operates at a level that demonstrates at least a core level of engagement each year.

Other forms of more public-facing engagement by the UNFCCC regime also flourished on the UNFCCC's belated awakening to gender issues. One prominent

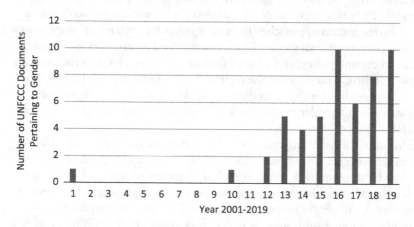

Figure 2.4 Numbers of UNFCCC documents pertaining per gender

development was the addition of a regular 'Gender Day' event to the meetings of the conference of parties from 2012 onwards (UN Women, 2012). A second innovation saw the regular inclusion of practically oriented gender-based workshops at regime meetings (UN FCCC, 2013). While these developments could have been fairly superficial, they have not only generated headlines but have also raised the profile of gender issues in the main business processes of the UNFCCC, at the same time as providing opportunities to interrogate the regime's shortcomings and attempt to build capacity to address them. Gender days have, for example, highlighted progress on gender equality and climate change (2014), the economic case for gender-responsive climate action (2017) and gender and national adaptation plans (2019). Gender workshops (reports on which are subsequently fed into the regime by its Secretariat) have addressed core topics such as gender and mitigation (2015), gender and adaptation (2016) and the gender-differentiated impacts of climate change and gender-responsive climate policy and action (2018). It is an indication of continuing institutional commitment that workshops on integrating gender into national climate actions have not been side-lined by the COVID-19 pandemic but instead moved online in a development which should help to retain their momentum (UN FCCC, 2020).

The Paris Agreement and gender

Much is made by the UNFCCC on its website of the fact that gender attained preambular status in the Paris Agreement (PA), which acknowledges that, as

> climate change is a common concern of humankind, Parties should, when taking action to address climate change, respect, promote and consider their respective obligations on human rights, the right to health, the rights of indigenous peoples, local communities, migrants, children, persons with disabilities and people in vulnerable situations and the right to development, *as well as gender equality, empowerment of women and intergenerational equity.*
> (UN FCCC, 2015b, emphasis added)

However, while the PA could have been a springboard for a more balanced, socially directed approach to climate change coming to the fore, in many ways it missed the mark, not least with regard to gender, which garnered a scant three mentions in the Agreement's main text. While the preamble to the PA adopts a more socially contextualised approach to climate change than hitherto, its role in the regime is comparatively limited, largely serving to provide an interpretative context for the substantive articles included in the main body of the Agreement. Gender had been a live issue during the negotiating process and prominent in the draft agreement – it was mentioned nine times in the bracketed negotiating text (UN FCCC, 2015a). Many were, however, hugely disappointed by what the PA actually delivered on gender, as the inter-state negotiating process took effect. One result was that elements perceived by some states as controversial, including, as Rochette observes, most of those on gender, were excised from the text (Rochette, 2017, p. 254). In the Agreement as adopted, gender was siloed in coverage of adaptation (UN FCCC, 2015b, Article 7) and mitigation (UN FCCC, 2015b, Article 11).

The use of language in the preamble to the PA, calling on states to 'respect, promote and consider' their human rights obligations, is remarkable and arguably signals a great deal in attitudinal terms. The usual parlance with regard to states' human rights obligations requires them to 'respect, protect and *fulfil*' (emphasis added) them. In short, under the guise of 'respect', states are expected to refrain from interfering with human rights. To 'protect' requires that states address human rights issues for individuals and groups – 'promotion' is a much looser, hortatory term and not a like-for-like substitute. The obligation to 'fulfil' requires states to take positive action to deliver on the human rights that they have committed to (UN OHCHR, n.d.). An obligation merely to 'consider' (i.e. to take into account) is not commensurate with an obligation to fulfil (i.e. deliver) on human rights. The orientation of this preambular clause of the PA is significant in real terms, as it is now seen as representing part of the institutionally accepted context for actions under the UNFCCC (UN FCCC, 2019c). This careful and deliberate use of language is crucially important – in effect, it seeks to pay lip-service to human rights, while at the same time diluting state obligations to act to secure those rights that are increasingly recognised by the UN's human rights machinery to be infringed/infringeable by climate change (UN OHCHR, 2019). This is particularly significant with regard to gender, as women are recognised as a particularly vulnerable group in the context of climate change and as having protected human rights in this regard (UN OHCHR, 2019, paras. 45–47 and 64, respectively). Such gendered effects of climate change have long been recognised by many UN entities (Morrow, 2017a, p. 34), not least CEDAW (most recently in CEDAW, 2018).

The Sustainable Development Goals, Agenda 2030 and gender

The inclusion of reference to human rights in the PA itself is ambiguous, recognising their relevance to climate change in principle while distancing the regime from their implementation in practice. Nevertheless, it is evident that, despite the evasive approach adopted, the very act of inclusion has affected the context of the climate change debate, serving to further fuel the active engagement of the UN human rights and gender machinery on climate change where it crosses into their respective remits. A similar situation arose in the cross-cutting of the gender/human rights/environment nexus (including climate change) expressed in the 2030 Agenda for Sustainable Development (Agenda 2030) (UN GA, 2015) and the Sustainable Development Goals (SDGs) (UNDP, 2015). This is significant as there is clearly acknowledged crossover between the UNFCCC and the SDGs and Agenda 2030, not least in respect of Goal 13, climate action, and its supporting targets (UN GA, 2015, para. 14). While gender is covered in the SDGs themselves, both discretely (under Goal 5) and as a cross-cutting concern in a number of other goals and appears in many of the regime's supporting targets

(Morrow, 2018), it is more prominent still in Agenda 2030. Nevertheless, in both contexts, rights-based coverage is again effectively side-stepped (Morrow, 2018). However, flawed as the coverage offered may be, the fact that gender is included in the SDGs provides an additional route to invoke gender equality, which is significant in the context of the increasingly regularised interplay between the climate change and SDG regimes. The latter is indicated, for example, by the inclusion of 'Action on Climate and the SDGs' under UNFCCC topics (UN FCCC Topics, n.d.).

A plethora of institutional activity – But to what avail?

The UNFCCC regime machinery has, belatedly, engaged prolifically with the gender agenda and its approach has matured considerably, yet its impact has been at best limited. This begs the question, what is needed to do better? On paper, the building of a sound foundation of accurate and transparent data, its dissemination within the UNFCCC, and tying it in, however symbolically, with the human rights and SDG agendas more broadly, demonstrates some promise for more thoroughgoing engagement with gender and climate change than hitherto. Were this all that had been done, it would, however, have been open to interpretation as a cynical exercise in ecomodernism, favouring form over substance. However, latterly it has been joined by the promotion of institutional reflection, debate and change, which ostensibly provides a more grounded basis for addressing substantive gender inequality issues. In practice, though, progress, even within the UNFCCC's own constituent bodies, has been frankly unimpressive. Self-evidently, more needs to be done – but what? Bringing gender into sharper focus than ever before has made the pervasive nature of gender inequality even more apparent. This is important in itself, as it lays bare the root of the failure of UNFCCC initiatives to gain much traction – state inaction on gender equality. The failure of states to live up to their international legal commitments is notionally a source of political embarrassment vis-à-vis other states and stakeholders and often described as a potent lever to motivate states to action in both international human rights law (Cassel, 2001) and international environmental law (Benedick, 1998). Yet slow progress and even regression in gender equality suggest that both in general (as outlined above) and in the context of the UNFCCC shame does not always serve as an effective driver of state behaviour. The UNFCCC has done much to set its own house in order, but, as is readily apparent from the discussion above, if it is confined to hortatory approaches to its state signatories, these will not suffice to bring about deeper change, as states are effectively free to ignore exhortations to act on existing gender equality commitments. In theory, the UNFCCC could do more; in recognition of the pervasive and deeply entrenched nature of gender inequality, a suite of tools has been identified that can help to forge progress, for example, temporary quotas to promote participation (CEDAW, 2004; UN OHCHR, 2014). Failure to address gender equality could also be subject to innovative corrective action within the UNFCCC's competence, for example, targeting gender-unequal state delegations by limiting participation rights to a guaranteed minimum but rewarding gender-equal delegations with support and increased opportunities to participate above that minimum. In practice, though, as an intergovernmental organisation, the UNFCCC is in a weak position to compel states to act, as it, in effect, relies on their goodwill to operate. The fundamental sovereign status of states in this area, as in all others, remains the ultimate brake on progress. Political buy-in and sustained commitment to the pursuit of gender equality by the UNFCCC's signatory states is crucial, but remains controversial for some states, standing in the way of consensus. Additionally, but relatedly, state financial buy-in is also necessary to further progress – if gender equality is not viewed as a priority, it is not adequately resourced and progress will be hampered. It is telling that most UNFCCC documents on gender equality repeat the refrain that action is subject to the availability of financial resources.

What is needed? Updating, expanding and fully realising UNFCCC engagement with gender issues

The UNFCCC began as a male-dominated, masculinist-oriented regime and despite concerted efforts to correct this, the continued paucity of women's participation and its limited impact underline that this remains largely the case in substance, if not in form. Insofar as equality between women and men within the global climate governance regime is concerned, we are dealing with a broadly improving picture, after a belated start, but the business in hand is very much unfinished and progress remains quite sporadic and slow overall. This is not the only area of gender concern where concerted action is needed. Our understanding of the complexity of sex and gender, extending far beyond the female/male binary, has been rediscovered (Independent Lens, 2015) and grown apace in recent years. The World Health Organisation (WHO) defines gender as referring to:

> the roles, behaviours, activities, attributes and opportunities that any society considers appropriate for girls and boys, and women and men. Gender interacts with, but is different from, the binary categories of biological sex.
>
> (WHO, n.d.)

Thus, an informed and current approach to gender must extend its reach beyond the conventional female/male binary and be genderqueered to embrace those of all sexes, genders and orientations and indeed those of none. The understanding of gender as expressed in our international institutions more generally (human rights bodies excepted, discussed briefly below) has now fallen far behind our social and scientific grasp of the issues that it raises and remains dominated by the female/male binary.

As Kenny has observed, feminist analysis has evolved beyond a simplistic sexed approach focused on 'women's issues' to embrace a broader perspective on gender as performance. Fully pursuing even this has, however, proven problematic when looking at issues of representation, 'where research continues to focus on female bodies as the "main vehicles" for institutional change and transformation' (Kenny, p. 94). It is also the case in global climate change governance that gender continues to be largely conflated with 'women's issues' and that broader gender concerns, notably those relating to LGBTQI+ persons, remain under-interrogated to the point of near invisibility. UN work in cognate areas, such as food security (UNEP, 2018, p. 33) suggests, in passing, that LGBTQI+ people face many of the same disadvantages as women, and as these issues intersect with climate change, it is therefore reasonable to assume that a broader notion of equality than that typically current should be applied to the treatment of gender-based vulnerability to climate change. This lacuna is fairly typical, as while the UN human rights machinery finally turned its attention to LGBTQI+ rights in the last few years, as was the case with women's issues for many years, other UN institutions remain relatively unengaged. It is also the case that the human rights activity in this area thus far remains fairly narrow, focusing primarily on first-generation (civil and political) rights (UN OHCHR, 2019). Second-generation (social and economic) and third-generation (solidarity) rights also need to be invoked, not least in the context of climate change, if we are to give full meaning to the universality of human rights and dignity in this regard; equal human rights are the entitlement of all, gender/no gender notwithstanding. Ebbs and flows in the respect accorded to human rights are, given the tide of human affairs, entirely to be expected. The core international protections owe their origins to reactions to the horrors of World War Two – and it speaks volumes that it was deemed necessary to articulate and garner state support for them in the context of vast infringements of the most basic entitlements of human beings. While arguments on the efficacy of international human rights law are commonplace (Cassel, 2001), what it does provide is an articulation of core protections that serves as a yardstick to evaluate state (and international institutional) behaviour. As the boundaries that human rights protection delineate are tested, not least by climate change, the willingness – or not – of the international community to deploy rights rhetoric as prevention and/or cure is immensely revealing. Human rights commitments, then, have an important role to play in the conduct of human affairs, albeit as part of a broad, complex and multi-layered framework of legal, political and social processes (Cassel, 2001). Where gender and human rights are concerned, a global backlash fuelled by some religious and ultra-conservative groupings is currently observable in relation to both women's (UN OHCHR, 2014; Lilja and Johansson, 2018) and LGBTQI+ rights (UN OHCHR, 2019). This backlash demonstrates the precarity of even the limited progress that we have made in this regard and that there is a real danger of regression that will have broad ramifications across our societies and act to the detriment of all.

Conclusion – Process is all very well, but progress must be prioritised

While there has been some progress in addressing gender in the global climate regime, an economics-driven, technocratic stance toward climate change continues to dominate. Insofar as gender, narrowly defined across the female/male binary, is concerned, advancement has been uneven and achingly slow. As Caroline Criado Perez puts it:

> we have to close the female representation gap. When women are involved in decision-making, in research, in knowledge production, women do not get forgotten. Female lives and perspectives are brought out of the shadows.
>
> (Perez, 2019, p. 318)

It is now also readily apparent that a largely binary focus is too narrow to fully address gender issues (UNEP, 2018, p. 196). Furthermore, broader gender identity and sexuality–related diverse perspectives share many of the hallmarks of exclusion that have long been identified with regard to the basic gender binary. Gender equality is not a novel claim, nor should it be regarded as a radical one, though given the paucity of progress that we have made in its realisation, one could be excused for thinking that it was both. It stands to our shame that the constant refrain that commentators must adopt in this field is slow progress; more needs to be done. It remains the case that, in an international legal system founded on state sovereignty (Dixon, 2013) and hallmarked by the limited enforcement capacity of a regime based on consent and consensus (Dixon, 2013), bodies such as the UNFCCC, however committed they are to advancing gender equality, are politically constrained in getting states to live up to their commitments. While more compelling approaches are notionally available, political realism in the context of the UNFCCC suggests that persuading states to act will continue to be the standard approach. Limited progress to date, however, suggests that this is incapable of delivering the paradigm shift required. It remains the case that representation – both in presence (which is necessary but not sufficient) and, in turn, in influence – is the *sine qua non* of progress on gender equality. Self-interest supports this; if climate change has taught us nothing else, it is that it will require the whole range of human ingenuity and agency to tackle it. It requires an inclusive approach that brings all human perspectives, female, LGBTQI+ and male, to the table – to proceed otherwise is a further act of self-sabotage, adding an avoidable additional challenge to the already obdurate problems that we face.

Notes

1 Initially there were 11 constituted bodies in total: the Executive Board of the clean development mechanism (CDM Executive Board); the Joint Implementation Supervisory Committee (JISC); the Compliance Committee facilitative branch; the Compliance Committee enforcement branch; the Least Developed Countries Expert Group (LEG); the Consultative Group of Experts on National Communications from Parties not included in Annex I to the Convention (CEG); the Adaptation Fund Board (AFB); the Technology Executive Committee (TEC); the Adaptation Committee (AC); the Standing Committee on Finance (SCF); and the Advisory Board of the Climate Technology Centre and Network (CTCN Advisory Board). The Executive Committee of the Warsaw International Mechanism for Loss and Damage associated with Climate Change Impacts (Executive Committee of the WIM) was added from

2014 and included in analysis of the report statistics from 2015. The Paris Committee on Capacity Building (PCCB) was added from 2017. The Facilitative Working Group of the Local Communities and Indigenous Peoples Platform (figures analysed relate to government members only) and the Katowice Committee of Experts on the Impacts of the Implementation of Response Measures were added in 2019.

2 Namely the AC, AFB, CDM Executive Board, CGE, CTCN Advisory Board, JISC, LEG, PCCB, SCF, TEC and Executive Committee of the WIM. Exceptions were task- or time-based.

References

Benedick, R. E. (1998) *Ozone Diplomacy: New Directions in Safeguarding the Planet.* Enlarged Edition. Cambridge, MA: Harvard University Press.

Cassel, D. (2001) 'Does International Human Rights Law Make a Difference?', *Chi. J. Int'l L.*, 2, pp. 121–135.

CBD, UNCCD, UNFCCC (2012) The Rio Conventions: Action on Gender [online]. Available at: https://unfccc.int/resource/docs/publications/roi_20_gender_brochure.pdf (Accessed: 23 July 2020).

CEDAW (2004) General Recommendation No. 25, on Article 4, Paragraph 1, of the Convention on the Elimination of All Forms of Discrimination against Women, on Temporary Special Measures [online]. Available at: https://www.refworld.org/docid/453882a7e0.html (Accessed: 05 August 2018).

CEDAW (2018) General Recommendation No. 37 on Gender-Related Dimensions of Disaster Risk Reduction in the Context of Climate Change, CEDAW/C/GC/37 [online]. Available at: https://tbinternet.ohchr.org/Treaties/CEDAW/Shared%20Documents/1_Global/CEDAW_C_GC_37_8642_E.pdf (Accessed: 31 July 2020).

Charlesworth, H. (2002) 'The Hidden Gender of International Law', *Temp. Int'l & Comp. L.J.*, 16, pp. 93–102.

Criado Perez, C. (2019) *Invisible Women: Exposing Data Bias in a World Designed for Men.* London: Chatto and Windus.

Dixon, M. (2013) *Textbook on International Law.* 7th Edition. Oxford: Oxford University Press.

Independent Lens (2015) A Map of Gender-Diverse Cultures [online]. Available at: https://www.pbs.org/independentlens/content/two-spirits_map-html/ (Accessed: 07 August 2020).

Kenny, M. (2007) 'Gender, Institutions and Power: A Critical Review', *Politics*, 27(2), pp. 91–100.

Lilja, M. and Johansson, E. (2018) 'Feminism as Power and Resistance: An Inquiry into Different Forms of Swedish Feminist Resistance and Anti-Genderist Reactions', *Soc. Incl.*, 6(4), pp. 82–94.

Morrow, K. (2006) 'Not so Much a Meeting of Minds as a Coincidence of Means: Ecofeminism, Gender Mainstreaming and the UN', *Thomas Jefferson Law Journal*, 28(2), pp. 185–204.

Morrow, K. (2017a) 'Integrating Gender Issues into the Global Climate Change Regime', in Buckingham, S. and Le Masson, V. (eds.) *Understanding Climate Change through Gender Relations.* London: Routledge, pp. 31–44.

Morrow, K. (2017b) 'Changing the Climate of Participation: The Gender Constituency in the Global Climate Change Regime', in Macgregor, S. (ed.) *Routledge Handbook on Gender and Environment.* London: Routledge, pp. 398–411.

Morrow, K. (2018) 'Gender and the Sustainable Development Goals', in French, D., Louis, J., and Kotzé, L. J. (eds.) *Sustainable Development Goals Law, Theory and Implementation*. Cheltenham: Edward Elgar, pp. 149–172.

Meyer, M. K. and Prugl, E. (eds.) (1999) *Gender Politics in Global Governance*. Lanham: Rowland & Littlefield.

Rochette, A. (2017) 'The Integration of Gender in Climate Change Mitigation and Adaptation in Quebec', in Griffin Cohen, M. (ed.) *Climate Change and Gender in Rich Countries: Work, Public Policy and Action*. London: Earthscan from Routledge, pp. 250–265.

UN BDPA (1995) Beijing Declaration and Platform for Action [online]. Available at: https://www.unwomen.org/-/media/headquarters/attachments/sections/csw/pfa_e_fina l_web.pdf?la=en&vs=800 (Accessed: 29 July 2020).

UN CEDAW (1979) Convention on the Elimination of All Forms of Discrimination against Women UNGA Res 34/180 of 18 December 1979 [online]. Available at: https ://www.un.org/womenwatch/daw/cedaw/text/econvention.htm (Accessed: 29 July 2020).

UNDP (2015) Sustainable Development Goals [online]. Available at: http://www .undp.org/content/undp/en/home/sustainable-development-goals.html (Accessed: 04 August 2020).

UNEP (2018) Global Gender and Environment Outlook [online]. Available at: https://we docs.unep.org/bitstream/handle/20.500.11822/14764/Gender_and_environment_ou tlook_HIGH_res.pdf?sequence=1&isAllowed=y (Accessed: 22 July 2020).

UN FCCC (1992) United Nationals Framework Convention on Climate Change [online]. Available at: https://unfccc.int/resource/docs/convkp/conveng.pdf (Accessed: 23 July 2020).

UN FCCC (2001) FCCC/CP/2001/13/Add.4 Decision 36/CP.7 Improving the Participation of Women in the Representation of Parties in Bodies Established under the United Nations Framework Convention on Climate Change or the Kyoto Protocol [online]. Available at: https://unfccc.int/resource/docs/cop7/13a04.pdf (Accessed: 23 July 2020).

UN FCCC (2012) Decision 23/CP.18 on 'Promoting Gender Balance and Improving the Participation of Women in UNFCCC Negotiations and in the Representation of Parties in Bodies Established Pursuant to the Convention or the Kyoto Protocol' FCCC/ CP/2012/8/Add.3 [online]. Available at: http://unfccc.nt/files/bodies/election_and_me mbership/application/pdf/cop18_gender_balance.pdf (Accessed: 23 July 2020).

UN FCCC (2013) Workshop on Gender, Climate Change and the UNFCCC Workshop on Gender, Climate Change and the UNFCCC [online]. Available at: http://unfccc .int/files/adaptation/application/pdf/in_session_workshop_agenda_web.pdf (Accessed: 23 July 2020).

UN FCCC (2014) Decision 18/CP.20 The Lima Work Programme on Gender Online FCCC/CP/2014/10/Add 3 [online]. Available at: https://unfccc.int/sites/default/files/r esource/docs/2014/cop20/eng/10a03.pdf (Accessed: 04 August 2014).

UN FCCC (2015a) Draft Paris Agreement FCCC/ADP/2015/L.6 [online]. Available at: https://unfccc.int/files/bodies/awg/application/pdf/draft_paris_agreement_5dec15.pdf (Accessed: 28 July 2020).

UN FCCC (2015b) The Paris Agreement FCCC/CP/2015/10/Add.1 [online]. Available at: https://unfccc.int/resource/docs/2015/cop21/eng/10a01.pdf (Accessed:28 July 2020)

UN FCCC (2016) Decision 21/CP. 22 Gender and Climate Change [online]. Available at: https://unfccc.int/sites/default/files/resource/docs/2016/cop22/eng/10a02.pdf (Accessed: 04 August 2017).

UN FCCC (2017) Decision 3/CP.23 Establishment of a Gender Action Programme FCCC/CP/2017/11/Add.1 [online]. Available at: https://unfccc.int/resource/docs/20 17/cop23/eng/11a01.pdf#page=13 (Accessed: 04 August 2020).

UN FCCC TP (2018) Entry Points for Integrating Gender Considerations into UNFCCC Workstreams FCCC/TP/2018/1 [online]. Available at: https://unfccc.int/sites/default/files/resource/01.pdf (Accessed: 31July 2020).

UN FCCC (2019a) Progress in Integrating a Gender Perspective in Constituted Body Processes (Synthesis Report) FCCC/CP/2019/8 [online]. Available at: https://unfccc.int/sites/default/files/resource/cp2019_08E.pdf (Accessed: 03 August 2019).

UN FCCC (2019b) Implementation of the Lima Work Programme on Gender and Its Gender Action Plan FCCC/SBI/2019/15/Add.1 [online]. Available at: https://unfccc.int/sites/default/files/resource/sbi2019_15E.pdf (Accessed: 04 August 2020).

UN FCCC (2019c) Decision 3/CP.25 Enhanced Lima Work Programme on Gender and Its Gender Action Plan FCCC/CP/2019/3/Add.1 ADVANCE VERSION [online]. Available at: https://unfccc.int/sites/default/files/resource/cp2019_13a01_adv.pdf (Accessed: 05 August 2020).

UN FCCC (2020) Integrating Gender into National Climate Actions—Online Regional Workshops [online]. Available at: https://unfccc.int/news/integrating-gender-into-na tional-climate-actions-online-regional-workshops (Accessed: 31 July 2020).

UN FCCC (undated) UNFCCC Topics [online]. Available at https://unfccc.int/topics (Accessed: 28 July 2020).

UN FCCC (undated) Gender and Climate Change Documents [online]. Available at: https://unfccc.int/topics/gender/resources/documentation-on-gender-and-climate-c hange (Accessed: 12 October 2020).

UN GA (2015) A/Res/70/1 Transforming Our World: The 2030 Agenda for Sustainable Development, 21 October 2015 [online]. Available at: http://www.un.org/ga/search/view :doc.asp?symbol=A/RES/70/1&referer=/english/&Lang=E (Accessed: 04 August 2020).

UN OHCHR (2019) Born Free and Equal. 2nd Edition [online]. Available at: https://www.ohchr.org/Documents/Publications/Born_Free_and_Equal_WEB.pdf (Accessed: 22 July 2020).

UN OHCHR (2014) Women's Rights are Human Rights [online]. Available at: https://www.ohchr.org/Documents/Events/WHRD/WomenRightsAreHR.pdf (Accessed: 22 July 2020).

UN OHCHR (2019) Human Rights Obligations Relating to the Enjoyment of a Safe, Clean, Healthy and Sustainable Environment A/74/161 [online]. Available at: https://undocs.org/A/74/161 (Accessed: 31 July 2020).

UN OHCHR (undated) International Human Rights Law [online]. Available at: https://www.ohchr.org/en/professionalinterest/Pages/InternationalLaw.aspx (Accessed: 31 July 2020).

UN SPFII Et al (2010) Gender and Indigenous Peoples [online]. Available at: https://www.un.org/esa/socdev/unpfii/documents/Briefing%20Notes%20Gender%20and%20Indi genous%20Women.pdf (Accessed: 05 August 2020).

UN Women (2012) 8th Conference of the Parties to the United Nations Framework Convention on Climate Change [online]. Available at: https://www.unwomen.org/e n/news/stories/2012/12/cop18-landmark-decision-adopted (Accessed: 23 July 2020).

Wängnerud, L. (2009) 'Women in Parliaments: Descriptive and Substantive Representation', *Annu. Rev. Polit. Sci.*, 12, pp. 51–69.

WHO (undated) Gender [online]. Available at: https://www.who.int/health-topics/gender (Accessed: 22 July 2020).

3 EU external climate policy

Gill Allwood

Introduction

Climate change has arrived centre-stage on the EU's policy agenda. The new President of the Commission, Ursula von der Leyen, announced in 2019 that Europe would be the first climate-neutral continent by 2050 and introduced the European Green Deal and the first EU climate laws as part of the project to achieve this goal. Externally, the EU plays a major role in international climate negotiations, partly to meet its global leadership aspirations, and partly to ensure that European manufacturing is not undercut by producers with lower environmental standards. Climate change also features prominently in EU foreign policy, where it is framed as part of a series of nexuses, connecting it with migration, security and conflict. The need to address climate change rests against a backdrop of crisis – economic, environmental, health and political – building a sense of urgency and strategic prioritisation. One of the consequences of this is the sidelining of gender equality, despite the EU's repeated assertions of its commitment to mainstreaming the goal of gender equality throughout all its internal and external activities.

This chapter draws on feminist institutionalism and the literature on policy integration, including gender mainstreaming, to show how and why gender is excluded from EU external climate policy. It asks, firstly, where, in external climate policy, do we find references to gender and what, if anything, do they contribute to achieving gender-just climate policy. Secondly, it asks how feminist institutionalism and policy integration studies can help us understand why gender equality is not mainstreamed in EU external policy and what institutional obstacles prevent its integration. I argue that gender has been excluded from EU external climate policy by a combination of institutional power struggles; a discourse of crisis and security, which pushes gender into the background; and a proliferation of nexuses and mainstreaming imperatives in which the treaty obligation to mainstream gender is pushed to one side.

EU climate policy

EU climate policy has three main components: mitigation, adaptation and climate diplomacy. Mitigation refers to strategies for reducing climate change,

DOI: 10.4324/9781003052821-3

largely through the reduction of greenhouse gas emissions. Adaptation refers to strategies for adapting to the effects of climate change, such as increased flooding, droughts and unpredictable weather patterns. Climate diplomacy refers to the EU's role in international climate negotiations and agreements, an area in which it has sought to assert leadership. However, financial, economic, and political crises from 2008 slowed internal climate policy, and internal opposition to strong climate action has grown, particularly from central and eastern European member states, including Poland (Dupont and Oberthür, 2015, p. 229). While initially, the EU's ambitions exceeded its own internal practice, since the mid-2000s it has tried to 'lead by example'. The EU, along with its member states, is a signatory to the United Nations Framework Convention on Climate Change (UNFCCC) and plays a key role in trying to reach agreements on global targets (Biedenkopf and Dupont, 2013). The EU was an influential player in the 2015 Paris Agreement, the first universal, legally binding climate agreement and continues to try to push global targets upwards (Oberthür and Groen, 2018).

Key policy frameworks are the Climate and Energy Package for the period 2020–2030 (COM [2014] 15 final), which sets out targets for greenhouse gas emissions reductions, renewable energy and energy efficiency; and the Environmental Action Programme, which provides an overarching framework for all environmental and climate policy. The European Green Deal (2019) sets out a strategy for achieving net-zero greenhouse gas emissions by 2050 while sustaining economic growth. The first 'climate law' (COM [2020] 80 final) was proposed in early 2020, aiming to enshrine in legislation the objective of climate neutrality by 2050.

Since 2013, internal EU climate policy has also included an adaptation strategy, in recognition that climate change is having an impact within the EU, as well as, more obviously, elsewhere. While climate change mitigation was more readily framed as an issue to be dealt with at the EU level, adaptation to the effects of climate change appeared, until recently, to require local responses or to be of concern only in countries hit most severely by the impact of climate change, which are concentrated in the Global South. The floods and heatwaves of the early 2000s raised awareness of the impact of climate change within the EU and of its cross-border nature (Rayner and Jordan, 2010). As a consequence, the EU's adaptation strategy was adopted in 2013 (COM(2018) 738 final). Member states are encouraged to produce national adaptation strategies, setting out, for example, how they will climate-proof their transport, energy and agriculture sectors, and protect their populations from flooding, droughts and heatwaves. As part of EU external relations, however, adaptation has a longer history. The visible impact of climate change in developing countries, and the use of development aid for adaptation purposes, means that climate change has long been prominent in EU development policy. A Commission Communication in 2003 (COM [2003] 85 final) declared climate change a development, as well as an environmental, problem. This has implications for the study of gender and EU climate policy, as will be demonstrated later. The well-established nature of gender and development within the European Commission and the European

Parliament (especially its FEMM Committee on Women's Rights and Gender Equality and the Committee on Development) means that adaptation has been the core concern of gender and climate change analyses and policy proposals emanating from these institutions.

The EU's internal and external climate policies are dealt with by different decision-making institutions and processes. The European Commission is responsible for drafting EU legislative proposals. The newly appointed President, Ursula von der Leyen, has declared climate change one of her top priorities, and the European Green Deal (2019) was introduced as one of the new Commission's first actions. Internal EU climate policy is proposed by the Commission Directorate-General for Climate Action, DG CLIMA, which was created in 2010. However, climate change cannot be addressed by discrete policy measures. Energy, environment, agriculture, transport and trade all play a crucial role. Successfully meeting climate targets requires other policy sectors to integrate climate change into their work, as will be discussed later in this chapter.

Externally, climate change appears in foreign, security, external migration and international development policy. Since the Paris Agreement (2015), external climate policy has included supporting partner countries with the formulation and implementation of their Nationally Determined Contributions (European Commission, 2019, p. 62). The EU's Global Climate Change Alliance Plus initiative (GCCA+) has fostered policy dialogue with, and support for climate action in partner countries, mainly Least Developed Countries (LDCs) and Small Island Developing States (SIDS) (European Commission, 2019, p. 64). Before the Lisbon Treaty (2009), climate change, like other EU policy issues, was expected to align with the principle of Policy Coherence for Development. Policy Coherence for Development is a treaty obligation and is intended to prevent EU policies in areas other than international development undermining development objectives. This applies to all EU activities, including climate action. Impact assessments are supposed to be conducted to ensure that proposed actions will not have adverse effects on developing countries. However, the relative weakness of development policy institutions within the EU; the hierarchy of political priorities; and the poor implementation of the impact assessment obligation mean that Policy Coherence for Development has been more of a rhetorical commitment than an accomplishment. The Lisbon Treaty (2009) brought about a shift in emphasis, with development objectives being set alongside new strategic priorities, such as security and migration, and then gradually losing place to them. Development policy is increasingly required to serve the interests of a security-focused foreign policy. According to the Global Strategy (High Representative of the Union for Foreign Affairs and Security Policy 2016), which is the overarching statement of EU foreign policy, 'Development policy needs to become more flexible and aligned with our strategic priorities'. EU foreign and security policy frames climate change as a security threat and a root cause of migration (High Representative of the Union for Foreign Affairs and Security Policy 2016, p. 27). The Council Conclusions on Climate Diplomacy of 26 February 2018 state that: 'without decisive action [climate change and environmental degradation] will become an even greater source of global risk, including forced displacement and migration'. Development policy is now expected to mainstream climate change adaptation to render partner countries more resilient when coping with climate change consequences in order to address the root causes of the migration crisis. The European Consensus on Development (2017) calls for climate change integration across all sectors of development cooperation. It commits to addressing the root causes of migration, including climate change. This is a reversal of the expectation prior to the Lisbon Treaty that climate action would be coherent with development policy objectives.

The European Commission's framing of the EU's global role stresses the importance of access to

> partner countries' markets, infrastructure and critical raw materials. This starts with enhancing the EU's energy and climate diplomacy, and with further mainstreaming climate change objectives and considerations in political dialogues, including in the area of migration, security and development cooperation.
>
> (COM [2018]773 final)

Climate mainstreaming is seen as a way to achieve the EU's trade, security and migration priorities, and is, therefore, a core priority for external action.

Where does gender fit into EU external climate policy?

This section of the chapter gives an overview of EU gender equality policy, showing where it intersects with climate policy. Focusing on external climate policy, it then asks where we can find references to gender and what, if anything, they contribute to achieving gender-just climate policy.

The EU is committed to mainstreaming gender throughout all of its internal and external activities. Article 8 of the TFEU states: 'in all its activities, the Union shall aim to eliminate inequalities, and to promote equality, between men and women'. EU institutions still tend to frame gender equality as 'equality between men and women', ignoring the heterogeneity of these two categories, the power relations within, as well as between, them, and the intersection of other structural inequalities with gender. There are institutional pockets in which a more nuanced gender analysis emerges. These will be discussed where they relate to external climate policy. The EU's approach to gender equality is contained in its Gender Equality Strategy, the most recent version of which was launched in March 2020. The strategy makes some move away from a focus on equality between women and men, aiming to achieve

> a Europe where women and men, girls and boys, in all their diversity, are equal ... The Commission will enhance gender mainstreaming by systematically including a gender perspective in all stages of policy design in all EU

policy areas, internal and external. The Strategy will be implemented using intersectionality as a cross-cutting principle.

(p. 2)

Although there is no detail about how this will be done, the recognition of intersectionality is an important step away from equating gender equality solely with a men–women binary. An intersectional analytical lens has been used by feminist scholars to make important contributions to our understanding of the impact of climate change and responses to it (Nagel, 2012; Tschakert and Machado, 2012; Kaijser and Kronsell, 2014; MacGregor, 2014; Sultana, 2014). Intersectionality can help us understand individual and group-based differences in relation to climate change. Rather than designating women as vulnerable victims of climate change, an intersectional approach demonstrates that social structures based on characteristics such as gender, socio-economic status, ethnicity, nationality, health, sexual orientation, age and place influence the responsibility, vulnerability and decision-making power of individuals and groups. For example, Kronsell's (2013) study of Sweden shows that there are gendered differences in energy consumption and transportation. She argues, however, that it is important to recognise that gender is not the only relevant factor; sometimes class matters more than gender and there are considerable differences within the Global North and Global South. An intersectional analysis asks which social categories are included in, and excluded from, the cases in question, what assumptions are made about social categories, and what type of knowledge is privileged (Kaijser and Kronsell, 2014). Buckingham and Le Masson (2017, pp. 2–3) argue that,

> If gender, and gender equality, is to be a meaningful policy objective, it must be recognised that it comprises relations between women and men, and between and among different groups of women and men, not to mention between different conceptualisations of masculinity and femininity, which can each be practised by either, and both, women and men … Understanding gender inequality also requires a simultaneous understanding of social and economic divisions based on class, ethnicity, age, disability, religion, sexuality, parenthood, among others, and how these divisions intersect to compound particular disadvantages and inequalities between and within social groups.

Internally, the EU's concern with gender equality has focused primarily on equal pay for equal work, a principle enshrined in the treaties since the Treaty of Rome (1957). The EU has been committed to gender mainstreaming since 1996, meaning that gender equality should be integrated into all areas and at all stages of policy-making. However, key areas of activity remain untouched, including energy, trade and transport. The European Parliament's FEMM Committee has highlighted many of these gaps in a series of reports and resolutions, as has the European Institute for Gender Equality (EIGE). For example, EIGE (2016), finds that the European Commission has only just begun to recognise the links between gender and transport, and according to the European Parliament (2018), 'current EU trade policy and its "Trade for All" strategy … lack a gender equality perspective'.

Gender equality and women's empowerment are declared core objectives of EU external action (Gender Equality Strategy 2020–2025). Since 2015, the second Gender Action Plan (GAPII), has required all actors in EU external action to implement and report on gender mainstreaming, specific actions and political dialogue in relation to gender equality. GAPII also introduced a cross-cutting priority: Institutional Cultural Shift. This was based on the recognition that gender equality will only be achieved once there is a significant change in institutional culture. It will be discussed in the next section of this chapter. The original Gender Action Plan (2010) was created within the Directorate-General (DG) for Development and aimed to translate gender and development commitments into effective policy outcomes. This illustrates the importance of EU development policy institutions in raising awareness of gender equality, gender mainstreaming and gender analysis. The strong influence within DG Development of Gender and Development experts and advocates has resulted in development cooperation being the most gender-aware area of EU policy. In the wake of the Lisbon Treaty, the creation of the European External Action Service, and the restructuring of DG DEVCO, the second Gender Action Plan, GAPII (2015–2019), applied to all areas of EU external action, not just development policy, and to all actors, including EU headquarters in Brussels, EU delegations in partner countries and member states. This shift from development to external action has diluted the influence of gender and development activists and advocates on the framing of gender equality issues, and gender equality remains less well integrated into areas of external action outside international development.

Development policy documents continue to enshrine a long-standing commitment to gender equality. For example, the European Consensus on Development (2017, para 15) states that:

> gender equality is at the core of the EU's values and is enshrined in its legal and political framework. It is vital for achieving the SDGs and cuts across the whole 2030 Agenda. The EU and its Member States will promote women's and girls' rights, gender equality, the empowerment of women and girls and their protection as a priority across all areas of action.

In line with the UN's 2030 Agenda and with growing awareness of diversity and intersectionality, the European Consensus on Development (2017, para 16) moves beyond a focus on gender alone:

> the EU and its Member States will continue to play a key role in ensuring that no-one is left behind, wherever people live and regardless of ethnicity, gender, age, disability, religion or beliefs, sexual orientation and gender identity, migration status or other factors. This approach includes addressing the multiple discriminations faced by vulnerable people and marginalised groups.

Statements of commitment to gender mainstreaming are pervasive in EU foreign and security policy, including external climate policy. The Global Strategy (High Representative of the Union for Foreign Affairs and Security Policy 2016, 11 and 51), states that: 'we must systematically mainstream human rights and gender issues across policy sectors and institutions'. The Council Conclusions on Climate Diplomacy of 26 February 2018, state that: 'gender equality, women's empowerment and women's full and equal participation and leadership are vital to achieving sustainable development, including climate change adaptation'. The Council Conclusions on Climate Diplomacy of 18 February 2019 state that: 'the EU will continue to uphold, promote and protect human rights, gender equality and women's empowerment in the context of climate action'. The European Council Strategic Agenda 2019–2024, sets out as one of its four main priorities a climate-neutral, green, fair and social Europe, stating that 'Europe needs inclusiveness and sustainability, embracing the changes brought about by the green transition, technological evolution and globalisation, while making sure no-one is left behind'. In this section, it adds: 'we need to do more to ensure equality between women and men, as well as rights and opportunities for all. This is both a societal imperative and an economic asset'.

These statements can be read as all-embracing or as add-ons to unchanged gender-blind policy. They are now systemic in EU external policy and can provide leverage for gender equality advocates, both inside and outside EU institutions, but they can also suggest that the problem has been resolved, removing incentives to address it.

In contrast, the European Green Deal (2019) is completely gender blind. It makes no reference to gender/women, although it does say that the transition to a climate-neutral economy must be 'just and inclusive' and must 'put people first' (p. 2). The European Green Deal mostly concerns internal policy, but a section on 'the EU as a Global Leader' stresses the importance of increasing climate resilience in partner countries to avoid 'conflict, food insecurity, population displacement and forced migration'. The EU's efforts to 'lead by example' in climate action are important, but the gender blindness of climate policy means that any norms it exports will also be gender blind.

One of the reasons why gender is absent from the European Green Deal is that (gender) equality and climate action are presented as two of the new Commission's top priorities, but they are kept separate from one another. Elements of gender-awareness in relation to climate change exist in fragments of EU policy, but they lack coherence. The Gender Equality Strategy has a short section on climate change which points out some of the ways in which climate change is gendered and argues that, 'Addressing the gender dimension can therefore have a key role in leveraging the full potential of these policies'. GAPII refers to climate issues only twice. The first reference states that the Commission's services and European External Action Service will contribute to 'women's increased participation in decision-making processes on climate and environmental issues' and Objective 20 in Annex 1 is: 'equal rights enjoyed by women to participate in and influence decision-making processes on climate and environmental issues'.

This is the objective least often reported upon. There are also reports on gender and climate change by EIGE and the European Parliament, some of which are limited to counting the numbers of women in climate decision-making, and some of which contain sophisticated gender analyses of climate issues and responses. The EIGE (2019) report, for example, reiterates the EU's obligations to integrate gender equality into its climate change policies, in line with the UNFCCC Lima Work Programme on Gender and the 2017 Gender Action Plan, as well as with the EU's own gender mainstreaming commitments. However, EIGE argues that these have not been translated into concrete actions, and EU climate policy has remained largely gender blind.

To summarise, climate change is mentioned briefly in gender equality documents, such as GAPII and the Gender Equality Strategy, but the major climate framework document of 2019, the European Green Deal, is gender blind. The Social Development Goals (SDGs) offer the potential to bring gender equality and climate change together, but this has not yet been realised, including within EU policy-making.

A feminist institutionalist analysis of gender and EU external climate policy

The previous section has shown that gender equality is not mainstreamed in EU external climate policy. Can feminist institutionalism help us to understand why this is the case? Feminist institutionalism tells us that institutions matter, that they are ridden with gendered power relations, and that they are resistant to change. Feminist institutionalism (Mackay et al., 2009; Krook and Mackay, 2011; Chappell and Waylen, 2013; Waylen, 2013) can help explain the gap between formal commitments to gender mainstreaming and gender equality in all policy areas and at all stages of policy-making and, on the other hand, persistently gender-blind policy in particular areas, in this case, climate change. It enables us to examine the institutional constraints, opportunities and resistances that affect gender mainstreaming within climate change policy-making. Feminist institutionalism emphasises the importance of informal practices, norms and values, exposing the ways in which they can constrain or distort formal rules (Chappell and Waylen, 2013; Waylen, 2013). A focus on the informal rules of the game provides clues that can contribute to explaining the gap between the rhetoric and reality of gender mainstreaming in specific areas of EU policy. Focusing on the relation between formal rules and informal practices can help us understand why gender mainstreaming – which is formally compulsory in all policy areas and all stages of policy-making – is ignored, overlooked, pushed down the agenda or out to the margins, while the main business of climate change policy-making continues unperturbed. Drawing on sociological institutionalism, feminist institutionalism suggests that actors are constrained by cultural conventions, norms and cognitive frames of reference which privilege a certain way of thinking about a policy problem and ensure that other perspectives remain submerged from view (Lowndes and Roberts, 2013, p. 30). Feminist discursive institutionalism also enables us to focus on the construction and contestation of meaning in the interaction between gender mainstreaming and climate change policies (Schmidt, 2012). Gender mainstreaming is interpreted and re-interpreted in day-to-day institutional inter-actions. Individual and collective actors engage in struggles to impose their under-standings of gender mainstreaming, and this is affected by the broader context of institutional power imbalances that push issues such as gender equality to the centre or the margins of particular policy debates. It can reveal ways in which gender mainstreaming is imbued with new meanings in day-to-day policy-making practices. It can highlight the ways in which issues are constructed as certain types of problem requiring certain types of solution. This can act as a constraint on those pushing other meanings. The special issue of *Political Studies Review* on gender and external action shows that in times of crisis, gender equality is pushed off the agenda (Muehlenhoff et al., 2020). Studies on the gendered impact of COVID-19 have revealed the same effect (John et al., 2020). Crisis discourse wrongly pro-claims that 'we are all in this together', whereas Cynthia Enloe (2020) insists that

> we are not all in this together. We're on the same rough seas, but we're in very different boats. And some of those boats are very leaky. And some of those boats were never given oars. And some of those boats have high-powered motors on them. We are not all in the same boat.

The EU's institutions have very different institutional cultures. The European Parliament, for example, has traditionally been greener and more gender equal-ity-friendly than the Council, but can be squeezed out of decision-making, par-ticularly in areas where member states seek to retain control. The Council has retained exceptional control of climate policy-making, meaning that the mem-ber states have more influence and the European Parliament has less influence than is set out in the treaties (Dupont, 2019). Climate and gender equality lag-gards, often from central and eastern Europe, can water down policy in both of these areas. EU decision-making is famously sectoral, although some cross-sectoral structures exist, for example, the European Parliament's gender main-streaming group and the promised Task Force for Equality in the Commission. Power relations between external climate institutions affect their relative ability to frame climate change. For example, the European Parliament Development Committee and the European Commission's Directorate-General for develop-ment are relatively weak. In contrast, security and migration institutions and frames are dominating the external climate policy agenda (Youngs, 2014). The Lisbon Treaty enhanced external action, but reduced the profile of development within it. There is also a gap between policy formulation in Brussels and imple-mentation by the EUDs in the partner countries. As Debusscher and Manners (2020) point out, it is not just about what happens in the institutions in Brussels. External climate policy is also made and implemented in the delegations in the partner countries and in partnership with the partner countries. Studies have shown that gender mainstreaming is not consistently present in these institutions and processes (Allwood, 2018).

Institutional resistance to gender-sensitive change was recognised in GAPII which applies to all external action and contains an important cross-cutting priority: Institutional Cultural Shift. This means 'shifting the Commission services and the European External Action Service's institutional culture to more effectively deliver on EU commitments'. Institutional Cultural Shift was based on the idea that change will only be brought about once a systemic shift has taken place. This aligns with Rao and Kelleher's (2005) work on gender and organisations, which argues that effective gender mainstreaming needs transformational institutional change. It is a response to the recognition that organisations matter and that they are 'sticky'. Path dependency might explain why climate change continues to be framed as a scientific, elitist, technical and masculinised issue. As Cornwall and Rivas (2015, p. 400) argue, 'Ultimately, a paradigm transformation is needed to reclaim the gender agenda and address the underlying structures of constraint that give these inequalities the systemic character and the persistence over time'.

Institutional Cultural Shift and changing mindsets are essential for gender transformation, but will not be effective without political will. Gender composition at management level in the European External Action Service is worse than in the European Parliament, European Commission and Council. Chappell and Guerrina (2020, p. 8) find that,

> A common normative stance within the European External Action Service has not yet fully evolved, indicating that where a gender perspective does occur, it is unlikely to have dispersed across the European External Action Service. Hence, there may well be 'pockets' of gender actorness, but not full gender mainstreaming. From an institutional perspective, this relates not only to how many women are working within the European External Action Service but also whether officials are looking at EU foreign policy through a gender lens.

Efforts to integrate gender equality in the institutions continue. The European External Action Service Gender and Equal Opportunities Strategy 2018–2023,

> aims to achieve gender equality in the European External Action Service in the broadest sense, whereby women and men enjoy equal rights, equal obligations and equal opportunities across the Service ... through accelerated progress towards gender balance and accelerated Institutional Cultural Shift in the European External Action Service.
>
> (p. 6)

This is justified in terms of productivity and effectiveness:

> measures will be put in place for the sustainable transformation of institutional reflexes and individual mindsets. Transformational measures will aim to counteract stereotypes, unintended bias and prejudices while at the same

time fostering inclusivity, diversity and collaboration through training, mentoring, coaching, networking and innovative learning and through providing equal development opportunities for women and men.

(p. 8)

However, the European Parliament draft report on Gender Equality in the EU's Foreign and Security Policy 2019/2167 (INI), of 27 March 2020, calls for more precision and specific targets on gender and diversity in this strategy. It also calls for member states to create a formal Council working group on gender equality, argues that the European External Action Service Principal Advisor on gender equality requires more staff and resources, and calls on the Vice President/High Representative to ensure that heads of European Union Delegations (EUDs) abroad ensure that gender equality is mainstreamed throughout all of the delegation's work and that there is a full-time gender focal point in each delegation.

An additional obstacle to mainstreaming gender throughout all EU external climate policy is that the latter is itself constructed as a cross-cutting issue. Gender mainstreaming is implemented more easily in discrete policy sectors and is more difficult to apply to other cross-cutting issues. There is a long-standing and widespread understanding that some objectives cannot be reached by treating them as standalone goals, but that they need to be woven into all areas of decision-making and at all stages. This is referred to as mainstreaming (as in gender mainstreaming), policy integration (as in environmental policy integration), or policy coherence (as in policy coherence for development). Collectively, they can be referred to as horizontal policy coordination. Gender mainstreaming, environmental policy integration and policy coherence for development are all treaty-based obligations. As policy-makers increasingly refer to cross-cutting issues and to policy nexuses, we need a way to understand and improve how they intersect.

We can draw on the literature on horizontal policy coordination and policy nexuses to try to understand how policy issues interconnect and what happens when they do. This literature considers how cross-cutting issues are integrated into policy sectors. It shows where this works and where it fails, and suggests explanations for these outcomes (De Roeck et al., 2018). It suggests that successful policy coherence requires a strong shared vision which acts as a strategic goal and maintains focus on the objective, and not on the procedural tools and instruments. The substantial literature on gender mainstreaming shows that, despite repeated rhetorical commitments by EU actors, it is still absent from key policy areas and is often treated as procedural, rather than substantive (Meier and Celis, 2011; Allwood, 2013; Guerrina and Wright, 2016). Kok and de Corninck's (2007, pp. 587–599) study of climate change mainstreaming shows that organisational structures were not designed for cooperation, coordination and joint decision-making on different levels. There are power imbalances between different Commission DGs; between different configurations of the Council of Ministers; and between the Council, the European Parliament and the Commission. The European Parliament, and in particular its various committees on the environment, development and gender equality, have been increasingly active in advocating the mainstreaming of these issues throughout all European Parliament decision-making, but the European Parliament can be excluded from forms of decision-making dominated by intergovernmentalism, and this applies to most of the Union's climate change policy. Power imbalances and inter-institutional rivalries mean that issues such as environmental protection can struggle to impinge on policies shored up by powerful economic interests such as trade and agriculture. Institutional resistance, often based on powerful economic interests, is identified as the main obstacle by Gupta and van der Grijp (2010), in their study of climate change mainstreaming. Climate change mainstreaming threatens the status quo and unsettles the vested interests of industry and the energy lobby. Resistance is therefore strong. Any policy competition or struggle for scarce resources will expose these imbalances, and rhetorical commitment to mainstreaming may lack underlying substance, particularly in times of economic crisis.

EU external policy increasingly uses the term 'nexus' to describe the intersection between two or more policy areas (Lavenex and Kunz, 2009; Carbone, 2013; Furness and Gänzle, 2017; De Roeck et al., 2018). Climate change is situated in a series of nexuses, including climate-security and climate-migration. The Council Conclusions on Climate Diplomacy of 26 February 2018 'resolve … to further mainstream the nexus between climate change and security in political dialogue, conflict prevention, development and humanitarian action and disaster risk strategies'. While this is an important recognition that policy issues are intersecting and cannot be addressed in isolation from each other, it raises substantial questions about how gender can be mainstreamed throughout other cross-cutting issues. Allwood (2020a) found that gender equality is absent from the migration-security-climate nexuses which are driving development policy priorities. This makes mainstreaming even more challenging. Not only are climate change, migration and other cross-cutting issues to be mainstreamed in development cooperation and policy dialogue, but the nexus between them must also be mainstreamed. This raises questions about the practicalities of addressing complex webs of intersecting issues, especially when some of them are accorded priority status. It also creates a context in which the mainstreaming of gender becomes even more difficult.

Conclusion

This feminist institutionalist analysis of EU external climate policy shows that the framing of climate change as a policy problem affects which institutions address it and which solutions are proposed. When climate change was framed as a development issue, decisions were made by development institutions in the context of policy coherence for development. The relatively strong influence of gender and development in DG DEV meant that there was some chance of external climate policy being tinged with gender awareness. Following the Lisbon Treaty, the creation of the European External Action Service, and the restructuring of DG

DEVCO, foreign and security policy gained prominence, and development policy became subservient to its priorities. Given that gender was relatively prominent in development policy, this contributed to gender slipping down the agenda. Crisis discourse accentuated this trend (Muehlenhoff et al., 2020).

External relations gained prominence on the EU's political agenda with the adoption of the Lisbon Treaty. The creation of the role of High Representative and the European External Action Service marked a new era in foreign and security policy. These new institutional arrangements had consequences for both climate policy and gender equality. The new external agenda prioritised migration and security, and climate change was constructed as an external issue in relation to these priorities. The EU makes much of its foundational myth of gender equality (MacRae, 2010), and of the fact that gender equality is a fundamental value of the Union. Commitments to gender equality and to gender mainstreaming as a means of achieving it, abound in EU policy documents, both internal and external. Internal gender equality policy is closely linked to equality between women and men in the workplace. Present in the treaties since the foundation of the Common Market in 1957, equal pay for equal work (and later, work of equal value) has been at the heart of the EU's gender equality narrative, even as the principle has extended to other areas of EU policy and has slowly begun to include other forms of inequality, discrimination and exclusion. Gender equality in external relations was, for many years, most visible in relation to development cooperation, which was influenced by gender and development theory and practice. Gender awareness was higher than in other parts of EU external action and the Gender Action Plan of 2010, demonstrated a desire to implement commitments to gender equality on the ground. This constituted an institutional context in which climate change as an issue affecting developing countries and the EU's relations with them could potentially begin to be addressed in a gender-just fashion. However, the reframing of climate change as a migration and security issue has acted as an obstacle to this.

The construction of climate change as a problem which can be solved with market, technological and security solutions has, until recently, excluded a people-centred approach, which could favour a gender-sensitive policy. There are signs that this is beginning to change very slowly. EU climate policy is edging away from an exclusive focus on technological solutions towards a recognition that climate change affects people, and that people are part of the solution. However, integrating diversity and intersectionality into the analysis of climate change and proposed responses to it is still a marginal concern (Allwood, 2020b). Efforts to address gender inequality and efforts to address climate change continue to exist in parallel, rather than being fully integrated into each other. Gender equality is not integrated into all aspects of decision-making and at all stages. Instead, it is tagged on or addressed in separate documents and debates, in what Acosta et al. (2019, p. 15) refer to as a 'stale reproduction of set pieces of text [pointing to] significant levels of inertia in thinking and practice around gender mainstreaming issues'. This is a result of institutional stickiness, which makes it difficult to bring about a much-needed Institutional Cultural Shift.

In addition to contributing to our understanding of why gender has not been integrated into climate decision-making, this chapter also opens up the broader question of how we can integrate gender throughout all (intersecting) policy in a way that is transformative and sustainable (i.e. not just inserting the word gender into policy documents, as Sherilyn MacGregor fears (2014, p. 624). If, as the SDGs suggest, gender equality is a prerequisite for achieving sustainable development, peace and security, and climate change mitigation and adaptation, then it must be integrated into all of these policy areas, including the intersections between them.

References

Acosta M., van Bommel S. and van Wessel M. (2019) 'Discursive Translations of Gender Mainstreaming Norms: The Case of Agricultural and Climate Change Policies in Uganda', *Women's Studies International Forum*, 74, pp. 9–19.

Allwood G. (2013) 'Gender Mainstreaming and Policy Coherence for Development: Unintended Gender Consequences and EU Policy', *Women's Studies International Forum*, 39, pp. 42–52.

Allwood G. (2018) *Transforming Lives? CONCORD Report EU Gender Action Plan II: From Implementation to Impact*. Brussels: CONCORD.

Allwood G. (2020a) 'Gender Equality in European Union Development Policy in Times of Crisis', *Political Studies Review*, 18(3), pp. 329–345.

Allwood G. (2020b) 'Mainstreaming Gender and Climate Change to Achieve a Just Transition to a Climate-Neutral Europe', *Journal of Common Market Studies*. https://doi.org/10.1111/jcms.13082

Biedenkopf K. and Dupont C. (2013) 'A Toolbox Approach to the EU's External Climate Governance', in Boening A., Kremer J.-F., and van Loon A. (eds) *Global Power Europe Volume 1*. Berlin: Springer-Verlag, pp.181–199.

Buckingham S. and Le Masson V. (2017) 'Introduction', in Buckingham S. and Le Masson V. (eds) *Understanding Climate Change Through Gender Relations*. London and New York: Routledge, pp. 1–12.

Carbone M. (2013) 'International Development and the European Union's External Policies: Changing Contexts, Problematic Nexuses, Contested Partnerships', *Cambridge Review of International Affairs*, 26(3), pp. 483–496.

Chappell L. and Guerrina R. (2020) 'Understanding the Gender Regime in the European External Action Service', *Cooperation and Conflict*, 55(2), pp. 261–280.

Chappell L. and Waylen G. (2013) 'Gender and the Hidden Life of Institutions', *Public Administration* 91(3), pp. 599–615.

Cornwall A. and Rivas A. (2015) 'From "Gender Equality" and "Women's Empowerment" to Global Justice: Reclaiming a Transformative Agenda for Gender and Development', *Third World Quarterly*, 36(2), pp. 396–415.

De Roeck F., Orbie J. and Delputte S. (2018) 'Mainstreaming Climate Change Adaptation into the European Union's Development Assistance', *Environmental Science & Policy*, 81, pp. 36–45.

Debusscher P. and Manners I. (2020) 'Understanding the European Union as a Global Gender Actor: The Holistic Intersectional and Inclusive Study of Gender+ in External Actions', *Political Studies Review*, 18(3), pp. 410–425.

Dupont C. (2019) 'The EU's Collective Securitisation of Climate Change', *West European Politics*, 42(2), pp. 369–390.

Dupont C. and Oberthür S. (2015) 'The European Union', in Bäckstrand K. and Lövbrand E. (eds) *Research Handbook on Climate Governance*. Cheltenham: Edward Elgar, pp. 224–236.

European Commission (2019) *Supporting the Sustainable Development Goals across the World. Joint Synthesis Report of the European Union and its Member States 2019*.

Enloe, C. (2020) 'Why the Use of "War" Narratives are Dangerous in Times of COVID-19', Interview with Quaker Council for European Affairs. Brussels: QCEA, https://www.youtube.com/watch?v=Gaif6mTwFw8, accessed 18 February 2021.

European Institute for Gender Equality (EIGE) (2016) *Gender in Transport*. Vilnius: EIGE.

European Parliament (2018) *Report on Gender Equality in EU Trade Agreements, (2017/2015(INI)*, https://oeil.secure.europarl.europa.eu/oeil/popups/ficheprocedure.do?lang=en&reference=2017/2015(INI), accessed 18 February 2021.

Furness M. and Gänzle S. (2017) 'The Security-Development Nexus in EU Foreign Relations after Lisbon: Policy Coherence at Last?', *Development Policy Review*, 35(4), pp. 475–92.

Guerrina R. and Wright K. (2016) 'Gendering Normative Power Europe: Lessons of the Women, Peace and Security Agenda', *International Affairs*, 92(2), pp. 293–312.

Gupta J. and van der Grijp N. (2010) *Mainstreaming Climate Change in Development Cooperation: Theory, Practice and Implications for the European Union*, Cambridge: Cambridge University Press.

High Representative of the Union for Foreign Affairs and Security Policy (2016) *Shared Vision, Common Action: A Stronger Europe. A Global Strategy for the European Union's Foreign and Security Policy*, http://europa.eu/globalstrategy/en, accessed 18 February 2021.

John N., Casey S.E. and Carino G., (2020) 'Lessons Never Learned: Crisis and Gender-Based Violence', *Developing World Bioethics*, 20(2), pp. 65–68.

Kaijser A. and Kronsell A. (2014) 'Climate Change through the Lens of Intersectionality', *Environmental Politics*, 23(3), pp. 417–433.

Kok M. and de Coninck H. (2007) 'Widening the Scope of Policies to Address Climate Change: Directions for Mainstreaming', *Environmental Science and Policy*, 10, pp. 587-599.

Kronsell A. (2013) 'Gender and Transition in Climate Governance', *Environmental Innovation and Societal Transitions*, 7, pp. 1–15.

Krook M.L. and Mackay F. (2011) 'Introduction', in Krook Mona L and Mackay F (eds). *Gender, Politics and Institutions: Towards a Feminist Institutionalism*. Basingstoke: Palgrave, pp. 1–20.

Lavenex S. and Kunz R. (2009) 'The Migration-Development Nexus in EU External Rleations', in Carbone M. (ed). *European Integration. Policy Coherence and EU Development Policy*. Abingdon and New York: Routledge, pp. 439–457.

Lowndes V. and Roberts M. (2013) *Why Institutions Matter. The New Institutionalism in Political Science*. Basingstoke: Palgrave Macmillan.

MacGregor S. (2014). 'Only Resist: Feminist Ecological Citizenship and the Post-Politics of Climate Change', *Hypatia*, 29(3), pp. 617–633.

Mackay F., Monro S. and Waylen G. (2009) 'The Feminist Potential of Sociological Institutionalism', *Politics and Gender*, 5(2), pp. 253–262.

MacRae H. (2010) 'The EU as a Gender Equality Polity: Myths and Realities', *Journal of Common Market Studies*, 48(1), pp. 153–172.

Meier P and Celis K (2011) 'Sowing the Seeds of its Own Failure: Implementing the Concept of Gender Mainstreaming', *Social Politics*, 18(4), pp. 469–489.

Muehlenhoff H.L., van der Vleuten A. and Welfens N. (2020) 'Slipping Off or Turning the Tide? Gender Equality in European Union's External Relations in Times of Crisis', *Political Studies Review*, 18(3), pp. 322–328.

Nagel J. (2012) 'Intersecting Identities and Global Climate Change', *Identities*, 19(4), pp. 467–476.

Oberthür S. and Groen L. (2018) 'Explaining Goal Achievement in International Negotiations: The EU and the Paris Agreement on Climate Change', *Journal of European Public Policy*, 25(5), pp. 708–727.

Rao A. and Kelleher D. (2005) 'Is there Life after Gender Mainstreaming?', *Gender and Development*, 13(2), pp. 57–69.

Rayner T. and Jordan A. (2010) 'Adapting to a Changing Climate: An Emerging EU Policy?', in Jordan A, Huitema D, Van Asselt H. (eds). *Climate Change Policy in the European Union*. Cambridge: Cambridge University Press, pp. 145–166.

Schmidt V. (2012) 'Foreword', in Lombardo E and Forest M (eds). *The Europeanization of Gender Equality Policies. A Discursive-Sociological Approach*. Basingstoke: Palgrave Macmillan, pp. xiii–xvi.

Sultana F. (2014) 'Gendering Climate Change: Geographical Insights', *The Professional Geographer*, 66(3), pp. 372–381.

Tschakert P. and Machado M. (2012) 'Gender Justice and Rights in Climate Change Adaptation: Opportunities and Pitfalls', *Ethics and Social Welfare*, 6(3), pp. 275–289.

Waylen G. (2013) 'Informal Institutions, Institutional Change, and Gender Equality', *Political Research Quarterly*, 61(1), pp. 212–223.

Youngs R. (2014) *Climate Change and EU Security Policy: An Unmet Challenge*. Brussels: Carnegie Europe.

4 How to make Germany's climate policy gender-responsive

Experiences from research and advocacy

Gotelind Alber, Diana Hummel, Ulrike Röhr, and Immanuel Stieß

Introduction

A cornerstone to open the political perspective of the German climate policy to gender and gender justice was a research project titled: 'The contribution of gender justice to successful climate policy: impact assessment, interdependencies with other social categories, methodological issues and options for shaping climate policy' (Gender and Climate research project) commissioned by the German Federal Environment Agency (UBA). It was carried out from November 2016 to October 2019 by three partners: Wuppertal Institute Environment, Climate, Energy, ISOE-Institute for Social-ecological Research and GenderCC-Women for Climate Justice. The project aimed at reviewing existing findings on the gender and climate change nexus in industrialised countries, examining the added value of the gender perspective for climate policies and developing recommendations for action for the German national framework, in line with the UNFCCC provisions on gender. Moreover, the knowledge of obstacles towards a gender perspective in climate protection and adaptation was to be enhanced (Röhr et al., 2018; Spitzner et al., 2020). The study was based on the assumptions that, a) a gender perspective supports the development of policies tailored for different target groups, thereby improving the effectiveness of climate policy in particular in terms of acceptance; b) a gender-responsive climate policy can counteract gender inequality; and c) in this way, synergies can be created between climate protection/adaption and gender justice. In the first step, a systematic literature review of national and international literature on sex and gender aspects in climate-relevant areas of action in countries of the Global North, as well as an analysis of gender provisions and mandates accruing from the UNFCCC process, have been conducted (Röhr et al., 2018). In a second step, a 'Gender Impact Assessment' (GIA) has been conceptualised for climate policies and measures, and developed in a participatory process. As a third step, the prerequisites for gender analyses and requirements for gender-responsive climate policies have been determined in order to derive action and policy recommendations at the federal level. The outcomes of the research project resulted in some advocacy activities within the ministry (Spitzner et al., 2020).

DOI: 10.4324/9781003052821-4

The research project as well as the developments in the German Ministry for the Environment, Nature Protection and Nuclear Safety (BMU) were backed by the latest gender-outcomes of the UNFCCC, which urge all governments to develop gender-responsive climate policies.

In the following section, we provide an overview of how gender considerations evolved in German climate policy. Then, in the next section, we look at two potential paths to integrate gender into climate policy. Finally, we discuss these options and, in conclusion, suggest next steps for working towards gender-responsive climate policy.

The starting point: Climate policy and gender in Germany

German governments have generally considered themselves as forerunners in climate policy. Germany has been successful in presenting and advocating for ambitious climate goals and thereby pushing the climate agenda forward at EU and international levels, but it has not, however, been as successful in achieving these goals. It should, nevertheless, be noted that Germany is currently striving for emissions reductions of 55% by the year 2030 while at the same time phasing out nuclear power and also – in the longer term – the use of coal energy.

The Federal Ministry for the Environment (BMU), run by a Minister from the Social-Democratic Party, has the lead in national climate policy. Yet, it is not in charge of central fields of action for implementing the climate protection goals. Therefore, even if the BMU was a pioneer in terms of attempts to integrate gender into climate policy approaches and goals, it has little influence on the implementation in the various sectors. Energy policy, for instance, is located in the Ministry of Economic Affairs, while transport, agriculture, housing and construction are in the hands of three other ministries. All these ministries are currently run by the conservative Christian Democratic Union (CDU) or Christian Social Union of Bavaria (CSU) and are working very closely with the private sector in the respective areas. They are not yet ready for a transformative policy in general, let alone for a gender-equitable policy. Thus, this fragmentation of climate policy is difficult in terms of effective climate change mitigation and it also hampers the integration of gender and intersectional approaches.

Mainstreaming gender into Germany's climate policy

The road to mainstreaming gender into climate policy in Germany started to look promising in 1995, when the first Conference of the Parties to the Framework Convention on Climate Change (COP1) took place in Berlin. It was chaired by the (at that time) German Environment Minister and former Minister for Women Affairs, Dr Angela Merkel, who later on, in 2005, became Chancellor of the Federal Republic of Germany. Theoretically, this could have been the ideal condition for advancing gender-equitable climate policy.

Furthermore, in 1998, the first red-green Federal Government agreed to implement gender mainstreaming and to recognise the realisation of equality of men and women as a universal guiding principle for policy planning in all ministries. These regulations are still in place. Since then, governmental coalitions have changed several times, and the strong efforts for an organisational anchoring of gender mainstreaming in the BMU have had only very limited success (Sauer, 2014).

With the exception of a few selective measures, gender-responsive climate policy has not been an issue in the German Environment Ministry in the past 20 years. Particularly, it is still striking that the gender mainstreaming process had not found its way into climate policy. So, despite the mandate to mainstream gender into all policy areas, it was only recently and in connection with and supported by UNFCCC decisions on gender, that it gained more attention in German national climate policy.

So, while the importance of gender relations for an effective climate policy was largely underestimated by the government for a long time, some progress has been made in the last few years, giving a reason to hope for a move towards a gender-equitable climate policy in the near future. After the federal elections in 2017, at the beginning of the current legislative period, Svenja Schulze came into office as the new Environment Minister. She placed greater emphasis on gender equality, both in terms of gender balance and of gender-sensitive or gender-responsive policy. She set up, or reactivated, internal networks and training for the staff, particularly at decision-making levels, installed a new department dealing with gender aspects of environmental policy, and announced the development of a gender strategy for environment, nature protection and climate policy (BMFSFJ, 2020). Moreover, the current Minister Schulze holds regular meetings with female leaders of environmental non-governmental organisations to discuss difficulties in, and solutions for, addressing gender and gender equality in the organisations, as well as to present and discuss up-to-date environmental policies and programmes. Considerable steps have been taken to make staff aware of the added value of integrating gender into the climate policy field, namely due to more effective emission reductions measures that address the needs and preferences of all sections of society and, thereby, might gain a higher acceptance by citizens.

The updated UNFCCC Gender Action Plan (GAP), adopted in 2019 (UNFCCC, 2019), added momentum to these efforts. While the vast number of earlier UNFCCC decisions (Röhr et al., 2018, p. 90ff) that call for the consideration of gender equality and the broad gender mandate provided by the Paris Agreement (PA) of 2015 (UNFCCC, 2015) were considered as applying mainly to developing countries, the enhanced GAP functioned as an eye-opener for the relevance of gender mandates for industrialised countries. Internal meetings in the BMU were held in order to identify core fields of action to implement GAP provisions. This involves, for example, capacity-building on the gender dimensions of climate policy which plays a prominent role in a number of UNFCCC decisions and the GAP (Röhr et al., 2018, p. 90ff; UNFCCC, 2019).

The advancing introduction of gender in the UNFCCC process was the context in which the research project 'Interdependent gender aspects of climate policy' was commissioned by the UBA. The project underpinned the new move at the BMU towards gender-responsive policies and it provided evidence to the importance of integrating gender into climate policy and the need for relevant data. As a result, at present (October 2020) there is a growing openness towards gender equality in environment and climate policy in the ministry, as well as in environmental organisations that adopt the minister's perception regarding gender equality in climate policy – yet, do not necessarily point out gender issues actively.

Two potential paths for climate institutions to bring gender into climate policy

As a first option, in the next section, we present and discuss our experience with raising gender issues in participatory processes, organised by the BMU to involve a broad range of stakeholders and citizens in the development of German climate protection programmes. With these extensive participation processes, new ground was broken in Germany. The following evaluation is based on personal involvement in the processes and – where available – evaluation reports from other organisations and researchers.

Thereafter, we present, as a second option, gender analysis as an attempt to strengthen a gender perspective within administrative processes. On an institutional level, gender mainstreaming requires governments to assess 'the implications for women and men of any planned action, including legislation, policies or programmes, in all areas and at all levels' (UN, 1997). This is usually done by integrating gender analysis in the policy process in order to analyse and evaluate the impact of political measures on gender justice. As a cornerstone, the 'Gender and Climate' research project developed a Gender Impact Assessment (GIA) tool to strengthen a gender perspective in the German Climate Protection Programmes. Hence, the second path we describe is based on our experiences in adapting and improving a GIA to the field of climate policy.

Implementing gender-responsive climate policy in participatory processes

'Stating a policy objective is no guarantee that it will be realised in practice' (Rose, 1984, cited from Engeli and Mazur, 2018, p. 117). Thus, how is the goal of gender equality operationalised in climate policy in Germany? The following section will shed light on the participation processes that support the development of German climate change programmes. Since 2015, the federal government has launched two participation processes, one of which (initially) focused on achieving the short-term emission reduction targets for 2020, the other focused on the more long-term target of 2050. Both processes were led by the BMU, and are very different, which is also due to the time horizon of emissions reduction targets for 2020/2030 and 2050. Such extensive participation processes previously have not taken place in Germany and can therefore be described as innovative.

The Action Alliance for Climate Protection (Aktionsbündnis Klimaschutz)

The first participation process, the Action Alliance, had its first meeting in spring 2015 upon invitation of the BMU. It is organised in the 'Vienna setting', which means that the various stakeholder groups are divided into so-called thematic 'benches'. Initially comprised of 15 benches in number, in 2019 'youth' was added as the 16th bench. Each bench consists of various organisations that appoint speakers for the meetings to represent the positions of the bench. All other representatives of the bench take part as listeners. The 'members of a bench exchange views on their respective climate protection positions, strive to understand the positions of other bench members and try to find synergies and similarities in their positions'.[1]

Gender issues do not have a separate bench, but some gender expertise is involved in the process. Currently, there are three women/gender-focused organisations regularly participating, all of them are part of the Environment, Nature and Climate Protection bench.

The Action Alliance meets twice a year for one-day meetings. At the first session, one of the women/gender-focused organisations, GenderCC-Women for Climate Justice, representing the bench for the Environment, Nature and Climate Protection, had the opportunity to explain fundamentally how gender relations and climate protection are linked. Since then, there have only been opportunities for very brief statements. As the speaking time per subject (usually two or three per session) is limited to five minutes per bench and about 30 organisations are represented on the environmental bench, the chance to introduce gender aspects to the agenda items is extremely limited.

In terms of content, until 2017 the Action Alliance dealt with short-term measures aimed at achieving the CO_2 reduction targets for 2020. Before the sharp emission decline due to the 2020 coronavirus pandemic, Germany was far from meeting its target, thus there was a high pressure to achieve as much emission reduction as possible in a very short time. This might have been one reason why the measures presented, and later on decided by the German Bundestag, focused mainly on technical solutions, respectively, solutions the government coalition could agree on. A transformative (gender) perspective, questioning the economic system and growth paradigm, criticising the focus on technical efficiency, and demanding actions for addressing sufficiency, was brought into discussion by gender representatives and some environmental organisations, but had no chance of being recognised, neither by most of the other benches, nor by the ministries responsible for the measures. This broader and more holistic perspective was postponed to the programmes dealing with the mid and long-term goals for 2030 and 2050.

Participation process for the development of the Climate Protection Plan 2050

The process for developing the climate protection programme for 2050 took place parallel to the Action Alliance for Climate Protection and its task to support the 2020 programme of measures for climate protection. This very complex process started with inviting 550 people as representatives of stakeholders. Representatives of women's associations, such as the German Women's Council – a major umbrella organisation which perceives itself as the lobby for women in Germany – was invited but, unfortunately, they did not participate in the process, and neither did other women's associations. Most German women's organisations focus on more traditional women and gender equality issues, like the gendered labour market, care economy, violence against women and health issues, etc. This might have been a reason why they didn't feel familiar with climate change issues and did not participate.

In addition to the stakeholder dialogue, 100 randomly selected citizens were each invited to a citizen dialogue in five cities, which later on resulted in two online dialogues to propose a number of climate protection measures and vote for the preferred actions. Without being able to go into the process in further detail here, the effort to participate was extensive. Particularly for civil society organisations relying on volunteer work, participation in this process was extremely challenging.

The process ended in March 2016 with the handover of a catalogue of 97 proposed measures to achieve the long-term climate protection goals to the BMU, but the impact was limited, as in the end, the Federal Government's climate protection plan was drawn up 'in 2016 on the basis of scientific studies and scenarios and in the light of the Paris Agreement. The results and recommendations of the broad dialogue **also informed the process**'[2] (accentuated by the authors).

Three evaluations of this process were undertaken. Greenpeace commissioned a rather critical assessment of the impact of the process (Rucht, 2016), the BMU commissioned the overall process evaluation (Bohn and Heinzelmann, 2017) and an evaluation of the citizen participation also took place (Faas and Huesmann, 2017). All three evaluations came to a cautiously positive assessment of the process. Despite this, Greenpeace (Rucht, 2016) criticises that there has been a significant shift of weight in favour of economic interests by clearly weakening and watering down a number of sectoral targets and proposed measures from the participation process (Rucht, 2016, p. 6). None of the three evaluations took notice of the lack of gender expertise in the whole process and in developing or assessing the measures. A content-orientated analysis of the Climate Protection Programme 2050 from a gender perspective can be found in the final report of the research project (Spitzner et al., 2020, Chapter 5.5.1).

In 2018, the two processes were merged in the Action Alliance for Climate Protection. The Action Alliance is now described on the BMU website as

> the central dialogue forum for the continuous discussion of climate protection policy positions between social groups and with the Federal Government. It supports the Federal Government in achieving the climate protection goals for Germany and recognises the shared responsibility of its members for the success of the transformation to a largely greenhouse gas-neutral society by 2050.[3]

In conclusion, both processes might have been important and innovative from the point of view of the German Government, as they led to the involvement of more stakeholders in defining Germany's climate policy. One expectation was to achieve acceptance for ambitious emission reduction measures and sectoral reduction targets. But these processes were, and still are, completely insufficient in terms of addressing the need for gender justice in climate policy and action.

In order to establish a connection between the climate policy process and the gender mainstreaming process, which run independently of one another, the research project 'Gender and Climate' has created a 'handout for gender-appropriate evaluation and updating of the 2030 climate protection program'[4] for BMU-internal use (Röhr and Stieß, 2019). This refers to the Climate Action Programme[5] finally adopted by the German Parliament at the end of 2019, its evaluation and update, as well as the envisaged (and meanwhile adopted) climate protection law. The handout recommends taking into account the social impacts including their gender differences, in the implementation, evaluation and updating of the programme, and where and how the corresponding expertise is to be included in the basic structures implemented by the law.

Gender Impact Assessment as an analytical tool in the area of climate policy in Germany

This section introduces Gender Impact Assessment (GIA) as a tool for researchers and practitioners in climate institutions for analysing the equality impacts of climate policy and discusses how GIA can be re-designed in order to introduce an intersectional and non-binary perspective on gender into climate policy-making. GIA is rooted in the field of impact assessment which is the 'process of identifying the future consequences of a current or proposed action' (IAIA, 2020, n.p.). In contrast to other impact assessment procedures (e.g., technology, environmental, or health impact assessment), a GIA takes up the basic idea of gender mainstreaming that gender equality should be promoted as a guiding principle through all policy measures. Gender is taken as the central category of analysis and the investigation is oriented towards the normative goal of gender equality.[6] The objective of GIA is to produce recommendations for political decisions, the outcome of GIA is usually binding and applied to a specific policy intervention.

According to this, a GIA is a structured application of procedures and methods aiming to assess whether a political project – for example, a draft law, regulation, programme, or concept – and the day-to-day administrative action takes adequate account of gender aspects and the command for equality. This assessment can take place at different points in the policy cycle: as ex-ante evaluation, but also as ex-post evaluation or during the implementation of a policy measure.

> The implementation of a GIA after the completion of a measure can reveal 'gender gaps', but in the case of measures that have yet to be developed or are ongoing, it is possible to identify gender aspects and then integrate them into the measure. In the case of the accompanying GIA, the possibility of

integrating gender aspects is far greater than in a retrospective GIA. If the GIA is integrated into the development of a measure, gender aspects can be systematically introduced into decision-making processes, negative impacts and unintended side-effects on gender equality can be identified before the political measure is implemented, and alternative solutions can be presented.
(Life e.V./FrauenUmweltNetz, 2004, p. 10f)

A GIA can be applied to different policy areas at the level of ministries, as well as subordinate authorities and administration.

Applying intersectionality and gender dimensions to a GIA guidance for climate policy

In Germany, a regulatory GIA baseline tool was adopted in 2006 to support gender mainstreaming by the federal government (BMFSFJ, 2007),[7] but the general practice of GIA has proven to be difficult in terms of practitioners' receptivity and the institutionalisation of the instrument. In the past 20 years, only a few policy GIAs were fully executed and documented in public administration (Veit, 2010; Lewalter, 2013; Sauer, 2018). There was a certain stickiness within the administration preventing them from including new ideas such as feminist concepts in the policy-making process (Sauer and Stieß, i.p.).

The GIA design for climate policy took the regulatory GIA baseline tool as the starting point. The re-design was based on a non-binary understanding of gender and also took intersectionality into account. The tool acknowledges different genders ('male, female, diverse'), in order to consider the diversity of gendered identities. Gender is conceived as an interdependent category – which is marked by the term 'Gender+'.[8]

The basic idea of the GIA is that gender and other social differences that generate inequality are to be considered on a case- and context-specific basis; the categories and their significance must be worked out in the process of analysis itself (cf. Hankivsky, 2014, p. 3). The GIA tool, therefore, applies an iterative procedure. Starting from the gender perspective, 'internal' intersections of gender with other categories of inequality are examined in an iterative process.[9] For the concrete subject of investigation, this does not claim to include all categories involved in the constitution of social inequality, but rather focuses on those that are most relevant to the respective area of action. In a first step, the impact on the hierarchical relations between the genders is examined. In a second step, further characteristics/categories of social inequality can be introduced into the analysis – for example, to examine the effect of a project with regard to further discrimination and exclusion due to socio-economic or socio-cultural characteristics such as age, income, religion, etc. – in the sense of 'where the difference done, matters the most' (Sauer, 2018, p. 146). Taking housing and energy poverty as an example, a first step of analysis could focus on differences between genders and then add income and place as further categories of analysis (Großmann, 2017).

As a second task, the GIA tool was thematically adapted to the policy area of climate protection and climate adaption. The tool's thematical structure is built on the so-called 'gender dimensions' which provide a framework for gender analysis. The gender dimensions are analytical categories for investigating the structurally unequal power relations between the genders and their causes and characteristics. They are linked to central areas or fields of societal structuring by which hierarchical gender relationships are established, maintained and reproduced (Verloo and Roggeband, 1996; Verloo and Lombardo, 2007). These interconnected structures include norms, values, institutions and organisations. Drawing on the seminal work of Mieke Verloo and Connie Roggeband and insights from GIAs in the field of environmental, research and transport policy (cf. Schultz et al., 2001; Turner, Hamilton and Spitzner, 2006; Alber and Röhr, 2011), the conceptual framework was elaborated. Finally, six gender dimensions and a seventh cross-sectional dimension ('symbolic order') were devised, representing areas of life in which a sex- and/or gender-specificity can still be discerned and which are relevant for climate policies. These gender dimensions describe the social areas of the care economy and caring work, employment, shaping of public infrastructures, institutionalised rationalities, participation in decision-making and physical integrity. The gender dimensions include:

- *Symbolic order* (as a cross-sectional dimension): hierarchisations, attributions and positions of meaning, gender-hierarchical narratives and modernisation strategies
- *Care economy/care work*: attribution, significance, distribution and instrumentalisation in societies' economies
- *Market-driven/labour economy*: horizontal and vertical segregation, economic job evaluation, poverty, property, financial capabilities
- *Public resources/infrastructures*: provision, alignment, prioritisation, accessibility, usability
- *Institutionalised androcentrism/power of definition*: masculinity models set as the standard and benchmark in action-field-specific rationalities, problem perceptions, methods, etc., institutionalised gender hierarchisations
- *Shaping power at the actors' level*: participation and consideration of gender expertise in decision-making in science, technology, and politics
- *Body/health/safety/privacy (intimacy)*: the social organisation of sexuality, health, freedom from violence, privacy, sexual self-determination

For each gender dimension, main gender aspects like the unequal distribution of care, income, wealth, or resources are presented. These general topics are linked to specific issues of climate change using the results of the literature analysis on gender and climate change carried out in the first step of the project as a reference (Röhr et al., 2018). For each gender dimension, specific questions are provided, linking the more general dimensions to specific topics of climate change (Stieß et al., 2019).

The GIA contains two-stages of evaluation:

- The preliminary or *relevance test* (screening) checks whether the analysed project has gender-relevant implications and decides whether a GIA is to be carried out or not. In the first step, the given measure or project and its target groups are identified. The relevance of gender equality is determined on the basis of key questions referring to the different gender dimensions. If gender equality relevance is noted, the main examination is carried out.
- The *main examination* builds upon the questions of the relevance check. Gender aspects that were identified in the first step are scrutinised further in an in-depth examination, comprising of both qualitative and quantitative methods such as expert interviews or the analysis of statistical data. For each gender dimension, a detailed analysis of the policy impacts is carried out on the basis of a set of questions. As a result of the assessment, concrete opportunities for improving the outcome on gender equality are identified. On this basis, recommendations are formulated.

The re-design of the GIA tool was carried out in a participatory process. A prototype of the tool was tested in a workshop in order to adapt it to the requirements and needs of the future users of the instrument. Employees from different departments of the UBA were invited to the testing. Ongoing projects of UBA were used as examples to examine how the set of questions can be applied to make the significance of gender for climate policy visible. The selected projects covered topics from the fields of climate protection and adaptation with a high gender relevance: scenarios for reducing the degree of motorisation in cities as a contribution to a compact mixed-function city (climate protection), and recommendations for behavioural heat protection in connection with the preparation of heat action plans (climate adaptation). Drawing on the literature review on gender and climate (Röhr et al., 2018), main findings in the respective fields were collected and structured with the help of the gender dimensions. Examples are the different mortality and morbidity rates of women and men in the case of heatwaves and the gendered distribution of car ownership in cities in the case of the motorisation scenario. These findings were presented in a test workshop. In a second step, the participants were encouraged to analyse the selected examples with the help of the questions of the GIA tool. The workshop yielded valuable insights and suggestions about how to improve the practical suitability of the tool and facilitate its application. These findings were taken into account in the revision of the tool as well as the recommendations gleaned from the comments of gender experts. Finally, a second test was carried out in a workshop with UBA employees and external gender experts.

In summary, the GIA tool proved to be useful to assess the gender impacts of climate policy. It has the potential to assess the impact of climate policy measures with regard to gender/'Gender+', and can support knowledge transfer between societal stakeholders and political and administrative stakeholders on gender topics.

Until now, no full GIA of a climate policy measure has been executed. However, the tool was applied in a consulting assignment. The assignment had the objective of identifying gender-relevant aspects in an early stage of climate policy-making. In the context of the implementation of the climate protection programme 2030 of the German Federal Government, selected measures of the package were analysed. The results were communicated in consultancy papers and presented and discussed in an internal expert hearing in 2019. These first experiences show that the tool is well suited for screening gender aspects and identifying relevant aspects/questions from a gender perspective.

However, a crucial prerequisite for the implementation of GIA in German climate policy would be conducive governance structures in the institutions. As Sauer (2018) states 'gendered analysis is rarely executed without strong institutionalised equality governance structures within and knowledgeable, partial actors to carry it' (Sauer, 2018, p. 510). In terms of governance structures required, Sauer suggests a helpful distinction between 'embedded', 'embodied' and 'entrenched' governance structures. 'Embedding' of gender or gender justice refers to processual aspects within institutions. 'Embodied' refers to the required gender expertise within relevant institutions. 'Entrenched' refers to the need for gender to be rooted in the knowledge cultures of institutions. 'Creating accountability is an effective method for entrenching gender equality' (Sauer, 2018, p. 516). However, such governance structures within institutions would have to be reflected in corresponding policy structures and processes of climate policy.

Discussion and outlook

Even if there is progress in mainstreaming gender into climate policy in Germany, there are still a lot of institutional challenges. The provisions adopted in the international process definitely support gender-responsive climate policy in Germany. But because of the limited scope and character of provisions in the GAP and 'the clash between the rhetorical commitments to gender mainstreaming and the actual gender norms embedded within an institution' (Mergaert and Lombardo, 2017, p. 106), the progress at national levels will continue to depend strongly on further factors, such as the engagement of individuals in decision-making positions and civil society interventions.

Which factors are decisive for a transition to a more gender-responsive climate policy? Based on studies in the field of institutional resistances (e.g. Waylen, 2013; Kaijser and Kronsell, 2014; Mergaert and Lombardo, 2017; Göldner, 2018) and experiences in our research and advocacy work, we identified the following drivers that could help to overcome resistance to change.

One of the institutional drivers is, of course, a certain tradition of addressing gender, and this is the case in Germany, although early efforts have been suspended for quite a while (Göldner, 2018). A fundamental requirement is a political will to implement a climate policy geared towards gender equality. This is currently the case in the Federal Ministry for the Environment and is shown, particularly, in the establishment of a ministerial unit on gender issues and in the planned development of a gender strategy. Furthermore, there is some awareness for this topic in the management levels of the BMU that supports the framing of climate change issues from a gender perspective. Also, there is an internal feminist agency in the ministry and the subordinate authority UBA, which can back the efforts. Another driver would be the involvement of ministries or other bodies in charge of gender equality, the so-called 'gender machinery' (Göldner, 2018).

For the sustainable, long-term implementation of gender-responsive policies, however, further structural changes are required. As stated by Arora-Jonsson (2014), it needs 'to remedy the "invisibility" of privilege – whether of men, class, caste that discriminates. The structures and relations that cause disadvantage to persist need to be challenged'. This includes 'making visible the mechanisms by which environmental governance takes place – the daily practices of knowledge production and action' (Arora-Jonsson, 2014, p. 306). Usually, it takes at least five to ten years after a policy has been put into place for it to make change visible. Thus, if the intended gender strategy of the BMU and, in particular, the GIAs are implemented in climate policy and if these actually lead to gender-just climate measures, it will only happen and be assessed in the long term (Engeli and Mazur, 2018, p. 120). Thus, it is important to follow up on the implementation with research in the next years.

Androcentrism is deeply embedded not only in the structures of institutions, but also in the framing of problems and solutions (Sauer, 2010; Spitzner u. a., 2020). A transition to a gender-just, low-carbon society would require challenging these norms (Mazur and McBride, 2010; Kronsell, 2013; Kaijser and Kronsell, 2014).

In the climate change process, the 'Fridays for Future' movement has shown how quickly external pressure can lead to changes, at least at the level of discourse. Whether these changes are reflected in real political effects or only result in including more youth representatives in participation processes is still an open question. Likewise, the advocacy of women and gender civil society groups would be key to emphasise the gender-climate-justice nexus. Yet, in Germany, there are only a handful of organisations and experts who deal with these topics. As noted, the German Women's Council – the most important actor at the national level – deals exclusively with more 'traditional' women's and gender equality issues. Previous attempts to get them on board with climate policy issues have so far failed. On the other hand, environmental organisations are showing increasing openness to gender issues, although this is still in its infancy.

Data and facts on the gender aspects of climate policy measures are fundamental to designing them in a gender-just way. For the first time, these were summarised for industrialised countries in the research project 'Gender and Climate' presented here, which also points to the large data gap and shows that there is still a long way to go to make climate policy sound and evidence-based in a gender-responsive manner (Spitzner et al., 2020).

Besides data, developing gender-responsive policies and measures – e. g. in the energy, transport, or agriculture sectors – requires challenging current priorities fundamentally, especially, but not only with regard to economic growth. A stronger focus on public services and common welfare is needed. If the political will is lacking to tackle these issues, the consideration of gender equality might decay into 'gender greenwashing'.

These restrictions, particularly the data gap, also hamper the application of the GIA in climate policy. Still, data, the expertise of the persons in charge, and above all resources (staff, time, money) are missing while climate protection specialists in the ministry are chronically overloaded. Even though there is a willingness to work on gender, it is desired that it 'should not take much time'. Hence, again, without providing resources for the gender mainstreaming process to the ministerial departments dealing with climate change (as well as with other environmental issues) – additional staff and money for external expertise for example – there is little room for implementing a gender-responsive climate policy.

A genuine intersectional approach would make it even more difficult. Hankivsky (2012) points out that the aim of an Intersectionality Based Analysis (IBA) is to identify how different policies address the inequality experiences of different social groups. An intersectional perspective 'changes the policy questions that are asked, the kind of data that is collected, how data is collected and how it is disaggregated and analysed to produce evidence-based policy making' (Hankivsky, 2012, p. 177). In view of the existing powerful political pragmatism, however, she sees difficulties in introducing even more complicated policy processes. Changes would therefore have to take place gradually (ibid., p. 178).

Next steps

In order to pave the way for gender mainstreaming in climate policy, the research project 'Gender and Climate' developed a wide range of recommendations (Spitzner et al., 2020). It included demands in the fields of methodology, data requirements and data collection, research needs and regulations. Moreover, the project's conclusions included principles for integrating gender equality into climate policy, general policy recommendations and recommendations for the implementation of the Gender Impact Assessment.

One of the demands is that the development and adoption of a national climate policy gender action plan that should define concrete steps for the coming years, including responsibilities, deadlines and deliverables, in order to strengthen commitment and clarity for all stakeholders. This national GAP might be included in the projected Gender Strategy of the BMU.

More generally, it is necessary for the planning and development of climate policy programmes and measures to be based on the state of gender-responsive specialist research. It is also advisable to draw on specialist gender expertise (in-house or external) when planning and implementing climate policy measures. In order to secure gender competence in the long term, it should be integrated into the training of civil servants in institutions and administration.

Binding and regular application of the GIA or similar instruments makes an important contribution to taking gender consistently into account in the development of climate policy projects. Appropriate resources should be available for this purpose. Accountability ensures that the instrument will be embedded within the institution.

And, finally, ongoing monitoring and evaluation with well-defined gender criteria and indicators to be developed from the gender dimensions elaborated in the research project are necessary (Spitzner et al., 2020, p. 39f). Particularly, the monitoring and evaluation might help to shed further light on the institutional and individual resistances and obstacles in the development of a climate policy, that is gender-just and addresses intersectional discriminations. Due to stakeholder participation and lobbying and a political will from the head of the administration, gender justice reappeared on the climate policy agenda of the German Federal Government. However, in order to make considerable progress towards gender-responsive climate policies, deeply embedded routines and knowledge cultures in the administration still need to be challenged. Experience so far has shown that this transformation will take time and appropriate governance structures and resources are necessary to make climate policy administrations more responsive and accountable for gender justice.

Notes

1 https://www.bmu.de/themen/klima-energy/climate protection/national-climate policy/action program-climate protection/action alliance-climate protection/
2 https://www.bmu.de/en/topics/climate-energy/climate/national-climate-policy/greenhouse-gas-neutral-germany-2050/
3 https://www.bmu.de/themen/klima-energie/klimaschutz/national-climate policy/action program-climate protection/action alliance-climate protection/
4 An important intermediate step to the 2050 climate protection programme.
5 The measures of the plan were developed in the ministries responsible for transportation, energy, housing, agriculture and forests, etc., according to their sectoral emission targets. They were presented and (critically) commented on in the meetings of the Actions Alliance, though with very little impact.
6 The concept of GIA originated in development cooperation and was applied mainly at a project level. Later, this concept was transferred to other policy areas and levels (cf. Verloo/Roggeband, 1996; EC 1997/98, p. 3). In the Netherlands, Mieke Verloo and Conny Roggeband (1996) developed a GIA for impact assessment of political programmes in the field of social policy and education. The authors based their work on existing instruments of impact assessment, especially environmental impact assessment, and regarded the GIA as 'an instrument designed to analyse potential effects of new government policies on gender relations' (Verloo/Roggeband, 1996, p. 3).
7 The predecessor of this instrument was developed originally by the Institute for Social-Ecological Research (ISOE), in a project for and with the UBA and the German Federal Ministry for Environment, Nature Conservation and Nuclear Safety (BMU) in 2002 and emulated environmental impact assessment (EIA). This early BMU GIA was successfully tested in the case of radiation (Hayn and Schultz, 2002); however, it was not routinely applied afterwards.
8 The term 'Gender+' (Gender plus) was first used in the European research project QUING ('Quality in Gender+ Equality Policies in Europe' [2006–2011]) (www.quing .eu). The authors of the QUING project apply the concept of intersectionality to the analysis of policy fields and use the concept of 'Gender+' to describe the relational and dynamic relationships between different forms of inequality. The Austrian project

'GIAClim' – Gender Impact Assessment in the context of climate change adaptation and natural hazards also uses the term 'Gender+' with reference to QUING: 'The term "Gender+" includes other individual factors such as age, origin or special needs in addition to gender in the analysis and thus makes multiple discrimination visible. Gender+ includes knowledge about the complexity of gender and structural inequalities' (Damyanovic et al., 2014, p. 23).

9 In addition to the differences, both concepts – intersectionality and interdependencies – also have significant commonalities: both are directed against an additive understanding of gender and inequality and the notion of a homogenous female or male subject. Both concepts consider it necessary to analyse the category of gender in its historically specific entanglement with other aspects of social and economic inequality.

References

Alber, Gotelind and Röhr, Ulrike. (2011) *Gender Analysis of the Policy Initiatives of the Member States in Relation to Climate Change in the Sectors of Transport and Energy.* Milieu Ltd/LIFE e.V. Brussels, Berlin (unpublished).

Arora-Jonsson, Seema. (2014) 'Forty Years of Gender Research and Environmental Policy: Where Do We Stand?' *Spec. Issue Gend. Mobil. Soc. Change—Guest Ed. Lena Nare Parveen Akhtar* 47(Part B), pp. 295–308.

Bohn, Axel and Heinzelmann, Susanne. (2017) *Evaluierung der Stakeholder-Beteiligung an der Erstellung des Klimaschutzplans 2050. Abschlussbericht.* Berlin: prognos.

BMFSFJ Bundesministerium für Familie, Senioren, Frauen und Jugend (2007) *Arbeitshilfe geschlechterdifferenzierte Gesetzesfolgenabschätzung. "Gender Mainstreaming bei der Vorbereitung von Rechtsvorschriften".* Berlin: BMFSFJ.

BMFSFJ Bundesministerium für Familie, Senioren, Frauen und Jugend (2020) *Gleichstellungsstrategie der Bundesregierung.* Berlin [online]. Available at: https://www.bmfsfj.de/blob/jump/158356/gleichstellungsstrategie-der-bundesregierung-data.pdf (Accessed: 19 July 2020).

Damyanovic, Doris et al. (2014) 'GIAKlim—Gender Impact Assessment im Kontext der Klimawandelanpassung und Naturgefahren. Endbericht von StartClim2013.F', in Startclim (ed.) StartClim2013: Anpassung an den Klimawandel in Österreich—Themenfeld Wasser. Wien.

Engeli, Isabelle and Mazur, Amy. (2018) 'Taking Implementation Seriously in Assessing Success: The Politics of Gender Equality Policy', *Eur. J. Polit. Gend.*, 1(1), pp. 111–129.

Faas, Thorsten and Huesmann, Christian. (2017) *Die Bürgerbeteiligung zum Klimaschutzplan 2050. Ergebnisse der Evaluation.* Gütersloh: Bertelsmann Stiftung.

Göldner, Lisa. (2018) *The Integration of Gender Considerations into National Climate Policies. Factors Driving the Integration of Gender Equality and Women's Empowerment into Nationally Determined Contributions (NDCs) to the Paris Agreement.* Master Thesis, FU Berlin.

Großmann, Karin. (2017) 'Energiearmut als multiple Deprivation vor dem Hintergrund diskriminierender Systeme', in Großmann, Karin, Schaffrin, André and Smigiel, Christian (eds.) *Energie und soziale Ungleichheit. Zur gesellschaftlichen Dimension der Energiewende in Deutschland und Europa.* Wiesbaden: Springer VS, pp. 55–78.

Hankivsky, Olena. (2012) 'The Lexicon of Mainstreaming Equality: Gender Based Analysis (GBA), Gender and Diversity Analysis (GDA) and Intersectionality Based Analysis (IBA)', *Can. Political Sci. Rev.*, 6(2–3), pp. 171–183.

Hankivsky, Olena et al. (2014) 'An Intersectionality-Based Policy Analysis Framework: Critical Reflections on a Methodology for Advancing Equity', *Int. J. Equity Health*, 13, pp. 1–16.

Hayn, Doris and Schultz, Irmgard. (2002) *Gender Impact Assessment im Bereich Strahlenschutz und Umwelt. Abschlussbericht im Auftrag des BMU.* Frankfurt am Main: ISOE [online]. Available at: http://www.genderkompetenz.info/w/files/gkompzpdf/gia_abschlussbericht_strahlenschutz.pdf (Accessed: 12 August 2020).

International Association for Impact Assessment (IAIA) (2020) *IAIA. The Leading Global Network on Impact Assessment.* IAIA. https://www.iaia.org/about.php (Accessed: 17 February 2021).

Kaijser, Anna and Kronsell, Annica. (2014) 'Climate Change Through the Lens of Intersectionality', *Environ. Polit.*, 23(3), pp. 417–433.

Kronsell, Annica. (2013) 'Gender and Transition in Climate Governance', *Environ. Innov. Soc. Transit.*, 7, pp. 1–15.

Lewalter, Sandra. (2013) *Gender in der Verwaltungswissenschaft konkret: Gleichstellungsorientierte Gesetzesfolgenabschätzung.* Speyer: Hochschule für Wirtschaft und Technik.

Life e.V./Frauenumweltnetz. (eds.) (2004) *Gender Mainstreaming in Deutschland. Auf dem Weg zu geschlechtergerechter Umweltpolitik in Deutschland.* Frankfurt am Main. Available at: http://www.genanet.de/fileadmin/user_upload/dokumente/Infopool_Publikationen/Broschuere_GMainstreaming_Deutschland_LIFE_FrauenUmweltN.pdf (Accessed: 28 February 2018).

Mazur, Amy G. and McBride, Dorothy E. (2010) Gendering New Institutionalism. in McBride, Dorothy E., Maruz, Amy G. (eds.) *The Politics of State Feminism.* Philadelphia: Temple University Press, pp. 217–237.

Mergaert, Lut and Lombardo, Emanuela. (2017) 'Resistance to Implement Gender Mainstreaming in EU Research Policy', in McRae, Heather, Weiner, Elaine (eds.) *Towards Gendering Institutionalism: Equality in Europe. Feminist Institutionalist Perspectives.* London: Rowman and Littlefield, pp. 101–120.

Röhr, Ulrike. (2017) *Geschlechterverhältnisse und Nachhaltigkeit. Abschlussbericht.* Dessau-Roßlau: Umweltbundesamt.

Röhr, Ulrike, Alber, Gotelind and Göldner, Lisa. (2018). *Gendergerechtigkeit als Beitrag zu einer erfolgreichen Klimapolitik. Forschungsreview, Analyse internationaler Vereinbarungen, Portfolioanalyse.* Dessau-Roßlau: Umweltbundesamt.

Röhr, Ulrike and Stieß, Immanuel. (2019) *Handreichung für eine geschlechtergerechte Evaluation und Fortschreibung des Klimaschutzprogramms 2030.* Berlin: Frankfurt a.M.

Rucht, Dieter. (2016) *Der Beteiligungsprozess am Klimaschutzplan 2050. Analyse und Bewertung.* Hamburg: Greenpeace.

Sauer, Arn. (2014) 'Gender und Nachhaltigkeit—Institutionalisierte Gleichstellungsarbeit und Gender Mainstreaming im Umweltbundesamt', *GENDER*, 1, pp. 26–43.

Sauer, Arn. (2018) *Equality Governance via Policy Analysis? The Implementation of Gender Impact Assessment in the European Union and Gender-based Analysis in Canada.* Dissertation/PhD thesis, Centre for Transdisciplinary Gender Studies Humboldt University, Bielefeld: transcript Verlag.

Sauer, Arn and Stieß, Immanuel. (i.p.) Accounting for Gender in Climate Policy Advice—Adapting a Gender Impact Assessment Tool to Issues of Climate Change. In: Impact Assessment and Project Appraisal. *IAPA.*

Sauer, Birgit. (2010) 'Framing and Gendering', in McBride, Dorothy E., Maruz, Amy G. (eds.) *The Politics of State Feminism.* Philadelphia: Temple University Press, pp. 193–216.

Schultz, Irmgard et al. (2001) *Gender in Research. Gender Impact Assessment of the Specific Programmes of the Fifth Framework Programme, Environment and Sustainable Development Sub-Programme.* Frankfurt am Main Bruxelles: ISOE und Directorate-General for Research, Energy, Environment and Sustainable Development.

Spitzner, Meike, Alber, Gotelind, Röhr, Ulrike, Hummel, Diana and Stieß, Immanuel. (2020) *Interdependente Genderaspekte der Klimapolitik. Gendergerechtigkeit als Beitrag zu einer erfolgreichen Klimapolitik: Wirkungsanalyse, Interdependenzen mit anderen sozialen Kategorien, methodische Aspekte und Gestaltungsoptionen. Abschlussbericht.* UBA-Texte 30/2020. Dessau-Roßlau: Umweltbundesamt. Available at: https://www.umweltbu ndesamt.de/sites/default/files/medien/1410/publikationen/2020-02-06_texte_30-202 0_genderaspekte-klimapolitik.pdf (Accessed: 22 February 2020).

Stieß, Immanuel, Hummel, Diana and Anna, Kirschner. (2019) *Arbeitshilfe zur gleichstellungsorientierten Folgenabschätzung für die Klimapolitik. Projekt Gendergerechtigkeit als Beitrag zu einer erfolgreichen Klimapolitik: Wirkungsanalyse, Interdependenzen mit anderen sozialen Kategorien, methodische Aspekte und Gestaltungsoptionen.* Frankfurt am Main: ISOE. Available at: https://www.umweltbundesamt.de/sites/default/files/medi en/1410/publikationen/2020-02-06_texte_30-2020_genderaspekte-klimapolitik.pdf (Accessed: 07 August 2020).

Turner, J., Hamilton, K. and Spitzner, Meike. (2006) *Women and Transport in Europe.* Strasbourg: European Parliament.

UNFCCC (2015) *Paris Agreement.* o.O. Available at: http://unfccc.int/files/essential _background/convention/application/pdf/english_paris_agreement.pdf   (Accessed: 27 October 2017).

UNFCCC (2019) *Enhanced Lima Work Programme on Gender and Its Gender Action Plan.* Bonn: United Nations Framework Convention on Climate Change. Available at: https ://unfccc.int/sites/default/files/resource/cp2019_13a01_adv.pdf. (Accessed: 10 October 2020).

[UN] United Nations General Assembly (1997) *Report on the Economic and Social Council for 1997.* New York: United Nations.

Veit, Sylvia. (2010) *Bessere Gesetze durch Folgenabschätzung?* Wiesbaden: VS Verlag für Sozialwissenschaften.

Verloo, M. and Lombardo, E. (2007) 'Contested Gender Equality and Policy Variety in Europe' in M. Verloo (ed.) *Multiple Meanings of Gender Equality. A Critical Frame Analysis of Gender Policies in Europe.* CEU Press, Budapest and New York, pp. 21–51.

Verloo, Mieke and Roggeband, Connie. (1996) 'Gender Impact Assessment: The Development of a New Instrument in the Netherlands', *Impact Assessment*, 14, pp. 3–20.

Walgenbach, Katharina, Dietze, Gabriele, Hornscheidt, Antje and Palm, Kerstin. (eds.) (2007) *Gender als interdependente Kategorie. Neue Perspektiven auf Intersektionalität, Diversität und Heterogenität.* 2nd Ed. Opladen: Leske + Budrich.

Waylen, Georgina. (2013) 'Informal Institutions, Institutional Change, and Gender Equality', *Polit. Res. Q.*, 67(1), pp. 212–223.

5 Promoting a gender agenda in climate and sustainable development

A civil servant's narrative

Gerd Johnsson-Latham and Annica Kronsell

Prologue

I, Annica Kronsell, became acquainted with Gerd's work on gender, sustainability and development when reading her report from 2007 (discussed below) and invited her to a conference on Gender, Power and Climate Change, at Lund University in 2010. Gerd's participation at the event was not in an official capacity; nevertheless Gerd's perseverance in pushing gender issues onto the governmental agenda on sustainability, climate and development issues truly impressed me. We have met many times since and Gerd continues to amaze me with her tireless work on gender and climate, whether it is in government or civil society organisations. Over the years, I have been particularly curious about her work in the institutional context of government offices, asking myself if Gerd can be seen as femocrat, as an outsider within, or an activist inside? Her experiences are important; thus, I asked Gerd to write them down as an input to this book. This being an academic text, we agreed that I would take her story and interpret it in relation to feminist institutional theory. The text presented in this chapter is, thus, the result of the encounter between a civil servant and an academic.

Introduction

This chapter discusses the promotion of gender perspectives in the sustainable development and climate agenda from the unique perspective of a civil servant with gender expertise, reflecting on a long career. Gerd Johnsson-Latham has been working with international (development) cooperation for the last four decades, primarily in the areas of gender equality, women's rights, climate justice and sustainable development in her role as a civil servant. Her main employer has been the Swedish Ministry for Foreign Affairs (MFA), but she has also worked with the Swedish Government Offices in other capacities, and served in the UN and the World Bank. Her knowledge of the multilateral context is solid and genuine and her competence broad as she has also pursued women's rights, gender and climate policies by virtue of her work in civil society: at the international level through the Association for Women's Rights in Development (AWID), and at the two Swedish NGOs: *Kvinna till Kvinna* and

DOI: 10.4324/9781003052821-5

Klimatriksdagen (The Climate Parliament). The narrative which constitutes the material for this analysis was provided by Gerd herself, then edited and framed by me, Annica Kronsell, in close dialogue with Gerd. The chapter aims to use Gerd's narrative to provide reflections on the institutional challenges that government officials face when promoting gender equality within government offices and, particularly, in international fora and how they navigate that space in order to gender mainstream the agenda. The reflections come from the perspective of feminist institutionalism (Chapter 1, this volume). The text which is indented is Gerd's own words, while the rest of the text is the analysis developed from the dialogue between us.

Feminist institutionalist analysis demonstrates that institutions – like ministries and multilateral organisations – organise power inequalities through formal as well as informal rules and practices (Mackay et al., 2010; Krook and Mackay, 2011; Ljungholm, 2017). Gerd's narrative sheds light on the power relations affecting the gender, sustainability and climate nexus of ministries and the multilateral institutional context. It tells a story of path dependencies (Kenny, 2007, p. 93; Waylen, 2014) in terms of the specific notions of what 'gender' tends to imply, or be understood as, basically concerning women – e.g. as *not* privileged men – and also how gender objectives are placed in relation to other competing objectives like trade, development and administrative principles. Gerd's narrative recognises that gender is nested in other power categories (Kaijser and Kronsell, 2014), such as economic status and place. Most importantly, Gerd's long experience as presented through her narrative also provides hope in that it shows that change is possible. Her story gives us valuable knowledge of how it is possible to affect the agenda, by pointing to critical acts that primarily female, gender-focused civil servants take. The critical acts that stand out in Gerd's narrative are: a) the reports that underscore the importance of going into depth, and generating new knowledge – on gender relations and privileges of males – as a way to push policy-making and action forward, b) the importance of being part of, and taking advantage of, networks of gender experts and political opportunities and c) the value of the everyday nitty gritty, hard institutional labour of analysing documents and finding ways to include gender in as many places as possible, because this can become a platform for action. Through the narrative, Gerd alerts us to the institutional inertia that affected her work and, finally, she concludes with visions for a transformative agenda, only really possible through political mobilisation.

The critical acts of a gender expert

Gerd describes her career in the following way:

> I started working at the Swedish Ministry for Foreign Affairs (MFA) in 1976. At the time, there were few female civil servants. In 2020, there was an equal proportion of female and male professionals in the MFA, at all levels within

the organisation and a female foreign minister, the sixth female to hold the title.

My different posts have taken me to numerous international conferences, organised by both the UN and the World Bank, including the Fourth UN World Conference on Women in Beijing in 1995. In my role as gender advisor and civil servant I have been part of negotiating teams on broader matters, at times with a focus on women, at times with a focus on economic development, on sustainability and climate, where the aim was to highlight effective strategies for how women in different segments of society can play a key role in such endeavours.

During her career, she has witnessed a dramatic increase in the representation of women in the MFA, both among civil servants and politicians. Although providing a critical mass (Chapter 1, this volume), according to Gerd this has not necessarily meant an increase in critical acts, that is, a gamechanging increased focus on women's rights or gender issues. She speaks of several challenges:

> In my professional experience I have faced various challenges when promoting gender equality and women's rights, primarily within international development cooperation, but also in areas regarding human rights, rule of law and disarmament. The same goes for my two-year assignment with the High Commissioner for Refugees (1980–1982), in Hanoi, and in my two years at the Nordic Office of the World Bank (WB), in Washington DC (1986–1988). In the 1980s, neither gender nor the environment were prominent in international cooperation but the WB had begun to address environmental issues (partly as a result of massive demonstrations against huge dam-building projects in Brazil and India financed by WB) and had just established a position to work on 'Women in Development' (WID) with Karen Mason as one of the first officers on the gender desk. However, WID's efforts must be considered largely insignificant in relation to the overall hardship of poor women, often aggravated by the harsh conditionality that typified the Structural Adjustment Lending (SAL) of the WB at that time.

In reflecting on the critical task of mainstreaming gender, Gerd compares her work with that of other gender advisors across the globe – i.e. the work involved in scrutinising policy documents, policy speeches and budget proposals from a gender perspective. It implied ardent work and long hours and was not always rewarding. The more satisfying undertakings, she explains, were those that lead to innovative thinking and empowered women in different areas and educational processes, including dialogues with colleagues in various departments of the MFA, looking for relevant entry points and processes in daily and strategic work where gender could easily be applied and be helpful to outcomes. One such example is how women could be actively involved in defining priority areas for disaster relief, or how women's rights could be put more to the forefront in multilateral negotiations, primarily within the UN but also in national legal reforms. As a gender expert, Gerd was also charged with a typical task of such a position, that of educating the staff:

> At the request of the political levels at the MFA (Pierre Shori and his political advisors), I organised and led the ministry's gender training efforts for a number of years. Through various seminars, I tried to disseminate gender awareness in different departments focusing on topics like multilateral and bilateral development cooperation, disaster relief, promotion of human rights, democracy, etc. A challenge in this context was – and still is – to get staff to prioritise gender training, keeping in mind their busy daily agendas and sharp deadlines.

Gerd's experience illustrates how gender mainstreaming can take many forms and encompass truly critical acts, but also be superficial; to be compared to "green-washing" in the environmental area where the concept "sustainability" often aims at obscuring unsustainable practices.

The importance of in-depth knowledge as critical acts

Gerd's demonstrated expertise on gender gave her the unique opportunity to engage in three in-depth studies on various aspects of gender equality/women's empowerment. These became critical acts – i.e. acts that civil servants can perform in order to elicit change. Gerd pushed the limits for each topic she undertook to study, not the least in terms of how she pinpointed male privilege and power and contrasted with and challenged the understanding of gender in the institutions, at least in the MFA and in the Ministry of Environment. However, in the Ministry of Justice, the framing was totally accepted, probably as a result of the practice within legal systems to explicitly pinpoint perpetrators. It should be highlighted that the studies Gerd presented tended not to be fed into the overall work of the ministry, but that was actually the case for many other strategic studies at the MFA. This is, of course, unfortunate but reflects the fact that the 'slow path' of these learning processes never really fit into the daily hurry at ministries, where speed often is prioritised. Yet, through these studies Gerd managed to convey her feminist perspective on gender power which she describes as follows:

> Male violence is crucial to gender power and often aims to subdue and control women. It is seldom a sign of pure rage, but rather a means to maintain power and privileges. The privileges are, of course, more evident among the richer and more influential men. Basically, all males are granted privileges in their capacity as males, for example in given roles as 'heads of family'. It follows the pattern of the feudal system, where every man, even at the bottom layers, gets rewards for obedience in preserving the overall power-structure. Thus, while poor men are deprived by more influential and richer males,

men have in return by tradition been entitled to control their own family members.

My framing stressed 'privileges' and not only 'power' and was not necessarily shared by other colleagues, who pointed to the fact that 'male privileges' was not a commonly used concept at the time. My first study 'Power and Privileges: On Gender-Based Discrimination and Poverty' was given considerable attention within civil society but failed to influence conventional approaches to pursuing gender equality within the MFA. For instance, the planned press-conference to launch the study was both questioned and delayed several times.

Hence, while Gerd's first study (2004) is an example of a critical act in that it pushed gender perceptions out of the comfort zone, according to Gerd its actual impact within Swedish Government Offices is doubtful, in contrast to Swedish civil society where Gerd was invited to present her findings in numerous areas such as within the Swedish Church community, Forum Syd (a major umbrella organisation for development cooperation). However, her work was evidently also recognised at ministerial level as she was asked to conduct a second study the following year, this time hosted at the Ministry of Justice.

The report was published in 2005 as 'Patriarchal Violence as a Threat to Human Security'. Again, my 'framing' of he problem did not focus on violence against women in armed conflict but in everyday life, an issue that was not regarded as a self-evident part of the human security agenda. Instead, a prevalent framing of that problem was in the family arena, generally labelled 'domestic' violence, or possibly 'violence against women' – but not language pointing out men as perpetrators. It reflects the fact that male violence against women was often socially and even legally accepted in many societies – though women, on the other hand, regularly pointed out male, often spousal violence as the major aspect of deprivation in their lives.

Her third study 'Gender Equality as a Prerequisite for Sustainable Development', connected to sustainability and the climate issue, had the purpose of demonstrating how, and to what extent, awareness of gender-based differences could inform and improve undertakings to enhance sustainable development. It was published in 2007 by the Swedish Environment Advisory Council (Johnsson-Latham, 2007).

At the time, there were few if any studies demonstrating how, and to what extent, awareness of gender-based differences pinpointing men could inform and improve undertakings to enhance sustainable development. Through gender-disaggregated statistics and data, the study showed how male behaviour affected carbon dioxide emissions, compared with emissions from groups with less economic power and resources. It highlighted how the lifestyles of well-to-do males tended to set a pattern for men (and some women) across the globe. As such is was one of the first studies to direct attention to the question of male privilege in relation to gender and connect it to climate change and sustainability issues. Further, it argued that lack of attention to these gender differences would jeopardise efforts to promote sustainable development.

This report has become a seminal text in the field of gender and climate change, and is widely cited among scholars, civil society actors and governments with 143 Google Scholar citations including key works in the field and 668,000 hits on Google including OECD, UNDP, UNFCC.

I was very pleased to also note considerable media interest in my study on gender as a prerequisite for sustainable development. In Sweden, the study was referred to on the front page of the second-largest daily newspaper, I was invited to a flagship news programme by the Swedish national television, and I was interviewed by at least a dozen smaller magazines covering gender and/or environmental issues. I also noted that uncountable articles highlighted my report and they emerged from all across the globe: Asia, Europe, Latin America, and the USA. I was further invited to present my findings in a paper for UNRISD (Johnsson-Latham, 2012). All these interventions pointed to the fact that this was the first time in history that a governmental publication had addressed male lifestyles as a cause of unsustainable development and carbon dioxide emissions.

Indeed, the report provided a new perspective and as Gerd explains:

In 2006, there were few – if any – policy papers within governmental offices (or research, for that matter) that pointed at males – especially wealthy males – as a problem. With my background in studies of male power and privileges in the context of poverty alleviation, I was given the task to study male lifestyles and consumption patterns as a key area for addressing sustainable development. Studies on gender and climate had, up until then, mostly studied women in the South as victims of drought and floods. I decided – with the experiences from my previous two studies – to shift the lens on the discourse on gender and sustainable development in a completely new direction and scrutinise male behaviour. (While I initiated this new framing, the idea was fully supported at political levels.)

Gerd's report (Johnsson-Latham, 2007) presents a transformative agenda for engendering climate politics and a convincing argument for why this is necessary. Due to assigned gender roles, women and men often have very different lifestyles and consumption patterns. This is evident in terms of mobility and transportation: a sector with some 25% of all global carbon emissions. In general, men enjoy greater freedom of movement, have more economic resources, work in prestigious, well-paid jobs, and are more likely to travel by more expensive means such as air and automobiles. Women – in every income segment – tend to focus more on their homes and families, earn and own less, and have less freedom of choice compared to men. Referring to studies in richer countries, the report shows that while men dominate decision-making fora, the tax resources societies spend on infrastructure primarily tend to benefit wealthier men, with the bulk of investments provided for aviation and highways. Fewer investments are provided to public transportation (trains, buses, metro, etc.), the use of which is more prevalent among women without driving licences and cars who often rely on more sustainable modes of transport.

While the trajectory of the report seems out of the league of a civil servant's action space, it can definitively be perceived as a critical activity that aims to transform conventional thinking on gender. Gerd could be considered to be a femocrat given the fact that her actions seem to fit into that description (Chapter 1, this volume). While Gerd's ideas sometimes were perceived as controversial, the reliance on in-depth knowledge in her pursuit often won acceptance.

> Of course, my position within the MFA was influenced by my strong stand on gender policies, and though I wanted my action to fit into the overall – often unspoken – cultural norms at the ministry, I have often been considered as head-strong; a person who does not avoid challenging prevailing policy options. In spite of this, I feel that I have by and large been respected, and even appreciated. Not least for a sharp pen and analysis.

The 2007 report's analysis and results were widely spread outside the government offices, but its precise paths of influence in the foreign ministry or its input to its feminist foreign policy from 2014 remain unclear or vague. Gerd worked at the MFA Department for Human Rights 2012–2017, as the Chief Editor for Swedish reports on adherence to human rights, rule of law and democracy across the globe. This included pursuing women's rights and pressing for sexual and reproductive health and rights (SRHR) in multilateral fora such as the UN Commission for Human Rights. It seems safe to say that Gerd has taken her role as a gender advisor seriously and has also accomplished important acts of gender politics.

Networks enable critical acts

Clearly, Gerd's ambition has been to realise goals set out at political levels to pursue gender equality. She was well aware all along that these goals sometimes conflicted with other goals *also set out at political levels*, particularly economic interests which, in general, indeed tend to be more decisive for government actions than gender equality. However, though she understood the difficulties, Gerd generally interpreted her role as gender advisor to be the one to push for options which were as beneficial as possible for gender equality. Critical acts of gender experts should not be perceived as individual pursuits as they are highly dependent on collaboration with others. A support network of gender advisors can be critical for any success or even the possibility to carry on work. This is strongly argued by Gerd and she exemplifies it with her experience in the Development Assistance Committee (OECD-DAC) and its gender group, in particular, providing a key source of support and inspiration for gender advisors across donor countries. Several members of this group played a leading role in the international negotiations on women's empowerment, notably the complicated and controversial deliberations on SRHR as evidenced in for instance, the UN World Conferences in Vienna, Cairo and Beijing in the 1990s. Facing similar institutional challenges the gender experts shared valuable experience – in coping with difficult processes – widely within the network.

An Italian colleague, Bianca Pomeranzi, later a member of the CEDAW-committee provided me with critical support in areas such as dominating epistemologies and the need to critically analyse 'knowledge' as knowledge at worst can be used for perpetuating power structures. Dr Carolyn Hannan Andersson – a world-leading gender advisor at SIDA and, later on, head of the UN Division for the Advancement of Women gave me most valuable insights on both advantages and shortcomings with gender mainstreaming. While conducting the study on poverty, I was also lucky to learn from many colleagues, including Amima Mama at the African Gender Institute in Cape Town on how power-relations within the Military often are replicated in civilian institutions. From IDS in the UK, notably researchers and experts on female poverty such as the professors Naila Kabeer, Sylvia Chant, Andrea Cornwall and Diane Elson, I learned about the similarities – regardless of geographical areas – in gender-based power-structures, in families as well as at national level. Diane Elson was, and is, one of many prominent experts within the Commonwealth who has developed tools for engendered budgeting – mechanisms applied by many, including the Swedish Ministry of Finance.

The importance of politics

Gerd's narrative also discloses how the potential of critical acts must be understood against the broader context of politics, i.e. beyond the institutions. Politics offer opportunities, but also throw up challenges. According to her, it was crucial for the work in the MFA that gender equality as a goal and gender mainstreaming as a means was furthered during Pierre Schori's time as Minister for Development Cooperation (1997–1998), when he and his political advisor, Lena Ag (now Director-General of the Swedish Gender Equality Agency), requested all departments of the MFA to produce gender equality plans:

Some of these plans lasted longer in the institutional memories of the departments, while others had shorter lifespans. By and large, many plans – notably the parts most easily put into practice such as disaster relief focusing among other things on SRHR – had a reasonably lasting effect.

It was also the political level – with Minister Mona Sahlin and her political advisor, Stefan Engström, both committed to gender equality (and Sahlin had played an active role at the Beijing Conference in 1995) – that commissioned Gerd to produce the study on gender and sustainable development discussed above. Mona Sahlin also opened a high-level, well-attended seminar when a draft of the study was presented in New York, in the context of the UN Commission on Sustainable Development in 2006. Gerd used that occasion to push a more transformative gender agenda and explains:

> During my years in development cooperation, the approach when addressing gender had been to strategically address women as victims in need of assistance. In general, that meant poor women were objects – but could also imply looking at women as agents, and even to apply an intersectional perspective. However, the gender approach very seldom meant that *men* and male behaviour were the focus of analysis and interventions with two exceptions to that rule: when men were the main perpetrators of violence against women and when highlighting men as fathers. I felt that the time was ripe and I seized the opportunity to introduce a new perspective in the sustainability discourse. To me, it was logical, based on my experiences from the poverty study, which had highlighted male privileges – and how these male privileges often deprived women of essential rights and resources.

The presence of political will – among leading politicians and gender experts – offers support and opportunities to push gender issues on agendas in a transformative way much beyond – albeit also important acts – of 'add women and stir' strategies or the inclusion of gender in important texts. Political will was declared strongly when the Swedish Government launched its Feminist Foreign Policy (FFP) in 2014. The FFP has attracted scholarly attention (e.g. Aggestam and Bergman-Rosamond, 2016; Aggestam et al., 2019; Robinson, 2019) and has been appreciated in government circles in many parts of the world (e.g. Rosén Sundström and Elgström, 2019). France and Canada (e.g. Thomson, 2020) have followed suit in establishing similar goals and policies. The launching of the FFP has given a strong and important signal – and remains a beacon – in a time when women's rights are contested, often violently, in very many parts of the world.

In the FFP, women in peace and conflict were a priority and SRHR was of paramount importance but climate issues were not. A notable change with the FFP is that the normal order of procedure would be that embassies and representations in multilateral arenas, such as the UN, would be requested to submit suggestions for gender action. In the 1990s, this standard was only applied to some departments; with the FFP it has become more all-encompassing, stretching across all departments. Gerd, however, has some strong objections to the FFP:

> The FFP is not a transformative agenda. For example, several Swedish NGOs have criticised the Swedish MFA for acting in manners inconsistent with the goals of the FFP. One example is the MFA allowing arms sales to countries

like Saudi Arabia and the United Arab Emirates where women are gravely discriminated and while both countries are also heavily engaged in warfare in Yemen.

The Swedish FFP emerged from a long engagement of the MFA on gender issues. Already, the Fourth UN World Conference on Women in Beijing 1995 was an example of Swedish leadership in the multilateral context. This implied work to include gender-relevant text in platforms and convention text. Gerd gives an account of her first-hand experience:

> Sweden (then a recent addition to the European Union), played a key role for attaining a ground-breaking and far-reaching Platform for Action (PFA). A major Swedish achievement was to include a recurrent paragraph in each and every section of the text, with the aim of ensuring that gender would be a key aspect in poverty reduction, education, health, violence, peace and conflict, economy, etc. The Swedish team – with gender expertise in a number of areas such as sexual and reproductive health and rights, international law, economics, conflict resolution and UN negotiations at large – was one of the stronger actors in Beijing in 1995, together with several other actors, including in the global south. It could be noted that one paragraph in the PFA gives special attention to women facing double or triple discrimination, such as migrant women, older women, women with disabilities and Indigenous women (though at that time the concept of intersectionality was not used).

Conservative politics as an obstacle to gender equity

There is much contestation on gender in the multilateral context in 2021. The Swedish position – stressing gender equality and later feminist foreign policy – stands in sharp contrast to an international system plagued by populist/national conservative winds and increased influence of anti-gender movements, e.g. in Russia, USA, Brazil, India and Eastern Europe.

> A reflection on the challenges I have faced when working multilaterally for the MFA is that there are radically diverging political views on gender, more than on most political matters. Over the years, the major obstacle to advancing gender equality has been the harsh resistance from several states, as well as the Vatican, a non-state observer within the UN, all endorsing conservative religious views. The most fierce political objection to gender equality relates primarily to SRHR; women's, young persons and LGBT-persons rights to their own bodies and sexuality. Many actors have worked closely together to resist secular approaches and all aspects of SRHR, regardless of faith (but often most vocal from the Russian Orthodox Church and also by the USA during Republican administrations). Their efforts have been particularly notable in the multilateral context, during numerous international

conferences organised by both the UN and the World Bank, and not least in fora set up to enhance human rights, such as the UN Commission for Human Rights.

Since 2000, gender equality has, according to Gerd, faced increasing challenges due to conservative political forces becoming more prevalent.

As I see it, reactionary forces were taken by surprise in Beijing 1995 regarding the achievements made in just two decades since the first UN World Conference on Women in terms of women's rights, in a direction that they did not favour. So, from then onwards, those who objected to gender equality and SRHR, have gathered support and collected huge amounts of money to pursue their agendas. This provides a stark contrast to the limited resources of women's rights groups. Even the EU has become less active in pursuing gender concerns since it expanded to include Hungary and Poland as member states in 2000. To include these states was part of a security agenda – but in reality, it became contrary to women's or gender security, notably in countries like Hungary and Poland. While it had previously been difficult to gain consensus within the EU on SRHR – due to objections from Malta and Ireland, supported by the Vatican – today with Poland, and Hungary as members, the EU has become a far weaker advocate for SRHR on key matters such as legal abortion, sexual education for adolescents, measures to reduce violence against women and permitting abortion for women and girls who have been raped, including during armed conflict.

Gerd also notes that misogyny has become more common as autocratic forces defend 'traditional' gender roles (cf. Chapter 12, this volume) which challenges Sweden's positions in multilateral fora. An example is the Geneva-based UN Commission for Human Rights, where since 2010, many states have given high political priority to stopping gains made in the previous 30–40 years related to advancing women's rights.

We have seen it with political leaders such as Orban, Kaczynski, Erdogan, Putin, Khamenei, bin Salman, Modi, Trump and Pence. We've seen it in countries across the globe, among populists and, nationalist groups, in political parties and on social media, and among 'lone wolves' and incels, who are prepared to kill to 'defend' their masculine privilege as well as – very often – their perceived white supremacy. While these leaders question and dismantle women's and LGBT rights within their own jurisdiction, they also collaborate internationally against those rights. Interestingly, many of these actors have also collaborated as climate change deniers including in negotiations in COP (Conference of the Parties) regarding the Paris Accord.

Thus, many men – and some women – who advocate a so-called traditional role for women also reject action to combat climate change and global warming. This fits well into my observations from efforts to mainstream and integrate *gender* in already established agendas such as economic development or legal reforms and in trying to integrate *climate awareness and sustainability* in relevant areas of concern. In doing so, I found that the same types of resistance surfaced in both areas – gender and global warming – when I and others tried to change the framing of problems. I think that the reason is that both gender and climate change is frequently perceived as a threat to many current male definitions of progress – and discourses which also challenge the expertise and status of well-established economists and policy-makers.

(cf. Chapter 12, this volume)

While the importance of political support for gender issues has been crucial for the work of gender advisors like Gerd, in the multilateral context, the strength and emergence of conservative politics is a counterpoint. It makes it not only difficult to advance gender equity, LGBT and SRH Rights but also implies that some of the victories won have been reversed or risk being reversed. In this context, pursuing a climate-related gender agenda and pointing to men and male privilege, may be particularly difficult.

Institutional path dependencies

As noted in the introductory chapter (Chapter 1, this volume), historically derived notions of gender tend to persist resulting in a path dependence that hampers opportunities for innovation and change (Kenny, 2007). Gerd reflects on her experiences of institutional path dependencies from the inside of the government offices and the foreign ministry:

The main institutional challenge government officials face when promoting gender equality within government offices is the odd constraint of insisting that the purpose is to address women in need of help – because this means it is not possible to address the real problem of masculine structures and the mostly male actors depriving women of means and rights.

The failure to recognise masculine structures is an institutional path dependence that works against transformation (e.g. Krook and Mackay, 2011; Mackay et al., 2010; Waylen, 2014) and instead leads to weaker, more superficial measures, such as increasing the representation of women, or 'mainstreaming'.

Gerd tells us that this applies not only in multilateral institutions but also within Sweden, where gender equality is generally accepted in society and SRHR has been a Swedish priority for decades, not the least in terms of pushing abortion rights and young people's access to sexual education. The Swedish Foreign Ministry has been pursuing women's rights, empowerment and (from the 1990s and onwards) LGBT interventions with considerable political and institutional backing. However, these issues remain prioritised only as long as desired projects and interventions fit within other overarching agendas, for example, if they fit with development efforts, peace-making, or conflict resolution and only as long as those agendas are not drastically changed by including gender concerns. Institutional inertia relating to the prominence of trade for Swedish welfare has strongly impacted sustainability issues. Gerd worked at the Ministry for Sustainable Development for two years (2006–2007), in a secretariat established to enhance sustainable development and experienced this firsthand:

> First, the secretariat had been placed in the prime minister's office but was later transferred to a less crucial part of the government offices. The secretariat had representatives from a handful of ministries, bringing together experts to create a multidisciplinary unit, in the hope that this would be conducive to creative thinking. While the secretariat gave a signal that Sweden continued to pursue the active role on environmental issues, a strong driving force behind Sweden's pursuit of sustainability was growth and trade-oriented and a way to demonstrate that Swedish technology and innovations can contribute to solving environmental problems and climate change.

> Our task was to mainstream sustainable development in key policy documents such as policies and speeches, at the national level, in the EU and globally in the UN and World Bank. Generally speaking, it was a question of adding some smartly formulated sentences and subject matters in order to place a reminder of the sustainability perspective – not really to transform the message as a whole.

Priority of trade

Sweden's economy relies heavily on international trade and its workforce is largely dependent on trade exports, thus the support for exports is robust and non-partisan. Swedish governments have strongly supported Swedish export industries, portraying Sweden as a modern country with technical solutions that are extremely conducive for the environment – i.e. the idea of ecological modernisation which has become an institutional path dependence. As an example, in the national country strategies of the MFA which are established to define the main features of Sweden's relations to other countries and facilitating exports, constitute the backbone and overarching concern in the strategies, while sustainability, the environment and climate issues are secondary. With an economy and labour force heavily oriented towards exports, Sweden often balances ambitions regarding gender equality with concern not to offend important trade partners, demonstrated in the past with the case of Saudi Arabia and the United Arab Emirates, countries which pursue gender apartheid.

Limiting frames

Other limitations had gender effects. In Gerd's view, gender-based discrimination constitutes examples of unsustainability in several areas, in terms of damage to both the physical and the social environment, something difficult to pursue fully due to the way the issues were framed and dealt with:

> While the aim of climate and sustainability efforts was to engage a multidisciplinary approach, the reality was that we often had to work with addressing forestry, water, energy and so on, in separate agendas. This gave very limited room for addressing gender and intersectionality in terms of *who has access* to sustainable forestry, to water resources, energy etc. The framing of the problems which conformed to a technical, rather than a social approach to sustainability failed to determine *who was causing* the problems. One consequence is that it gives the impression that *men* in their role as experts and engineers have the solutions to the problems; not the women. Thus, female experts are rarely called upon in this context, even when they could have given a voice to different segments of women and men, for example on the need for more public transportation.

Gerd demonstrates that how sustainability and climate change are framed can be limiting, as the frame prioritize specific topics and actors and excludes others.

> One of my suggestions is thus that a gender-equality perspective helps to enhance innovative thinking and can help us see that moving to carbon-free societies could for many – women as well as men – not mean a sacrifice but instead lead to emphasis on more crucial and sustainable aspects of well-being. One such example, could be to press for zero-tolerance regarding male or societal violence; to be secured by legal provisions, law enforcement, adequate funding etcetera.

In her work on sustainability and climate issues Gerd discovered that although she had support from political levels as we noted above, there was less support from what she calls the 'line' particularly to her report 'Gender Equality as a Prerequisite for Sustainable Development'. There was a trio of female senior civil servants at the Ministry for Environment who decided that the report should not – unlike her previous reports on poverty and violence – be published within the government offices, but was instead diverted to a government subsidiary agency (*Miljövårdsberedningen*). It seems as if the ministries were not ready for this new framing of gender issues proposed by Gerd.

> The 'line' of civil servants at the ministry – primarily females, at all levels – were far less enthusiastic about the study. Instead, they prioritised time within the secretariat to focus on daily tasks such as commenting on EU or UN statements to be held in Brussels and New York on matters relating to sustainable development. There, they would concentrate on a sentence or two to be inserted in texts that could, for instance, address poor women as victims of environmental degradation – but certainly not to address males as causing degradation. At the time of my study on gender and sustainability,

the Ministry of Sustainability was, among many things, involved in several working groups under the UN Commission on Sustainable Development to promote sustainable lifestyles. Generally, however, these working-groups – a part of the so-called Marrakesh-process – followed the overall more cautious approach of mainly dealing with innovative techniques and women as victims who should be more involved in decision-making.

Administrative practices

Through Gerd's narrative, we learn that politicians may support a transformative perspective and in particular that women at the political level share a strong common commitment to advance gender equality and women's rights and to the extent possible seek to transform the political agenda to attain those goals. However, politicians come and go, and what prevails is an administrative logic in the workplace enacted by both males and females in senior, managerial levels. With the exemption of gender advisors, they do not necessarily support a transformative agenda. Gerd explains about this path dependence:

> According to my experiences, an impediment for more far-reaching transformation is often civil servants, mainly at senior, managerial levels, who are unwilling or afraid of being considered driven by personal agendas. They point – often rightfully – to the fact that the task of civil servants is precisely that: to be 'servants' of the politicians, not to be activists. That is, of course, a rule not followed by all, the breaking of which is often appreciated by some politicians who are prepared to promote gender equality even while such paths conflict with other political goals.
>
> Ministries are maybe more advantageous for people who go by the book and follow the rules, not for strong-headed enthusiasts. Mostly, staff prefer to go by the book. The same goes for many young civil servants, notably within the MFA, who do not want to risk their future and career by being considered headstrong. All in all, at least in the MFA, it is more important – and appreciated – among peers never to make a mistake, even at the expense of not being brilliant, and even at the expense of not saving lives.

To follow such administrative logic means that civil servants will not advocate anything that challenges power structures, even if civil servants in the MFA, for the most part and regardless of their gender, are committed to addressing the needs of poor women and enhancing women's rights. However, at best it means that women can be considered as actors, agents, addressed in all their multitude through advanced intersectional analysis. Such analysis would still be insufficient in addressing the real problem: male privileges, male 'rights' and unsustainable male consumption and lifestyles (Gore, 2020). In speculating why this is the case Gerd writes:

> I am convinced that the reason that senior-level officials within government offices do not advocate transformative approaches is that the overall agenda

is to always focus on helping those at the bottom – but never challenging those at the top. Further, this approach is also a means to promote economic growth – perceived as a prerequisite for providing resources for less advantaged groups – even if it is detrimental for the climate, whereas obliging the rich to share their resources with those less well off would be much better for both climate and biodiversity.

Conclusion

Gerd's narrative highlights the committed and zealous labour that gender advisors put into diverse institutional contexts of multilateral organisations and government agencies, work that consists of analysing documents and finding ways to include gender in relevant texts that can then become platforms for further action. Such activities are well served by expert knowledge, coming from commissioned studies and research often advanced through networks of gender experts. These are imperative and critical acts. Gerd suggests that to gender mainstream climate change issues effectively, more knowledge and data are needed it is imperative to address excess consumption of rich males. She points to the necessity of providing sex-disaggregated data and data on wealth and income to identify who is causing CO^2 emissions and specific data on the beneficiaries of the fossil fuel economy.

Through her narrative, Gerd alerts us to the institutional inertia of specific framings of gender in relation to economic priorities that affected her work and how gender equality in reality is just one of many political goals set out to be pursued by means such as 'gender mainstreaming'. She also pointed to the limitations of political and administrative institutions with embedded restrictions on civil servants in terms of their roles and tasks largely determining their scope of action. Nevertheless, Gerd's narrative also signposted that opportunities could arise for civil servants to take action, and to the necessity of being brave enough to seize those opportunities (as also argued in Chapter 6, this volume). While all acts that have been discussed in the chapter can potentially advance gender issues, politics plays a crucial role both in terms of supporting transformative gender concerns and in obstructing them. Gerd maintains that for real transformation, another kind of politics is necessary. Gerd's visions for a future transformative agenda for gender, intersectionality, climate and sustainability, is that it 'is only really possible through political mobilisation. Transformative ideas have to become part of the ideas supported by the dominant political parties and governments of democratic countries' and mobilise the needed broader support, she continues: 'It is important to identify those who promote and advocate transformation as to enable collaborations and alliances between actors both inside and outside government and across societies'.

References

Aggestam, K. and Bergman-Rosamond, A. (2016) 'Swedish feminist foreign policy in the making: Ethics, politics, and gender', *Ethics & International Affairs*, 30(3), pp. 323–334.

Aggestam, K., Bergman-Rosamond, A. and Kronsell, A. (2019) 'Theorising feminist foreign policy', *International Relations*, 33(1), pp. 23–39.

Gore, T. (2020) 'Confronting carbon inequality: Putting climate justice at the heart of the Covid-19 recovery', *Oxfam*. Available at: oxfamilibrary.openrepository.com (Accessed: October 28 2020).

Johnsson-Latham, G. (2004) *Power and Privileges—On Gender Discrimination and Poverty*. Stockholm: Swedish Ministry for Foreign Affairs.

Johnsson-Latham, G. (2007). *Gender Equality as a Prerequisite to Sustainable Development*. Report 2007:2 of the Environment Advisory Council, Sweden: Stockholm.

Johnsson-Latham, G. (2012) *Gender Equality as Key in Defining Human Well-being and Enhancing Sustainable Development*. UNRISD 30 March 2012.

Kaijser, A., & Kronsell, A. (2014). Climate change through the lens of intersectionality. *Environmental Politics*, 23(3), 417–433. https://doi.org/10.1080/09644016.2013.835203

Kenny, M. (2007) 'Gender, institutions and power. A critical review', *Politics*, 27(2), pp. 91–100.

Krook, M. L. and Mackay, F. (eds.) (2011) *Gender, Politics and Institutions. Towards a Feminist Institutionalism*. Palgrave Macmillan, Basingstoke.

Ljungholm, D. P. (2017) 'Feminist Institutionalism revisited: The gendered features of the norms, rules and routines operating within institutions', *Journal of Research in Gender Studies*, 7(1), pp. 248–254.

Mackay, F., Kenny, M. and Chappel, L. (2010) 'New institutionalism through a gender lens. Towards a feminist institutionalism?', *International Political Science Review*, 31(5), pp. 573–588.

Robinson, F. (2019) 'Feminist foreign policy as ethical foreign policy? A care ethics perspective', *Journal of International Political Theory*, 17(1), pp. 20–37, 1755088219828768.

Rosén Sundström, M. and Elgström, O. (2019) 'Praise or critique? Sweden's feminist foreign policy in the eyes of its fellow EU members', *European Politics and Society*, pp. 1–16.

Thomson, J. (2020) 'What's feminist about feminist foreign policy? Sweden's and Canada's Foreign Policy Agendas', *International Studies Perspectives*, 21(4), pp. 424–437.

Waylen, G. (2014) 'Informal institutions, institutional change, and gender equality', *Political Research Quarterly*, 67(1), pp. 212–223.

6 Take a ride into the danger zone?

Assessing path dependency and the possibilities for instituting change at two Swedish government agencies

Benedict E. Singleton and Gunnhildur Lily Magnusdottir

Introduction – The need for institutional change

With the exception of the community of climate change deniers, there is largely consensus that global human society needs to go through a variety of changes in order to ensure continued sustainability. The UN's sustainable development goals (SDGs) emphasise the urgency of these changes and provide an incentive to recognise the importance of social differences, including gender differences, in climate politics (see also Morrow, this volume). Furthermore, the necessary societal changes needed for a successful climate transition call for deeper understanding of the complex social, economic and environmental effects of any climate action on different social groups, defined by intersectional climate-relevant factors such as gender, age, class, education and location. This means that the societal changes needed for sustainable climate politics will have diverse, disparate effects within the population (Magnusdottir and Kronsell; Bipasha, this volume). Gender, nested with other intersecting social differences, has been shown to be of relevance for consequences of climate action (Alston and Whittenbury, 2013; Magnusdottir and Kronsell, 2015, 2016; Nagel, 2012; Resurrección, 2013). As such, how to transition to sustainable societies is a topic of considerable scientific debate as is the discussion of the importance of social justice in climate transition. Whilst 'Green Radicals' may argue for substantive, wholesale changes to the underpinnings of contemporary economies and societies, still 'environmental problem-solvers' argue for a more pragmatic stance, i.e. reorientation rather than revolution (Dryzek, 2013). In this context, we argue that it is important to explore the role and possibilities that climate institutions and the civil servants embedded in them can play in orienting society in a more sustainable and socially just direction. This chapter examines the possibilities for action among civil servants within two Swedish climate institutions, with a specific focus on their potential to recognise and work with climate-relevant social differences, such as gender, age, location and class, in their policy-making. The climate institutions in question are two government agencies: the Swedish Environmental Protection Agency (*Naturvårdsverket*, SEPA) and the Swedish Transport Administration (*Trafikverket*, STA). Based on interviews with agency staff, we draw on feminist

DOI: 10.4324/9781003052821-6

institutional theory to examine the extent that respondents consider themselves bound by the *path dependency* of their institution and the extent to which they report potential action towards change. The purpose of this analysis is to highlight the potential ways that agency staff may be empowered to make use of a feminist or intersectional understanding of social differences in order to develop socially inclusive and just climate policy and potentially overcome institutional path dependency.

To be successful, climate transition strategies need to acknowledge social differences and thus consider the needs of different social groups. Otherwise, they not only risk being deemed undemocratic but also ineffective in their attempts to drive forward the large-scale societal changes needed for climate transitions (Laestadius, 2018). Our intention is to contribute by signalling the steps climate institutions could take in order to achieve their transformational potential (cf. Eckersley, 2020).

The chapter is structured as follows. In the next section, we discuss our guiding theoretical tool – feminist institutionalism – and its relevance to our case, e.g. how a 'gendered logic of appropriateness' might prevent the full inclusion of gender into climate policy-making. We then discuss the cases in question and the material and methods employed. Thereafter, we analyse the collected data, with the help of feminist institutionalism, highlighting how respondents framed their activities in terms of following a top-down and historical trajectory, while at other times they would describe how they are able to act on their own initiative. We conclude by suggesting that efforts to orient Swedish Government agencies towards a just climate policy require tools designed both to 'convert' senior staff to adopt the approach and to demonstrate civil servant initiatives fall within the agencies' legitimate purview.

Following a well-beaten path: institutional changes, path dependency and the 'stickiness' of institutions

We make use of institutional literature, especially including feminist institutionalism in our analysis. We are inspired by intersectionality literature on climate change (Kaijser and Kronsell, 2014) and we expand the traditional gender lens of feminist institutionalism to include climate-relevant social factors, such as education, class, age and geography, which intersect with gender (see further in Magnusdottir and Kronsell, in this volume).

Institutionalist approaches study *institutions*. Separate strands of institutionalism define institutions differently and we define them broadly, including both formal and informal climate authorities. Thus, our definition of 'climate institutions' also includes: 'rules, norms and practices embedded in historical traditions and legacies, giving some stability and predictability to political and public life' (Hysing and Olsson, 2018, p. 9; cf. March and Olsen, 1996; Mackay et al., 2010). Put another way, as social beings, humans coproduce understanding of the world with institutions through which they live and act (Douglas, 1987). People perform social relationships concomitantly with narratives of the world, which together comprise institutional patterns (Thompson, 2008). As social organisations, what people embedded in institutions

> assume to be true shapes what they value. This shaping occurs through the processes of proactive manifestation through which assumptions provide expectations that influence perceptions, thoughts, and feelings about the world and the organization. These perceptions, thoughts, and feelings are then experienced as reflecting the world and the organization. Members recognize among these reflections, aspects they both like and dislike, and on this basis they become conscious of their values (without necessarily being conscious of the basic assumptions on which their experiences and values are based).
>
> (Hatch, 1993, p. 662)

A key concern of institutional theorising is how institutions change. In this, it echoes wider social scientific discussions about the relationship between social structures and individual agency or power (Mahoney and Thelen, 2009). Within feminist institutionalism, institutional practices are understood to be based on a logic that may make the inclusion of gender appear less appropriate and less desirable. This builds on March and Olsen's (1996, pp. 21–38, p. 161) new institutionalist ideas about how institutions are reproduced through patterns of action in a 'logic of appropriateness'. In the context of this chapter, this might, for example, mean that dominant economic and technical solutions to societal problems are 'sticky' or deemed more appropriate than solutions focusing on climate-relevant social differences such as gender. The fact that Swedish environmental politics have historically been framed in the context of 'ecological modernization' (Lundqvist, 2004), which prioritises economy over social concerns might lead to path dependency in Swedish climate institutions and limit the scope for inside activism. Thus, economic solutions might be reproduced and normalised within institutions, limiting space for innovation and civil servant activism. Put differently, individual policy-makers are assumed to follow embedded rules and routines, according to what is appropriate for their social and professional role and individual identity. What is considered appropriate in any given situation is not trivial. 'Actions are fitted to situations by their appropriateness within a conception of identity' (March and Olsen, 1996, p. 38).

'Well-established institutions are even taken for granted and tend to produce path-dependent action' (Hysing and Olsson, 2018, p. 9). What this means is that those embedded in institutions act based on past institutional actions, with positive feedback from successful earlier actions encouraging the development of a particular institutional 'path' or 'mission trajectory' (Goodsell, 2006). Put another way, an institutional worldview represents a paradigm within which particular actions are encouraged or constrained, providing a framing of how the world should be. They may also function to retroactively maintain or alter existing assumptions (Hatch, 1993, p. 664). In this chapter, we are interested in the extent that respondents frame their actions around socially just climate transition with reference to the institutions within which they are embedded (the government agencies).

Path dependency may potentially be challenged by actions in the wider world; however, endogenous change is also possible. Not all institutions proceed along a static trajectory until an external crisis precipitates change. Institutional theorists have highlighted that those within institutions, including civil servants, do not necessarily agree with an institution's worldview and are able to work overtly or covertly to engender institutional change (cf. Johnsson-Latham and Kronsell in this volume). As such, civil servants can be 'inside activists' seeking to further their values within an institution or, in contrast, toe the institutional line (Hysing and Olsson, 2018). Governing authorities are of particular interest, as the officials within them necessarily require adaptation and discretionary power to carry out their role in dynamic situations where official direction may be vague and/or conflicting (Applbaum, 1992). However, concomitant to this discretionary power is the need for a certain legitimacy. If this is lost then public confidence in government may be undermined (Hysing and Olsson, 2018, p. 141). Linked to this, as the rise of Trump in the US illustrates, politicians may make political capital by depicting public officials as part of a 'deep state' opposed to what they consider the will of the people. As such, public officials walk a fine line and become involved in various societal conflicts through the actions they take. This highlights two dimensions relevant to institutional change. Firstly, institutions must be mindful of their social legitimacy – their standing within broader society affects their actions (Hatch, 2018). Secondly, and linked to the first point, institutional worldviews will contain assumptions about their place and role in society. Both of these usually entail a certain path dependency as agencies are steered by their own and outsiders' ideas about what they can and should do.

At the individual level, civil servants do make decisions and periodically can act against dominant institutional structures, which suggests individual agency and room for 'inside activism' (Chapter 5, this volume). This is one of the ways that institutions change. Within dynamic contexts, actors and structure thus hang together within processes of structuration (Giddens, 1984). As such, there are times when actors embedded in institutions display path dependency whereas at others, they are able to act with limited reference to the organisation (Hysing and Olsson, 2018).

Identities and role expectations

Feminist institutionalism has almost exclusively focused on gender, but we suggest, as already stated, that the intersectional understanding of identities and social factors other than gender is needed for our analysis. This means that we expand the gender lens of feminist institutionalism to include factors such as education, class, age and geography, when exploring both the role of professional identities of policy-makers and their understanding of climate-relevant social differences.

Professional identities in climate institutions are important to study since climate actions and policies are not only the product of path-dependent, sticky institutions but also determined by how individual policy-makers make sense of their position and professional identities. Feminist institutionalism argues that institutions organise power inequalities in two main ways. Firstly, through formal as well as informal rules and practices. Secondly, through co-production of identities intertwined with the daily life and logic of institutions. In most Western states, including Sweden, climate policy-making primarily takes place in the energy, environment or transport sectors. As such, the norms and path dependencies expressed in the rules and culture of the institutions active in that sector will affect understandings of how social power relations work within policy-making. In addition, the professional identity of the policy-makers, perhaps educated as engineers or economists, might make technical and economic solutions seem most appropriate. In our analysis, we side with Arora-Jonsson and Sijapati (2018) and propose that professional identities of civil servants and institutions are of considerable significance when studying climate institutions. As outlined above, climate policy has been criticised for being overly reliant on technical and economic action instead of also acknowledging the social dimension of climate change. Further recognition of the social dimension of climate change would, for instance, mean that climate-relevant differences in behaviour, adaptation, views and vulnerability of different social groups are taken into further consideration in climate policy-making. This limited recognition of social difference arguably limits the possibilities for institutional action, with consequences for the possibility of just sustainable climate transitions.

Of particular relevance to this chapter is the notion of 'green inside activism' (Hysing and Olsson, 2018), not dissimilar to the notion of femocrat or gender activist, referred to in other chapters in this volume. Inside activists are individuals 'engaged in civil society networks and organizations, who hold a formal position within public administration, and who act strategically from inside public administration to change government policy and action in line with a personal value commitment' (Olsson and Hysing, 2012, p. 258). In our study, we look for such inside activism in terms of how these civil servants describe their possibilities for initiating climate acts – e.g. critical acts built on an intersectional or gendered understanding of climate change.

Cases and methods

We collected the data for this chapter through an interview study in the spring of 2020, of civil servants in Swedish climate institutions, and based it upon the responses of two government agencies, the Swedish Environmental Protection Agency (EPA) and the Swedish Transport Administration (STA). We briefly present these agencies in turn. The Swedish governmental system is characterised by relatively small government ministries (headed by elected politicians) and relatively large and rather autonomous government agencies (*myndigheter*, staffed by public officials). Agencies fall under the direction of the government, but have a certain amount of their own influence, dealing with actors at all social scales. Agencies typically have a sector of responsibility, within which they are influential. In their work, agencies provide evidence for policy, act to real-ise government policy and respond to Swedish and international climate goals (e.g. Agenda 2030) (Swedish Government, 2017). Notable to Swedish climate policy is the absence of a specific climate agency. Instead, climate action is dis-persed within different agencies as considered relevant. This structure is similar to how climate politics are organised in many other European states, including other Nordic states with the exception of Norway (Magnusdottir and Kronsell, 2015), the EU (Chapter 3, this volume) and the German Federal Government (Chapter 5, this volume). However, governmental agencies in Sweden might be considered more autonomous in their decision-making than many of their European counterparts.

The Swedish Environmental Protection Agency is the state agency for envi-ronmental issues, employing around 600 people across two offices in Stockholm and Östersund (Oscarsson, 2020). It supports Swedish governmental action plans every four years alongside other activities (Persson et al., 2019). This includes (but is not limited to) monitoring Sweden's environment in terms of pollutants, invasive species and so forth, and collating knowledge. It also carries out work maintaining Sweden's network of nature reserves. The SEPA also develops and implements environmental policy, cooperating with other agencies. The SEPA was selected for this research project due to it's heavy involvement in climate activities across the board. In contrast, the Swedish Transport Administration, (STA) is responsible for long-term transport infrastructure planning, construc-tion and maintenance, and as such works with climate policies in relation to transport. Transport is one of the heaviest contributing sectors to Swedish carbon emissions, rendering the STA relevant to this project (Naturvårdsverket, 2012). The STA is an extremely large agency, with around 9000 employees. Based in Borlänge it has regional offices in six other Swedish cities (Trafikverket, 2019).

Initially, we selected respondents purposively, with individuals at each selected agency previously signalling their interest in participating in this study. We then selected further respondents utilising snowball sampling. We continued to seek out new respondents until we reached a saturation point, with the same people repeatedly suggested to us (Bryman, 2004). In total, we conducted nine inter-views with SEPA staff (iEPA01-09, four women and five men) and six from the STA (iSTA01-06, two women and four men). Respondents held different levels of seniority within their respective organisations. SEPA respondents held vari-ous relevant roles, such as investigators and analysts around carbon emissions, as well as policy advisors. STA respondents included investigators, project manag-ers and goal directors. Because of the ongoing Covid-19 crisis, we conducted all interviews online using Zoom or Skype for Business software. Having acquired informed verbal consent, we recorded each interview. We organised each inter-view in a semi-structured manner around a research guide. Questions focused upon a) respondent background and experience; b) interpretations of social jus-tice within each agency; and c) the institutional contexts within which agen-cies operate. Following the recordings, we transcribed each interview and utilised ATLAS 8 qualitative analysis software to interrogate the data. The interviews were originally conducted in Swedish, but we have translated the extracts utilised here.

Several writers have tried to theorise the links between institutional world-view and institutional activity and change (cf. Hatch, 2018). For example, Hatch (1993) describes a dynamic system where assumptions and values of the world manifest in cultural symbols and artefacts. This research does not directly utilise such a framing but shares with it the basic ontological standpoint that respondents' interview discourses are reflective of their organisational cultures, coproduced with their institutional worldviews. We, thus, pay attention to the assumptions and values evidenced by respondents.

Top to bottom, bottom to top? Change possibilities at the SEPA and STA

Respondents from both the SEPA and STA made various statements about their ability to change their agencies' emphases around climate change action and social justice. The picture painted was, at times, seemingly contradictory. Respondents would frame dominant approaches towards climate and social justice as dependent on instructions laid down elsewhere, predominantly from the government. However, at other times, respondents would also assert that they had space for individual and institution-instituted action. The following paragraphs discuss this dissonance. We then turn to the implications this picture presents for efforts to overcome institutional path dependency and incorporate social justice into the SEPA and STA's climate change action.

When questioned on the possibilities for incorporating broader conceptualisa-tions of social justice in their climate change work many respondents asserted that a key factor was their agencies' remit and the tasks (*uppdrag*) that they received from the government. Respondents would characterise the historical trajectory of their agencies' involvement with climate change work as a product of Swedish governmental decisions. This could be in the form of direct assign-ments of tasks, or it could be through the governments' undertakings regarding international agreements (such as Agenda 2030) and climate targets, which may express social sustainability goals:

> Either you get your assignments via the instruction [Swedish government rules about each agencies remit], or you get it every year when you receive a regulation letter [from the government] or you get different types of govern-ment assignments during the year. ... We get [assignments] not only from the Ministry of the Environment, but we get it from ... the Ministry of Finance or... [The orders] That we together with other authorities should do things.
>
> (iEPA04)

> We have taken Agenda 2030 to our hearts ... It was a government assign-ment that was about what can [STA] do to achieve Agenda 2030? We have

interpreted this as partly what should the transport system do. We call it Target 2030, ten aspects and fourteen goals. … we call it which piece of the cake we should make, what is it that the Swedish Transport Administration can contribute to, for example, achieving the climate goal or traffic safety, or accessibility throughout the country or accessibility for everyone … [the] more social sustainability goals.

(iSTA04)

This also meant that respondents voiced an unwillingness to act on things that they perceived as outside of the agencies' purview. This presented a particular challenge when it came to issues such as social justice that cut across multiple actors' jurisdictions within the Swedish governance system.

The Swedish Transport Administration is not allowed to work to influence behaviour, so this means that [social justice] requires close cooperation with other actors, such as municipalities. So that it is they who, as it were, are allowed to carry out these first steps to influence behaviour …. yes, talking about norms and values and so on. It is not a mission we have.

(iSTA01)

Personally, I try to achieve a transparent basis for decision-making. We are not politicians, we are not decision-makers, we are civil servants, so we must, as it were, show politics what the consequences will be of their decisions.

(iSTA04)

This also meant that when social justice did come in the fact that it was neither agency's main focus meant that it was dealt with in an ad hoc fashion. Thus, for example, some SEPA respondents described social justice as more of an add-on to their broader environmental role.

We do some analysis; we look at distributional effects as far as possible … Most recently, I looked at tax deductions for work journeys, looked at how it affected different geographical areas. However, I must say that it is something that we need to develop, we discuss it a lot: how we should incorporate additional distributional effects into our analysis … *We look primarily at the emissions effect.*

(iSEPA02, our emphasis)

Repeatedly, respondents would thus paint a picture of top-down decision-making directing their activities. A distinct normative assumption emerged that agencies work under the government, and concomitantly values around maintaining this relationship were in evidence (which is expanded on later).

However, it would be inaccurate to state that agency staff do not have a role in shaping and carrying out policy (Svara, 2001). At certain other times, respondents from both the SEPA and the STA would also indicate that they had a certain amount of innovative freedom. This would involve the formation of internal networks to discuss issues beyond the official remit as well as anticipating the agencies' future roles.

> But the government has never really had to tell us directly … 'we want you to make [climate-based] demands in your procurement'. Without it, we have ourselves [made the decision]. However, [the government] was very interested when they saw that we started making such demands. But they have never said anything about 'you should not do that, or you should do that'.
>
> (iSTA02)

> But we also do some self-initiated analysis where we try to be a little proactive and see what the things are that we will need to work on in the future. And do not just sit and wait for the assignments to come. Because they can have a fairly short response time, and then it is good that you have done as much analysis of the problem as possible.
>
> (iEPA07)

Furthermore, while the agencies' roles include responding to requests for information from the government, they also had a role in screening what information the government receives from them. Thus, respondents also described making judgements about the amount and nature of information policy-makers need as well as consideration about what issues they need to raise to prominence. Respondents, thus, highlighted that agency staff had a certain discretion in their action (Applbaum, 1992).

> What we do is that we work with regard to our taskmaster, the government, and then give them the analysis that they need to have, or *what we judge they need doing*.
>
> (iSEPA02, our emphasis)

> But then I think this pressure to work with social sustainability now, it probably comes both from above and below, I think. A bit like the environmental issue has been, that there are also interested people, *it is enthusiasts in the organisation who think it is important*.
>
> (iSTA05, our emphasis)

Likewise, this involved some awareness of the way broader political winds were blowing, and at times respondents voiced an unwillingness to becoming embroiled in broader left-wing/right-wing politics. In these cases, issues of social justice and indeed, environmental issues, were framed as part of left-wing politics. However, this implied tying the hands of the agency staff. As employees of expert-driven, theoretically non-political agencies, such questions were not theirs to discuss. There was also a further element to this; respondents admitted that agency workers could and would try to anticipate future information gaps and needs in their work. So, for example, one respondent described the SEPA preparing for future information requirements around European climate action by starting analysis early, before requests for information started coming in. This is suggestive of institutional values encouraging creativity rather than simply rote bureaucratic obeisance to established routines.

Such behaviour speaks of the potential of agency staff to be 'inside activists' in the sense that they can overtly and covertly act to further their own political agendas to a certain extent, independent of the directions of their taskmasters (cf. Hysing and Olsson, 2018). However, in this ability to make judgements and act independently of government direction, respondents were also open and reflexive about their own limitations in this regard. Respondents described how their own professional and academic backgrounds played an important role in their approach to their work and the need to incorporate social justice issues. Thus, this contributed to a tendency to follow extant ways of doing things.

> When I started there in the 00s, you already have a form of ... policy instruments ... in place. And we have that as a starting point when we write our report, we describe that we have a carbon dioxide tax in Sweden, because we already had that then. ... So, it is probably some form of path dependency, that we already from the beginning have ... certain instruments in place.
>
> (iEPA05)

Several respondents characterised agency employees as often coming from natural scientific, economic, or engineering backgrounds and this affected how they understood their role, which is in line with the institutional literature (Arora-Jonsson and Sijapati, 2018, Hysinge, 2018). Thus, an agency's own backgrounds and interpersonal networks would affect how they approached particular issues or assignments.

> I think there are many at the Swedish Transport Administration who are engineers and thus, the culture is like engineering culture. We probably have many other professions as well, but the culture is engineering culture. Then you want to measure, you want it a little simple, you want little neat tables.
>
> (iSTA04)

> There are still large parts of the Swedish Transport Administration who believe very much in technological development as the solution to the climate issue, for example. Little boys' toys, like this [laughs].
>
> (iSTA05)

The question of 'activist behaviour' within government agencies was, to a certain extent, considered ambivalently. When queried on their action, respondents would voice a respect for their role following the government's direction. Indeed, one respondent in particular articulated that there were risks to agency workers following their own personal passions rather than doing their job.

> [T]he most common disease is civil servant activism. That you want so much with your own agenda so that you push something very hard yourself, more than what you actually have a mandate for, either from a political or civic point of view. And to be honest, it is actually very common in the environmental field, and that's probably because it is so important.
>
> (iSTA06)

The same respondent asserted that different agencies seemed to have different organisational cultures in this regard, with some agencies' staff being in the respondent's eyes 'too activist', which brings with it the risks that they base their work solely upon their own experiences/preconceptions, rather than decisions in a rational manner upon a scientific basis. One of the stated risks of such an approach was losing touch with the reality of policy effects upon people's lives, in turn causing iniquities and/or fomenting public resistance (iSTA06). This echoes concerns of unchecked civil servant activism making their actions more manipulative than simply facilitating the smooth running of Sweden's systems (Bellone and Goerl, 1992; Irvin and Stansbury, 2004). Thus, while a certain independence was valued, this was also shackled by concerns about the legitimacy of agency action and its relationship to democracy.

This combination of top-down and bottom-up governance had direct implications for respondents' views on the possibility of organisational change. Thus, when asked to consider ways of incorporating the understanding of intersectionality as a way to develop socially just climate policies, the respondents repeatedly asserted that change should come from above, either in the form of government direction, or from senior agency staff. At the same time, respondents' were also interested to learn more about an intersectionality approach in climate policymaking and expressed a preference for concrete examples having direct relevance for their work. Indeed, several respondents felt that there was a need to find ways to better incorporate social justice into their agencies' climate work.

Discussion: implications for institutional change

These data suggest an immediate first interpretation, in the case of the SEPA and STA, that institutional change is dependent on top-down guidance from the Swedish Government. Respondents framed institutional values around the legitimacy of government direction and their limitations in acting independently. Whilst this top-down view of agency work is the formal framework for Swedish state agencies, it misses the undoubted action possibilities that government agencies have. The study also found that agency staff take an active position in interpreting their role and guiding the government by making judgements about what information the government needs and anticipating future work and research directions. Values around creativity and expert-based predictions of future needs were thus extant. Agency staff were seen to have the potential to push particular issues that they themselves were passionate about, highlighting the possibility for 'inside activism' (Hysing and Olsson, 2018). However, in this, staff are also constrained by the particular historical trajectories of the institutions that enfold them (cf. Hatch, 1993) and the extent that they may exercise discretion in their professional roles is a long-standing debate (Applbaum, 1992). These constraints, or alternatively, the agencies' institutional path dependency might make the inclusion of gender and other social factors appear less desirable or appropriate and thereby limit the room for inside activism (Mackay et al., 2010; Magnusdottir and Kronsell, this volume).

Agencies very seldom attempt to address problems in completely novel fashions; past experiences affect both interpretation of their various assignments by agency staff, as well as the way they innovate and institute institutional change (or not). As such, path dependency was evident on a number of levels. Firstly, the tendency to follow previous practice directly affected the types of information that agencies accessed and utilised in their work. In the Swedish context, where climate policy-making primarily takes place in the energy, environment, or transport agencies, the norms and path dependencies expressed in the rules and culture of those institutions might make technical and economic solutions seem most appropriate and, thereby, information and knowledge about the relevance of gender and other climate-relevant social factors might be excluded. Secondly, the institutional frames employed had a normative feature – how respondents framed Swedish society affected how they approached different tasks (cf. Singleton et al., forthcoming). Thus, path dependency is influential even in situations where respondents are able to take action towards institutional change. Put differently, respondents mentioned that, at times, it was hard to know what they missed because their epistemological positions were informed by their institutional embeddedness and experience.

All this has direct implications on any efforts to inculcate change. This section discusses these implications with particular reference to Swedish agencies as institutions. Particularly, we focus on what we have established as a particular identity that many respondents framed in their interviews. We call this their identities as civil servants (*ämbetsmänniskor*) that they perform (or aspire to perform) in their work roles. To varying degrees, many respondents framed an ideal that civil servants should be neutral in their work and this limited the extent to which they should be 'activists' for social justice within their climate activities. At times, they evidenced a certain pride in this role (cf. Olsen, 2006). Throughout the interviews, there was a distinct preoccupation with the *legitimacy* of respondents in their role as civil servants, which one could also interpret as the logic of appropriateness of the role. While the institutions that civil servants find themselves in are not totally and completely controlling (cf. Goffman, 1961), they do affect how those embedded in them go about action. For example, iSTA06 argued, in effect, that too much discretionary activism would push agencies into a danger zone with civil servant acting illegitimately. These institutions can likely affect several of the dilemmas that may arise for them in their work, for example how 'radical' they can be – for instance, do the institutional norms allow a gendered or an intersectional approach in their work? Or how much action they can take without explicit direction/collective agreement and how open any activist activities can be (Lowi, 1993; cf. Hysing and Olsson, 2018, pp. 110–114). Respondents articulated norms of agencies as apolitical units, seeing their legitimacy in part as linked to their separation from 'politics' (Svara, 2001; cf. Olsen, 2006). Fortunately, the literature on public administration suggests several ways in which civil servants may take legitimate action. These, combined with the data, provide potential avenues for engendering institutional change without endangering the social legitimacy of government agencies (cf. Hatch, 2018). It should, however, be noted that these suggestions are rooted in the collected data, which constitute respondents' representations of their agencies. As such, more in-depth (ideally field-based) work is required for more precise, targeted suggestions (e.g. along the lines of Hatch, 1993).

The first, and most obvious, strategy for institutional change suggested by the data is to enlist allies further up in the hierarchies of the agencies, in the form of politicians or senior agency figures. Legitimacy could thus be garnered by involving those with democratic legitimacy (Hysing and Olsson, 2018, p. 150). Respondents expressed a desire for clear direction on the integration of social justice in their climate change work. Practically, however, this may prove challenging and indeed, the idea of a neat separation of politicians and bureaucracy (i.e. change is simply a matter of top-down direction) is an inaccurate portrayal of the interwoven reality (Svara, 2001, 2006). Nevertheless, limited recognition of social justice or climate-relevant social differences could be perceived as a matter of a democratic deficit, potentially affecting the legitimacy of institutions.

Secondly, on a simpler level, respondents' views on integrating social justice into climate work stress the importance of following the direction of politicians. In effect, legitimacy in their eyes links to following the decisions of those with a democratic mandate. A first simple step, then, is to highlight how desired institutional change – rather than embodying revolutionary action – actually involves following the direction of the agencies' political masters. For example, in the case of our ongoing research on Swedish climate institutions,[1] we will seek to frame an intersectionality approach as following government policy, rather than agency activism. Going forward, we will highlight how intersectionality has direct relevance to Sweden's climate obligations (for example, Agenda 2030), in turn directly related to social justice issues and concepts (Swedish Government, 2017). Legitimacy is thus conferred by highlighting how apparent change conforms to established rules and norms of sticky institutions (March and Olsen, 1996; Beetham, 2013, p. 15). By establishing that change is evolutionary within an agency's legitimate 'mission trajectory', respondents' concerns that they are acting out of turn will be ameliorated by illustrating that it falls within established rules and norms (Goodsell, 2006). The feminist institutionalist literature, however, points out that institutions organise power inequalities through their formal as well as informal rules and practices. This might mean that efforts to institute 'legitimate' institutional change based on established norms and rules only reinforce social power inequalities, excluding new intersectional knowledge. Linked to the second point, one can also make appeals to arguments about how agencies taking action may be legitimate. Margaret Stout (2013) has identified three ways that bureaucratic actions may be shown to be legitimate. These are by '(a) decentralising authority and responsibility to the people; (b) defining public interest in relation to specific situations in dialogue with affected stake-holders; and (c) engaging in processes in which citizens can actively hold offi-cials accountable' (quoted in Hysing and Olsson, 2018, p. 155). As such, where appropriate, one can make arguments that a change will improve the institu-tional accountability and thus add legitimacy. Accountability has four dimen-sions 'Transparency, liability, controllability, responsibility and responsiveness' (Koppell, 2005, p. 94). Based on our intersectional and feminist institutionalist starting points, we can argue that greater nuance in framing Swedish society will allow for greater representation and acknowledgement of marginalised groups in agency work and thus improve institutional legitimacy (cf. Hysing and Lundberg, 2016). Public administrations also gain legitimacy through their roles as collators and providers of expert knowledge for government use (Goodsell, 2006, p. 630). As such, we can argue that by integrating intersectional understanding into their work, respondents will be developing needed-expertise and, thus, gain greater legitimacy as the experts on the various sectors of Swedish society. Along similar lines, we will also point to research highlighting the robust support within the Swedish population for multiculturalism, for example (Ahmadi et al., 2016). We can thus stress the democratic legitimacy for seeking broader framings of Swedish society.

Conclusion

In this chapter, we have presented data and analysis drawn from interviews with staff at the Swedish Environmental Protection and the Swedish Traffic Administrations. As a part of our research project seeking to find ways to integrate intersectional insight into agency work on climate change, we were interested in how institutional path dependency manifests in respondents' representations of their roles. Our findings reveal that, in varying degrees, respondents agreed that their possibilities for action were limited both in a top-down manner by the direc-tions of the Swedish Government and by sticky norms and values within their organisations. There was also strong emphasis on values of democracy, account-ability and legitimacy, which according to many of the civil servants limited their room for potentially critical action leading to further recognition of social differ-ences. This is interesting since one may interpret the limited representation of different social groups in climate policy-making, as well as lack of understanding about the intersectional nature of climate effects, as a form of democratic deficit. However, at the same time, many respondents also revealed that agency staff had certain discretionary powers to influence both research and policy and, at times, could play a steering role for governments by making decisions about the information policy-makers receive. In this description, respondents, thus, spoke positively around creativity and problem-solving. Both of these sets of values embody a certain path dependency. In the former, external prescriptions regarding agency roles set the path on which the institutional worldviews develop; in the case of the latter, staff drew upon their professional and academic backgrounds, favouring the types of initiatives and the policy research and implementation tools they were receptive to and used. This contradictory picture leads to suggestions about how to develop a more just and sustainable climate policy in a path-dependent institutional environment, which still allows for some critical initiatives. These are: firstly, gathering and including various types of knowledge and information on climate-relevant social factors early in the policy-making process. Secondly, increasing civil servants' awareness, with the help of feminist institutionalism, about challenges in institutional environment. Thirdly, enlisting high-level allies (top-down governance) to encourage particular directions. Finally, highlighting how the proposed tool of intersectionality can both sustain and even increase the legitimacy within the agencies' extant remits and within wider Swedish society as embodying widely held values. This would prevent civil servants from feeling that such action entails entering a 'danger zone' of illegitimate action. In future, we will seek to do this by designing educational materials and methods to illustrate that bringing a wider view of social justice into agency climate work is evolutionary and revelatory rather than revolutionary, with societal legitimacy.

Acknowledgements

We are most grateful to all respondents who participated in this study.

Note

1 The authors are involved in a four-year research project titled *Intersectionality and Climate Policy Making: Ways Forward to a Socially Inclusive and Sustainable Welfare State*. The project, financed by the Swedish Research Council for Sustainable Development, explores how climate policy-makers in government agencies work with social inclusion and how intersectional aspects relate to climate policy-making at the municipal level.

References

Ahmadi, F., Palm, I. and Ahmadi, N. (2016) *Mångfaldsbarometern 2016*. Gävle: Gävle University Press.

Alston, M. and Whittenbury, K. (2013) 'Does climatic crisis in Australia's food bowl create a basis for change in agricultural gender relations?', *Agriculture and Human Values*, 30(1), pp. 115–128.

Applbaum, A.I. (1992) 'Democratic legitimacy and official discretions', *Philosophy and Public Affairs*, 21(3), pp. 240–274.

Arora-Jonsson, S. and Sijapati, B.B. (2018) 'Disciplining gender in environmental organizations: the texts and practices of gender mainstreaming', *Gender Work and Organizations*, 25(3), pp. 309–325.

Beetham, D. (2013) *The Legitimation of Power*. New York: Palgrave Macmillan.
Bellone, C.J. and Goerl, G.F. (1992) 'Reconciling public entrepreneurship and democracy', *Public Administration Review*, 52(2), pp. 130–134.
Bryman, A. (2004) *Social Research Methods*. Oxford: Oxford University Press.
Douglas, M. (1987) *How Institutions Think*. London: Routledge and Kegan Paul.
Dryzek, J.S. (2013) *The Politics of the Earth*. Oxford: Oxford University Press.
Eckersley, R. (2020) 'Greening states and societies: from transitions to great transformations', *Environmental Politics*. Online 10.1080/09644016.2020.1810890.
Giddens, A. (1984) *The Constitution of Society*. Cambridge: Polity Press.
Goffman, E. (1961) *Asylums. Essays on the Social Situation of Mental Patients and Other Inmates*. Harmondsworth: Penguin Books.
Goodsell, C.T. (2006) 'A new vision for public administration', *Public Administration Review*, 66(4), pp. 623–635.
Hatch, M.J. (1993) 'The dynamics of organizational culture', *Academy of Management Review*, 18(4), pp. 657–93.
Hatch, M.J. (2018) *Organization Theory. Modern, Symbolic and Postmodern Perspectives*. 4th Edition. Oxford: Oxford University Press.
Hysing, E. and Lundberg, E. (2016) 'Making governance networks more democratic: lessons from the Swedish governmental commissions', *Critical Policy Studies*, 10(1), pp. 21–38.
Hysing, E. and Olsson, J. (2018) *Green Inside Activism for Sustainable Development: Political Agency and Institutional Change*. Cham: Palgrave Macmillan.
Irvin, R.A. and Stansbury, J. (2004) 'Citizen participation in decision making: is it worth the effort?', *Public Administration Review*, 64(1), pp. 55–65.
Kaijser, A. & Kronsell, A. (2014) Climate change through the lens of intersectionality. *Environmental Politics* 23(3): 417-433.
Koppell, J.G.S. (2005) 'Pathologies of accountability: ICANN and the challenge of "multiple accountabilities disorder"', *Public Administration Review*, 65(1), pp. 94–108.
Laestadius, S. (2018) *Klimatet och omställningen*. Stockholm: Borea.
Lowi, T.J. (1993) 'Legitimizing public administration: a disturbed dissent', *Public Administration Review*, 53(3), pp. 261–264.
Lundqvist, L. J. (2004) *Sweden and ecological governance. Straddling the fence*. Manchester: Manchester University Press
Mackay, F., Kenny, M. & Chappel, L.. 2010. New Institutionalism Through a Gender Lens. Towards a Feminist Institutionalism? *International Political Science Review* 31(5): 573-588.
Magnusdottir, G.L. and Kronsell, A. (2015) 'The (in)visibility of gender in Scandinavian climate policy-making', *International Feminist Journal of Politics*, 17(2), pp. 308–326.
Magnusdottir, G.L. and Kronsell, A. (2016) 'The double democratic deficit in climate policy-making by the EU commission', *Femina Politica*, 25(2), pp. 264–277.
Mahoney, J. and Thelen, K. (eds.) (2009) *Explaining Institutional Change: Ambiguity, Agency, and Power*. Cambridge. Cambridge University Press.
March, J. and Olsen, J. (1996) 'Institutional Perspectives on Political Institutions', *Governance*, 9(3), pp. 247–264.
Nagel, J. (2012) 'Intersecting identities and global climate change', *Identities* 19(4), pp. 467–476.
Naturvårdsverket (2012) *Underlag till en färdplan för ett Sverige utan klimatutsläpp 2050*. Stockholm: Naturvårdsverket.
Olsen, J.P. (2006) 'Maybe it is time to rediscover bureaucracy', *Journal of Public Administration Research and Theory*, 16(1), pp. 1–24.

Olsson, J. and Hysing, E. (2012) 'Theorizing inside activism: understanding policymaking and policy change from below', *Planning Theory & Practice*, 13(2), pp. 257–73.

Oscarsson, B. (2020) Organisation [online]. Available at: https://www.naturvardsverket.s e/Om-Naturvardsverket/Organisation/ (Accessed 1 October 2020).

Persson, T., Fogelström, E., Lundell, Y., Notter, M., Hansen, K., Jacobsen, K., Skovdal, A. and Dahlgren Axelsson, S. (2019) *Handlingsplan för Naturvårdsverkets arbete med klimatanpassning*. Stockholm: Naturvårdsverket.

Resurrección, B.P. (2013) 'Persistent women and environment linkages in climate change and sustainable development agendas', *Women's Studies International Forum*, 40, pp. 4033–43.

Singleton, B.E., Rask, N., Magnusdottir, G.L. and Kronsell, A. (Submitted) 'Intersectionality and climate policy making: the inclusion of social difference by three Swedish government agencies', *Environment and Planning C: Politics and Space*.

Stout, M. (2013) *Logics of Legitimacy: Three Traditions of Public Administration Praxis*. Boca Raton: CRC Press.

Svara, J.H. (2001) 'The myth of the dichotomy: complementarity of politics and administration in the past and future of public administration', *Public Administration Review*, 61(2), pp. 176–183.

Svara, J.H. (2006) 'Introduction: politicians and administrators in the political process – a review of themes and issues in the literature', *International Journal of Public Administration*, 29(12), pp. 953–976.

Swedish Government (2017) *Ett klimatpolitiskt ramverk för Sverige (A Climate Policy Framework for Sweden)*. Bill 2016/17:146, approved by the Swedish Parliament, 15 June 2017 (Stockholm: Ministry of Environment and Energy, 14 March 2017).

Thompson, M. (2008) *Organising and Disorganising*. Axminster: Triarchy Press Limited.

Trafikverket. (2019) Trafikverket [online]. Available at: https://www.trafikverket.se/en/ startpage/about-us/Trafikverket/ (Accessed 1 October 2020).

Part 2

Sectoral climate institutions

7 Towards a climate-friendly turn?

Gender, culture and performativity in Danish transport policy[1]

Hilda Rømer Christensen and Michala Hvidt Breengaard

Introduction

Transport has become an explicit part of climate policy and CO_2 reductions, and in 2019 the Danish Government set up an ambitious target of reducing CO_2 emissions by 70% by 2030. Currently, we see transport policy and electric cars as vital components of broader environmental and climate policy agendas. A fresh example of how cars are still being prioritised in this new climate policy agenda can be seen in a recent report from the government-appointed Climate Council. In Denmark, transport has been explicitly moulded into the broader agenda of climate policy; yet with the potentials of electric cars still being considered as the sole avenue of transport change, in tandem with pressure from big energy suppliers and wind power companies (Cyklistforbundet, 2020; Klimarådet, 2020.[2] All in all, it is hard to see how the targets of 70% CO_2 reductions will be met. In Denmark, major political parties – from social democrats to liberals to conservatives – support the development and modernisation of 'climate-friendly', yet still car-centric, transport solutions (Venstre, 2016; Socialdemokratiet, 2019). Furthermore, many studies and surveys demonstrate that both non-motorised mobility, as well as gender and other social categories, are still by and large missing from the tone-setting agendas, councils, strategies and surveys (Christensen and Breengaard, 2019; TINNGO, https://www.tinngo.eu/).

This chapter will focus on how transport policy, perceived by many as a dull subject, is dramatically conveyed by the media. It includes the representation of certain roles and performances of a group of central agents, which are acting and enacting certain policies and preferences according to gendered norms of appropriateness (Krook and Mackay, 2011). They seem to connect to and reproduce an unquestioned matrix of soft and strong, climate-friendly or unfriendly transport practices and values. The omission of gender and other socio-cultural practices appear as impediments to change and climate-friendly initiatives.

As for the structure, we begin the chapter with a short prelude to feminist approaches to transport studies, ending with the introduction of useful concepts related to the cultural and performative turn of transport policy studies. In the next section, we present a mapping of the entangled gendered and social codes of transport policy – addressing the general relevance of gender representation,

DOI: 10.4324/9781003052821-7

the question of substantial gender policy interests and gender mainstreaming as a tool of limited success. In the final section, we conduct an empirical case study of communicative events connected to the first female Minister of Traffic in Denmark, Sonja Mikkelsen. She was a warm supporter of public transit and the analysis reveals that gender, power and environment/climate issues are hot topics that cut across policy at all levels, including language, bodies, food, festivals, etc. In the last section, we reflect on the usefulness of gendering the cultural and performative turn in policy analysis.

Theoretical and methodological departures

The gendered dimensions of transport and car culture is a vital but also neglected field, which is lagging behind more sustained scholarly and political equality concerns about equal opportunities in family life, workplace, health and education. Geographer Susan Hanson has pointed to significant gaps both in existing feminist studies and in mainstream transport research and has called for refined analysis in either field (Hanson, 2010). According to Hanson, feminist analysis has focused too heavily on gender and has underplayed the specificities of mobility and transport. Feminist research, she argues, based on qualitative analysis has mainly focused on how mobility shapes gender but lacks perspectives on how gender shapes mobility. Several studies along these lines have pointed to transport and mobility as producing gender stereotypes, indicating women and femininity as synonymous with immobility and aligned with home and domesticity. It is claimed that this is an enduring feature that surfaces in everyday practices as well as in multiple cultural and political forms (Edensor, 2004; Cresswell, 2006; Transgen, 2007). The neglect of the institutional gendered dimensions of transport policy-making comes to hide the unequal distribution of resources, for example, in prioritising of cars rather than public transportation and non-motorised transport, and hampers the recognition of different kinds of mobility. It is vital, therefore, to study the gendered and cross-political character of such cultures and hegemonies. As well as throwing light on how gendered norms have been shaped and nurtured within these institutional and hegemonic structures.

In the field of feminist transport research, new trends are underway, addressing both issues of equal access and transport justice as well as methodological challenges in transport studies. Caren Lévy, a UK transport researcher, for example, appeals for broader and more complex ideas of transport and social identities and for what she calls the 'deep integration' of social identities in all areas of transport (Lévy, 2013). The social identities of transport 'users', she argues, are deeply imbedded in social relations and urban practices, ranging from the everyday lives of people to urban policies and planning. Even in studies of multiple identities of urban residents, such categories are often referred to the margins as the 'social and distributional' aspects of transport. Transport, Lévy contends, is not an isolated and delimited field, but one which has critical implications for how diverse citizens – men and women, girls and boys, young and old and non-binary persons – are able to exercise and use 'travel choices' both as individuals and collectively.

As we will show later, it seems as if these more structural perspectives on transport use are only gaining very fragile grounds in and around Europe.

In this chapter, we widen such perspectives and apply them to transport policy research. In doing so, we connect to recent approaches in the field, which can be clustered under the 'cultural turn' in policy studies and, furthermore, we connect such perspectives with central concepts of gender and diversity developed in feminist studies.

Several scholars have pointed to the need for taking policy studies beyond established theories and methods, such as overcoming dualism between structure and individuality, and replacing ontologies of substance with more dynamic and processual ontologies, as well as rethinking power as productive and as acting in many fields. Also, methodological break-ups have been urged for, from the focus on text and talk to the inclusion of wider mundane materialised and embodied experiences and affects (Lissandrello, 2015, 2016; Adler-Nissen, 2016). In order to assist the analysis, we use the notion of transport policy as a performative act. More specifically, we would like to connect to the distinct ideas of gender performativity, which at its initial stage was developed by the American philosopher Judith Butler. According to Butler, gender can be seen as a particular type of process; a 'set of repeated acts within a highly rigid regulatory frame'. The subject is not free to choose which gender it is going to enact. Yet, the subject is already determined within a cultural regulatory frame and has, therefore, only access to a limited number of choices. The idea of choice has been compared with a wardrobe, which consists of a limited selection of outfits, depending on gender as well as class, ethnicity, geography, etc. (Salih, 2002, p. 63). Following this feminist approach, acts can – from a broader perspective – be understood as 'shared experience and collective actions' (Butler, 1988, p. 525 cited in Lissandrello, 2015). Translated into the field of transport policy, this means that there are nuanced and individual ways of executing policy. The acts of policy-making are performed by gendered individuals, who act in accordance with certain cultural, political and social apparatuses. Therefore, this is not a fully individual matter.[3]

The idea of performativity relates to core ideas in feminist institutionalism, which has to do with how power is reproduced through acts of normalization and the ways in which institutions are reproduced through a certain logic of 'appropriateness' (Kronsell, 2015; Magnusdottir and Kronsell, this volume). It is a certain logic that makes the inclusion and analytical reflection of gender less appropriate in policy-making and institutions. While at the same time, it makes certain masculinity norms and stereotypical gender notions the most accepted and hegemonic. We conclude the chapter with reflections on the potentials of institutional opportunities and challenges for opening up new windows of climate-friendly transport politics for all.

In terms of methods, we apply a critical discourse analysis in a broad and open-ended approach. Following Norman Fairclough, the approach suggests that the world is constituted by both discourse and other social practices. (Fairclough, 1995; Winther Jørgensen and Phillips, 1999). Along this line, studies of communicative events take centre stage in forms of representations, such as text, talk, images and other forms of social practices, such as meals and networking styles. This provides a methodology that complies with the analytical requirements of the broader and more complicated ideas, agendas and practices related to governance and public sensibilities today. We have combined qualitative and quantitative methodologies of data analysis. We have combined language with other cultural expressions including cultures of meetings and transport as part of broader lifestyles and preferences. In this manner, our methodology is characterised by a widening of analytical attention and tools in the representation which turns out to be highly relevant in transport and climate policy-making.

More precisely, this chapter builds on screenings of the Danish Infomedia's online archive (https://infomedia.org/). We have conducted searches on a range of person-related keywords connected to ministers and central civil servants, such as 'Sonja Mikkelsen transport minister' (123 hits), 'Sonja Mikkelsen Traffic Minister' (4063 hits), 'Ole Zacchi, head of traffic department' (235 hits), 'Jørgen Halck, head of traffic department' (27 hits).[4] In order to widen the scope of media data we also searched topical keywords such as alternative transport, congestion ring, climate panel, etc. From this voluminous material we identified approximately 100 articles which we used as key data for the qualitative study. A smaller selection of them will be referred to explicitly in the analysis, while the others repeated or confirmed the claims and trends in these articles. The material was extended with an online interview with Sonja Mikkelsen in spring 2020 where the media information was supplemented and sustained. Throughout the chapter, we have enhanced this material with general reports on transport and research as well as public lists of ministerial appointments and scholarly articles.

Gendered representations in transport policy – Do women in transport policy make a difference?

A feature running as a distinctive undercurrent in transport policy is that it is a 'male' policy area and has remained so even in recent decades. Transport and mobility count as an old policy area in Denmark as well as throughout Europe. The first Danish minister for traffic was appointed as early as 1900. Environmental policy came about in the 1970s, following growing mobilisations and concerns and gained its own ministry in 1971. Climate has had a ministry of its own since 2007, following rising concerns and political attention. The issue of gender representation and women's participation is important for discussions of inclusive climate-friendly transport strategies. From a gender representational perspective, transport from the very beginning has figured as a manifest policy area conducted by men as argued in the introduction to this volume (Magnusdottir and Kronsell, this volume). This somehow differs from environment and climate, which seem more open and have already presented more female ministers than the area of transport has over the last 120 years.[5]

Gender representation and the question of fairness in elected assemblages is a longstanding issue in feminist theory as well as in political practices. Anne Phillips in her path-breaking study *The Politics of Presence* challenged the dominant idea

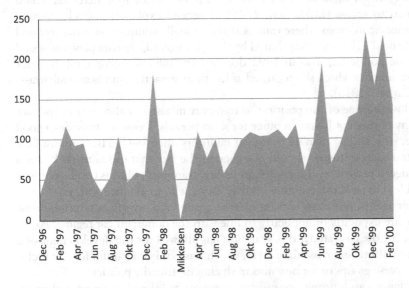

Figure 7.1 Number of articles about Westh and Mikkelsen during their time in office

of representation as only concerning ideas and not the people representing ideas (Phillips, 1998). We see the presence of diversity and gender as a participatory democratic dimension, which is currently aligned with the strategy of gender mainstreaming. Gender mainstreaming requires equal participation of women as well as the integration of other social groups in political and public life. Gender mainstreaming emphasises broad and mobilizing strategies, such as networking, dialogue, social mobilization and the involvement of NGOs in all stages of policy-making. The focus on equal representation responds to a deeper democratic deficit and the problem of gender imbalances in policy-making, which poses particular problems and gaps in transport policy. Since people tend to spot their own needs while not necessarily recognizing or knowing about the needs of other people, a lack of a wider representation in the various transport councils and ministries might easily end with the omission and ignorance of the diversity of travel patterns and needs (Christensen and Breengaard, 2019). Even though gender has been addressed in the EU since the mid-1980s, the gendered imbalances are still striking in current European transport councils and committees as demonstrated in the present EU project TiNNGO, focused on a game changer in European transport policy and practice. In the recent TiNNGO mapping of national parliamentary committees across the EU countries UK, Portugal, Romania, Germany, Spain, Sweden, France and Denmark, the proportion of women in 2019 varies between 7% and 40%. Similar mapping of transport ministers in the same countries shows that since 1945 only 23 out of a total of 245 ministers were women (Christensen and Breengaard, 2019). One explanation, provided by Sonja Mikkelsen, refers to the government transport committee as particular powerful e, which aligns more with masculine performance and, therefore, is hard for women to access (Mikkelsen, 2004). A related explanation circles around gender-specific interests where transport traditionally counts as unattractive and as a 'male field'. This view was shared by Ritt Bjerregaard, another powerful social democratic politician, who in 1992 declined the offer to become minister for transport; an offer which she regarded as less than attractive and as an embarrassment (Carlsen, 2018).

The low prestige of the position as transport minister is also seen in the fact that many of them only stay in office for short periods. Denmark provides a good example, with no less than 46 transport ministers since 1945. The constant flow of new transport ministers should not conceal the fact that the ministry handles huge budgets, projects and agencies. In reality, transport policy is conducted by a less visible, but powerful group of civil servants, who in alliance with politicians and other agents, represent various interest groups and private enterprises, most of whom are explicitly or implicitly in line with the dominant car regime. As we will show in the case study, this situation has allowed for the development of a particular culture and masculine style, which has not been gainful for the inclusion of diverse groups or for new and fresh climate-friendly politics.

Such figures and assumptions about transport as a low-key resort and male-dominated sector have been confirmed by a recent network study of the power elite in Denmark (Larsen, 2015). Transport, it is shown, does not figure as a prestigious or particular visible area in Danish societal and policy context. According to the extended mapping and network analysis of the Danish elite, the minister for transport figures at the margins of the elite network. Moreover, the network study shows that not only is the field of transport and infrastructure marked by an extremely high proportion of men, but it also stands out even in general male dominance in the elite in which only two out of the 100 most powerful persons are women (Larsen, 2015, pp. 62–63). From a sectorial perspective, transport, infrastructure and their related fields also proliferate as some of the most gender-divided sectors. In the topical elite network analysis, women make up under 10% of the subjects in the sectors of roads (7%), car sales (8%), shipping (9%), travel (9%) and cars (10%). In contrast, women dominated in the 'soft' areas of public institutions, such as health care, education, humanities and in terms of transport as cyclists. Following Bourdieu's interpretation of hard and soft subjects and of left and right wings of the state, this hardline gender division also places men in a favourite position related to the 'right hand' of the state, which is tied to the general patterns of dominance in the political field (Bourdieu, 2017, quoted in Larsen, 2015, p. 65).

The question is if and how such imbalances affect substantial policy and here specifically on the enhancements of sustainable and climate-friendly transport policy-making. In contrast to Sweden, available data is scarce and fragmented and points in several directions (Kronsell et al., 2015). In parallel with Swedish studies of transport policy, a fresh Danish analysis from 2019 shows how gender affects political priorities and representations in Danish municipalities. It turns out that there are clear gender differences – e.g. that women across party lines support more resources to childcare, healthcare and people with disabilities (Hermansen, 2019). What is more, female politicians in municipal councils favour collective transport higher than their male fellows, while policy areas such as 'roads' and 'environment' show a lower level of interest among women. While the study does not provide any explanations of this phenomena, we might assume that it relates to women's generally higher dependency on public transit as well as their transport responsibilities for children and elderly relatives. It is also shown in the study, that women only have a particular impact on their favourite policy issues in municipalities where they make up both a critical mass and where they are in powerful positions – for example, as mayors (Hermansen, 2019).

At the national level, the manifested gender gaps in transport are not only something of the past but seem to continue. In a recent list of appointments to councils and institutions associated with the ministry of transport, there have only been minor changes in gender representation. A count in 1998–1999 showed that approximately 10% of all 300 appointments were women (Mikkelsen, 2004, pp. 72–73). While a count in 2020 showed approximately 20% or 32 women out of a total of 154 members appointed to all committees affiliated to the transport and housing ministry. This conveys the message of enduring male dominance (Mikkelsen, 2004, pp. 72–73; Trafikministeriet, 2020).[6]

The strategy of gender mainstreaming has been the answer to address such imbalances both at European and global levels. Gender mainstreaming has been developed as an institutional method and consists of a range of tools to identify imbalances and inequalities in processes where gender, so far, has been invisible or regarded as not important.[7] The strategy helps to uncover how resources are used, how some groups benefit more from efforts than others, or how approaches can hide gendered imbalances. Besides, gender mainstreaming has been suggested as a tool for mobilisation and participation and for addressing gender and diversity in new and forward-looking ways (Squires, 2005; Christensen and Breengaard, 2011). So far, however, it does not seem as if gender mainstreaming and associated knowledge production has had any substantial impact on integrating gender and diversity in Danish or European transport planning and policy (Transgen, 2007; Christensen and Breengaard, 2019; Magnusdottir and Kronsell, this volume). Again, this refers to the reluctance in institutions and policy-making to further alternative priorities of non-motorised mobility and to go beyond the norms of addressing other than genderless (male) car drivers in the political priorities and institutions.

In the following section, we will explore how these complex configurations of gender and transport policy-making have unfolded in Denmark by examining a range of communicative events related to gender and the enhancement of more climate-friendly transport practices. A particular focus of attention is Sonja Mikkelsen's arrival and time as the first female traffic minister in Denmark. Her time as minister embodied the contested intersections of female gender- and climate-friendly politics. While the time she held office turned out to be very turbulent, it also marked a soft paradigmatic shift in transport policy in Denmark.

In general, Danish transport policy does not figure as a well-researched or comprehensive theme in policy or historical studies (Toft et al., 2000). This means that we had to collect evidence and establish both the context and the specificities and interpretations of the cases in question.

Sonja Mikkelsen: 'The first lady minister'

Sonja Mikkelsen was appointed as minister for traffic in Denmark during the social democratic–social liberal government transformation in 1998. Her appointment was subject to keen media interest. Not only was she the first female minister for traffic, she was also a social pattern breaker with a background in a working-class family from the rural area of Thy in Jutland, known for its scenic beauty and remoteness from the 'capital culture' in Copenhagen. At the time of appointment, her hometown and constituency were Aarhus in Jutland and she left behind a husband and two young children to work in Copenhagen – something that, at the time, was unusual for a female minister (Mikkelsen, 2020). By her time of appointment in 1998, she was already an experienced politician, and had spent around 14 years in parliament and as a former member of the parliamentary research and traffic committees. Besides this, Sonja Mikkelsen was known as a left-winged and determined social democrat, a warm supporter of public transit. She had voted against one of the big transport compromises, as well as her own party line, when highways in Northern Jutland were exchanged for support to the Great Belt Bridge by some of the politicians from Jutland.

The Ministry of Traffic routinely figured among the low-prestige institutions – also in terms of media attention. Yet Sonja Mikkelsen's entrance was followed with a keen interest in the print media. 'Highway hater' was an informal quote from one of the nervous civil servants, at her inaugural reception. Other informal claims were circulated such as: 'She is a solid left winged social democrat, she hates highways, anyway in Northern Jutland' and 'She seems to be guided by higher (political) powers – she needs to moderate that' (Rehling, 1998). The press immediately noticed the masculine atmosphere of the reception and the well-known fact that the transport sector was dominated by men:

> Surrounded by men, men and men and men again, the first female transport minister was welcomed in the ministry ... among the guests were the chiefs of the DSB (Danish State Railways, the Rail Council, Post Denmark, the Road Directory, the Traffic Council, the State Car inspection and A/S Great Belt ... all males.
>
> (Ritzau, 1998)

The head of the traffic department in 1998 was experienced and influential. He was known as both 'colourful' and powerful as well as as 'cheeky as a butcher's dog'. When he left his position in 2000, he was called 'the oldest male baboon' in parliamentary circles (Dahlin, 2000; Politikken, 2000). He was unusually outspoken as a civil servant and was, as chief head of the department, said to have seriously outmaneuvered a number of both fellow civil servants and ministers. A style which was also demonstrated in the appointment of Sonja Mikkelsen where he welcomed her as minister in gendered and patronising terms:

> You are the first lady [sic] as minister in the long history of the ministry. We are looking forward to seeing if you prioritise other values. We know that you have the courage and dare to express your opinions opinion unreserved. This can make us all sweat. Please remember that you are member of a government.
>
> (Rehling, 1998)

He also reminded Sonja Mikkelsen that the Ministry of Traffic was an empire that included 20 councils and ten companies, counting also the recently established Øresund company 'and often you will read in the newspapers what happen in them' (Rehling, 1998).

The Department of the Traffic ministry was known as a particular world of its own. The ministry and various governments had for decades granted their chief heads of department a level of autonomy which was unseen in other ministerial departments. Chief heads of departments or permanent secretaries, according to Danish democratic administrative rules, are routinely held under parliamentary and committee control. In this particular ministry, however, they seemed to have established their own rules and value regimes.[8]

On this occasion, the Head of the Traffic Department in the said goodbye to the tenth minister and hello to the eleventh minister.

> The Ministers forgo, but the ministry survives', was a standing joke in the ministry and he was in the routine of welcoming and saying goodbye to ministers. He himself had been the head of the department in the traffic ministry since 1993 after 11 years as head of the department in the ministry of housing.
>
> (Kjær, 1999; Nielsen, 1999; Politiken, 1999a)

An early goal of his as head of the department was to 'become master in my own house' which was an easy task due to the ministry he was serving in, which was seen by many as weak. Another reason for the weakness of the ministry was its role as a 'swing door' to more attractive political posts and resorts (Politiken, 1999a). He was a self-proclaimed social democrat but differed from Sonja Mikkelsen in his commitment to the conservative car-loving segment. In spite of growing attention from the social-liberal and conservative governments during the 1980s and 1990s to CO_2 emissions and environmental problems caused by private cars, in 1996 he bluntly claimed: 'We love private car-drivers in the Traffic Ministry. We regard cars as the best friends of humans. One cannot work in the ministry of traffic and be opposed to (car) traffic' (Det Fri Aktuelt, 1996, here quoted from Politiken, 1999a).

The central position of private car-driving was a clear indication that in 1996 traffic policy and the atmosphere of the Ministry of Traffic was still closely equated with cars. This horizon was, however, challenged during Sonja Mikkelsen's two-year reign as a minister. Sonja Mikkelsen was described as a 'warm supporter of public transit' and her clear aim as minister was to strengthen public transit and to give access and provide mobility for groups without cars in order to protect the environment (Toft et al., 2000).

Disruption of privileges

Many people find transport policy a dull subject. In general, the Danish population has only a vague feeling of who holds the power when it comes to transport and daily mobility. According to the media representations in Denmark, the power of transport politics around the year 2000 seems to have been identical with the parliamentary transport committee, a group of middle-aged or older white men. They were criticised for primarily feeding their own interests as well as the interests of a rather unspecified 'car lobby' consisting of various interest organisations in the field.[9]

The traffirc committee/members in question, in the media bluntly nicknamed as a mafia and as 'ronkedors' ('old male elephants'), consisted of a small group of male politicians with a long track of first-hand knowledge of major traffic policies in the 1980s and 1990s. They were not only used to special treatment but also special catering when they routinely attended their Friday meetings in the ministry. Meetings were – according to media accounts – held with high coated open sandwiches and accompanied by beer and 'at least one shot of aquavit' (Lautrup-Larsen, 2017), an 'old boys' network', which the new minister, Sonja Mikkelsen, thought she could put an end to. When she invited the committee to the first meeting in the ministry, the menu had changed to carrot sticks and cauliflower florets with dip. Sonja Mikkelsen later recalled the conflicts with the committee, saying that:

> They acted as a couple of fools, and in my experience, they never have fostered their own ideas. I think they became very offended when they found out that I removed their sand box and that I did not want to continue their food club.
>
> (Aarhus Stiftstidende, 2005)

There were several occasions where the head of department and the department did not provide adequate professional support for Traffic Minister Mikkelsen, which explains one of the reasons why she as minister ended up in one political storm after another. Several members of the parliamentary traffic committee criticised the ministry and the department for lack of damage control as well as lack of adequate briefing (Abild, 1999b). Mikkelsen recalls it as a relief when in 2000 the head of department was catapulted into a new position as head of a new research institute for transport. The idea of the new research institute was, ironically, enough to galvanise and qualify the knowledge basis of transport policy in

Denmark, something which up until this point he found no need for as head of the department (Mikkelsen, 2020).

The particular institutional landscape, which Mikkelsen entered as a minister, was both epitomised and disrupted during her two years as transport minister. Her presence from day one was perceived as a disturbance of the traditional order, both because she embodied a difference to the dominant image of the minister as male and due to her broader take on traffic politics and support of social concerns and public transit. By the late 1990s, the main actors in the landscape consisted of a strong triangle of males in the parliamentary traffic committees, a strong and independent group of civil servants in the ministry, and those in transport councils, according to Mikkelsen, were all car driving, middle-aged or older men, who did not understand the importance or significance of public transit (Mikkelsen, 2004, 2020). The media, headed by an influential press, eagerly covered every conflict and contributed to the image construction of Mikkelsen as a muddler.

Traffic policy and endless conflicts

Traffic policy was, when Mikkelsen became minister, far from being conducted as transparent policy processes, based on qualified knowledge. Traffic policy-making, as demonstrated, was to a great extent part of a special culture of informal meetings and particular close ties between some ministers and members of the parliamentary policy committee. Policy decisions seem to have taken place at the informal weekly meetings between tone-setting liberal and conservative politicians and various members of the 'blue block' and shifting ministers. Those informal arrangements made policy discussions at the parliamentary level a kind of political window dressing, since decisions and agreements had been horse-traded beforehand. What is more, Danish transport policy lacked a coherent strategy based on knowledge and research. A tradition, which differed from the other Nordic countries where transport policy was guided by major political and long-term strategies (Sørensen and Gudmonsson, 2010). In the Danish context, transport policies, therefore, were marked by random and short-sighted projects and decisions. One case in point here was the Great Belt Bridge against new highways in Northern Jutland and other compromises, which contributed to the general mistrust and disillusion about transport policy.[10]

Sonja Mikkelsen arrived with the whole package of antidotes to this male-dominated Maverick-like culture. Not only in terms of policy but also in terms of lifestyle – a contrast which was immediately represented and pinpointed in the media as entertainment. The situation in the ministry – and her contrast with it – made her time as a minister very turbulent. In the media, she was nicknamed 'muddling Sonja' and with a range of other degrading labels.

The day-to-day political initiatives and compromises turned out to be troublesome during Mikkelsen's time as minister. It seemed as if one scandal and conflict followed after the other. Often, these scandals were related to the difficult transition of the public transport units from state-run to more independent companies, but they also related to internal competition and disagreements between powerful public stakeholders and the government. For example, DSB, the Danish state railway department, time and again played a counterproductive role. Besides this, Mikkelsen broke the routine of compromising with liberal and conservative representatives in the traffic committee and started to include the left-wing parties – to the extraordinary regret of the blue representatives.

Among the most attended media events, was the 'tunnel case', which was connected to the construction of a tunnel between the new Copenhagen Metro and the old S-trains at Nørreport Station, one of the busiest stations in Copenhagen with 60,000 daily commuters. The metro was constructed and run by a new company, independent of the old national rail company, DSB. Not only did the price double during the construction process from 5 billion to 10 billion DKK, but DSB opposed the tunnel and positioned the new metro and the old S-train systems as simply incompatible. The issue was subjected to tense media attention and lengthy and humiliating accounts of the minister as powerless and unable to solve such a simple case. In the end, however, the conflict was overturned by Mikkelsen as the minister in charge and the two rail systems were connected by a tunnel, an intervention she recalls as one of her successful results (Mikkelsen, 28.4.20). Another spectacular case concerned Combus – a reconstructed bus company and a successor to the old state-run busses by DSB. When Combus, in 1999, ended with a large deficit and fear of bankruptcy, Mikkelsen 'rescued' the company with a 300 million DKK grant without the consent of the parliamentary traffic committee (Juel, 1999; Politiken, 1999b).

Yet, the single most contested event in the time of Mikkelsen as Minister of Traffic was the negotiation of the Finance Act in 1999, which planned to include a major transport change and a long-term agreement of improving the DSB, the national rail provider. Politiken, one of the tone-setting capital newspapers referred to the dramatic negotiations as follows: 'First Sonja Mikkelsen shows the liberal and conservative members out of the door. Next she contributed to the creation of a proposal to finance improvements of public transport with input from the left winged and Christian people's parties' (Politiken, 1999b).

The idea of Mikkelsen and the left-wing members of the traffic committee was to create a red/green solution with taxes aimed at financing public transport. The proposal was regarded as a provocation both by the conservative members and tone-setting social democratic ministers, because they were against transfers between transport and tax policy. The incident was eagerly reported in detail in the press, that foresaw her dismissal. Both the minister of finance and the prime minister

> were very upset, when she on Wednesday night chose to do it alone and presented a catalogue of green charges, aimed at financing an extended rail network. In other words, she went far beyond her mandate. Shortly before both the Liberal and the Conservative representatives had left and slammed the door; they under no circumstances could accept the proposals from the 'greens'.
>
> (Havskov, 1999)[11]

The unfinished negotiations were then removed from Mikkelsen's roles and responsibilities and finalised by the powerful minister of finance. The agreement still implied a six billion DKK budget to improve the railway tracks, but not the longstanding investment in an electric railway (Buksti, 2005).

Sonja Mikkelsen was, during her entire period as minister for traffic, subjected to both verbal and parliamentary attacks – in particular from the 'blue' representatives of the traffic committee. The subtext was that the transfers made in favour of public transit were seen as an attack on private car owners. The opponents often addressed her and her political allies as children and for 'stealing money' from the car owners. As said by one of the representatives:

> If she continues her car hunt, we do not want to create compromises with her in other fields. We have reached the bottom level. We do not want to enter into compromises with her – when she consequently spent the savings on (public transit).
>
> (Abild and Westh, 1999)[12]

The car policy, which referred to higher charges for diesel cars, was seen as a witch-hunt on car owners – a perception which was echoed in readers' letters:

> As a car-owner one has become a target of the hunt issued by the current traffic minister Sonja Mikkelsen and others are in the process of limiting our right to motorised mobility in the cities … it is called environmental regards'.
>
> (Ekstra Bladet, 1999a)

The events from car charges to metro mistakes were placed not only as a political mistake, but as a national catastrophe:

> If this was the private sector she would have been sacked a long time ago. Sonja Mikkelsen might not be a burden for the government, but she is a catastrophe for the country – one needs to look for long after a bigger mistake.
>
> (Ekstra Bladet, 1999)

While Mikkelsen's visions of broader and more knowledge-based transport policy was acknowledged, both foes and friends regretted her lack of interest or abilities to form alliances. This tended to leave her as a lonely and contested rider for her policies.

Sonja Mikkelsen evaluated – A bracket or a new beginning for transport policy?

In general, the Danish media seems to have had an arrogant and convicting attitude towards the long and shifting row of transport ministers. 'The truth is that the shifting ministers for transport have never been the sharpest knives in the cupboard' (Palle, 2009) and 'the ministry of transport has been marked by an endless row of incompetent predecessors in this resort' (Jyllands Posten, 2011) are just a couple of examples of this harsh tune. Yet, the style – both in terms of vocabulary as well as the political assessments – was particularly sweeping towards Sonja Mikkelsen as minister. What is more, her political opponents in alliance with the media created an image of her time as minister as a bracket, a disruption in the otherwise smooth running of the ministry's car-friendly politics. In 2001, a fellow social democrat succeeded Sonja Mikkelsen not as traffic, but as transport minister; a change intended to mark a broader horizon for the policy resort. The new transport minister, Jakob Buksti, was immediately appreciated – by a businessman from Jutland – for promoting new highway plans in Jutland. The positive welcome to the new transport minister was seconded by the chair for Danish transport and logistics:

> We have a very positive cooperation in every respect … Even though he (the minister) does not make any promises, he seems to be a much more popular personality compared with Sonja Mikkelsen, his predecessor. She did not do well with the transport sector, but also here we now hear new tunes.
>
> (Erhvervsbladet, 2001)

Members of the parliamentary traffic committee also commented on their strained relation with Sonja Mikkelsen and appreciated the former social democratic ministers: 'The working climate in the committee was substantially better with ministers such as Bjørn West and in particular with Jan Trøjborg'. On the one hand, left-wing politicians and environmentalists were critical and from time to time found Sonja Mikkelsen too weak on environmental issues. On the other hand, they tended to appreciate the efforts of Sonja Mikkelsen as: 'The only one of a long row of transport ministers, who did a lot to handle transport as a whole. She fought courageously for more than roads and cars' (Politiken, 2000). Her climate-friendly politics were acknowledged. 'She is one of the best – she is the one who had prioritized public transit the most and also supported the repair of trains and tracks' (Politiken, 2000).

In Sonja Mikkelsen's entry to the Ministry of Traffic, her break with the tradition of forming alliances, the change in acts of policy-making, together with the large critical media coverage of her as minister of traffic, there are ongoing complex gender (im)balances at stake. Looking at the gender performativities and gendered positions in this particular policy culture, shows several issues. One is that the traditional dominance of male agents led to a culture in which groups of male politicians regarded transport policy as a playground and a game – one which belonged to them. Through talks and actions they enacted a suite of conservative and oppressive masculinities related to core virtues of traditional Danish culture. Second, their acts were closely entangled with the fight for privileges of private car owners and road transport. The older and powerful members of the traffic committee regarded the new female minister, Sonja Mikkelsen, as an intruder in this unspoken, but also very established culture, which included not only policy-making but a whole way of life both in the ministry and outside. In the aftermath of her time as Minister of Traffic, she positioned herself as the rational actor, who wanted to enhance rationality and transparency in this complicated policy area: 'I am not a jolly type of person, I am more after the content and not so keen on window dressing and trappings. That's my way' (B.T., 2000). An interpretation of herself, which echoes the analysis of masculinity and car culture by Catharina Landström, who argues that certain processes of interpellation initially invite men into an imagined homo-social community and a shared culture of cars and transport artefacts. Car culture, thus, becomes implied in the 'doing' of heterosexual masculinity and pleasure, and in such a stereotyped framework women are often constructed as practising a rational femininity as opposed to, or even as a threat to, this type of male sociality and pleasure (Landström, 2006).

The question remains how transport policy and transport as a culture can be opened up and made useful and a climate-friendly playground not only for certain men, but for other groups as well, not least for a broader group of women with various backgrounds.

Epilogue

There is, by 2020, ample evidence that policy and planning are still in a circuit of knowledge produced by a neo-classical growth-oriented paradigm or a genderless sustainable paradigm. The critical paradigm, which highlights the structural inequalities in terms of privilege and disadvantages related to gender and other categories in transport and mobility is weak and still absent from political agendas (Kębłowski and Bassens, 2018; NOAH, Denmark). This means that alternative transport such as non-motorised and climate-friendly transport policy is still contested, and supporters have recently been addressed as 'climate fools 'in the Danish media.[13]. Yet, cycling and more attention to climate-friendly motorised and non-motorised transport are also gaining currency. The ideas of sustainability and 'bikeability' in combination with a decent public transit system are not new but have persistently been advocated by NGOs such as the Danish Cyclist Association but with limited influence on the governmental transport committee. Also, and currently, following the COVID-19 crisis, various forms of biking – not least e-bikes – are enjoying a growing popularity both inside and outside of Copenhagen. Last but not least, climate change has lately become a hot policy topic partly generated by a younger generation and revitalisation of critical social movements where young women are upfront. This seems to show promise for a game changer in the circuit of gender and diversity in transport and mobility as well as fresh and more diverse knowledge production.

Notes

1 This chapter forms part of the TiNNGO project, which received funding from the European Union's Horizon 2020 research and innovation programme under grant agreement number 824349.

2 The Danish Climate Council is an 'independent' expert council which provides suggestions for cost-effective climate solutions.

3 The idea of transport policy and planning as a performative act has been applied by several scholars, but not reflecting agents as gendered, see also Lissandrello (2016).

4 The name of the ministry was changed from Ministry of Traffic to Ministry of Transport in 2005. – In this chapter we follow this chronology using traffic about the ministry, committee, department, etc., till 2005 and ministry of transport after 2005.

5 From the 1990s, the environment and climate have been placed in various constellations as political resorts, which also echoes political priorities of governments.

6 In 2020, the list includes committee members in both Transport and Housing which might bring women's participation levels up.

7 The current EU Gender Equality Strategy 2020–2025 combines gender mainstreaming and an intersectional politics approach. EU Gender Equality Strategy, 2020.

8 This is a particularity which also applied to the former head of the traffic department, who served in the ministry all his working life (38 years in total, with the last 20 years as chief head of the department). He was extremely unpopular among male politicians and was nicknamed with degrading labels such as 'Godfather', 'a snake' and 'an intriguing emperor' etc. The ideal public servant in Denmark is the 'unpolitical person', that at all times follows seated governments and ministers. 'It is a fundamental principle that public employees at all levels must both be and appear to be impartial and that they must make decisions and rulings based on objective grounds' – these principles were updated in Code of Conduct in the Public Sector in 2017.

9 Besides the interests of (e)automobility are catered for by, for example, the FDM (Association of Danish Motorowners, Danish car import/Danish Transport and Logistics (DLTL) and more lately by e-car associations: such as Association of Danish E-car Drivers and Danish E-car Alliance (from list of hearing partners related to the suggested climate law, 2.3. 2020).

10 Denmark with many separated islands – and waterborne borders – has made big bridge-building projects a main priority and a major issue of contestations. From the Great Belt Bridge in the 1980s and 1990s to the Oresund Bridge in 2000, up until today when big projects are also underway: with the Femern Belt Bridge linking Denmark with Germany and the projected Kattegat Bridge linking Copenhagen with Aarhus over water. The bridge projects are often backed onto other compromises – and therefore hard to change.

11 Politicians later recalled that the conflict also unfolded around the unspoken gender composition of the members of the transport committee/the team of negotiators consisting of four women.

12 14.2. 1999, Berlingske Tidende. (Lars Abild og Maria Westh).

13 A degrading label suggested by Pia Kjærsgaard, former chair of the Danish Peoples Party.   https://www.dr.dk/nyheder/politik/pia-kjaersgaard-jeg-staar-100-procent-ved-der-er-mange-klimatosser

Bibliography

Aarhus Stiftstidende (2005) Sonja Mikkelsen 50 år. 21 juni 2005.

Abild, L. (1999a) Omstridt trafikminister vil slå rekorden. Berlingske Tidende Sektion. 27 juni 1999 [online]. Available at: https://apps-infomedia-dk.ep.fjernadgang.kb.dk/mediearkiv/link?articles=AZ939472 (Accessed: 15 October 2020).

Abild, L. (1999b) Trafikministeriet anklages for svigt. Berlingske Tidende. 29 juni 1999 [online]. Available at: https://apps-infomedia-dk.ep.fjernadgang.kb.dk/mediearkiv/link?articles=AZ939083 (Accessed: 15 October 2020).

Abild, L. and Westh, M. (1999) M. Trafikministerens trængsler. Berlingske Tidende. 14 februar 1999 [online]. Available at: https://apps-infomedia-dk.ep.fjernadgang.kb.dk /mediearkiv/link?articles=AZ908370 (Accessed: 15 October 2020)

Adler-Nissen, R. (2016) 'Towards a Practice Turn in EU Studies: The Everyday of European Integration', *Journal of Common Market Studies*, 54(1), pp. 87–103. DOI: 10.1111/jcms.12329. (Accessed: 5 October 2020).

Buksti, J. (2005) Skønmaleri af Flemming Hansen. Berlingske Tidende. 26 maj 2005 [online]. Available at: https://apps-infomedia-dk.ep.fjernadgang.kb.dk/mediearkiv/li nk?articles=e046eaf3 (Accessed: 15 October 2020).

Carlsen, M. E. (2018) *Konger uden land.* Copenhagen: Lindhardt og Ringhof.

Carstensen, J. (2001) Trafikminister fik rosende ord på transportmesse. Erhvervsbladet 2 marts. 2001 [online]. Available at: https://apps-infomedia-dk.ep.fjernadgang.kb.dk/ mediearkiv/link?articles=Y3558717 (Accessed: 15 October 2020).

Christensen, H. R. and Breengaard, M. H. (2011) 'Mainstreaming Gender, Diversity and Citizenship: Concepts and Methodologies', WP7, Working Paper, no. 4, Femcit, Bergen.

Christensen, H. R. and Breengaard, M.H. (2019) Gender Smart Transport. Translating Theory into Practice. TINNGO Road Map. (Innovation Gender Observatory mobility. EU Horizon 2020 project) forthcoming, Routledge 2021.

Cresswell, T. (2006) 'On The Move', *Journal of Transport Geography*, 14(6), pp. 471–472.

Cyklistforbundet. [online]. Available at: https://www.cyklistforbundet.dk/om-os/presser um/ 9.3.20 (Accessed: 29 September 2020).

Dahlin, U. (2000) Slotsholmens ældste hanbavian. Information. 29 Juni [online]. Available at: https://www.information.dk/2000/06/slotsholmens-aeldste-hanbavian (Accessed: 28 September 2020).

Edensor, T. (2004) 'Automobility and National Identity: Representation, Geography and Driving Practice', *Theory, Culture & Society*, 21 (4–5), pp. 101–120.

Ekstra Bladet. (1999) Letter to the editor. Hader Sonja bilisterne? 9 februar 1999.

Ekstra Bladet. (2000) Letter to the editor. 14 februar 2000.

Ellegaard,C. and Boddum, I. (1999) Herre i eget hus, Jyllandsposten. 5 December 1999 [online]. Available at: https://apps-infomedia-dk.ep.fjernadgang.kb.dk/mediearkiv/link ?articles=Z5096776 (Accessed: 15 October 2020).

EU Gender Equality Strategy 2020. [online]. Available at: https://ec.europa.eu/info/poli cies/justice-and-fundamental-rights/gender-equality/gender-equality-strategy_

Fairclough, N. (1995) *Critical Discourse Analysis.* Boston: Addison Wesley.

Hanson, S. (2010) 'Gender and Mobility: New Approaches For Informing Sustainability', *Gender, Place and Culture: A Journal of Feminist Geography*, 17 (1), pp. 5–2.

Havskov, J.A. (1999) Trafikministeren risikerer fyring. BT. 6 November 1999 [online]. Available at: http://dk.ep.fjernadgang.kb.dk/mediearkiv/link?articles=AZ983037 (Accessed: 15 October 2020).

Havskov, J.A. (2000) Sonja slap for sablen. BT. 24 februar 2000 [online]. Available at: https://apps-infomedia-dk.ep.fjernadgang.kb.dk/mediearkiv/link?articles=AY000304.

Hermansen, D. A. (2019) 'Køns betydning for politisk repræsentation i danske kommuner', Aarhus: Institut for Statskundskab, Aarhus Universitet. (How Gender affects political representation in Danish municipalities. Dissertation (with English summary). Department of Political Science. University of Aarhus. 2019. Available at: https:// surveyselskab.dk/wp-content/uploads/2020/03/Speciale_Amalie-Dahlerup_2019.pdf.

Juel, F. M. (1999) Trafikminister løber fra ansvar. Erhvervsbladet. 2 juni 1999 [online]. Available at: https://apps-infomedia-dk.ep.fjernadgang.kb.dk/mediearkiv/link?articles =Y3574961 (Accessed: 15 October 2020).

Jyllands Posten. (2011) Editorial. 4 April 2011.

Kębłowski, W. and Bassens, D. (2018) 'All Transport Problems are Essentially Mathematical: The Uneven Resonance of Academic Transport and Mobility Knowledge in Brussels', *Urban Geography*, 39(3), pp. 413–337.

Kjær, F: 60 ÅR I DAG. Ole Zacchi kalder en spade for en spade. Jyllandsposten 24 september, 1999 [online]. Available at: https://apps-infomedia-dk.ep.fjernadgang .kb.dk/mediearkiv/link?articles=Z4842103 (Accessed: 15 October 2020).

Klimarådet. (2020) 'Kendte veje og nye spor til 70 procents reduktion', Klimarådet, Marts [online]. Available at: https://klimaraadet.dk/da/rapporter/kendte-veje-og-nye-spor-til -70-procents-reduktion (Accessed: 29 September 2020).

Kronsell, A. (2015) 'Sexed Bodies and Military Masculinities: Gender Path Dependence in EU's Common Security and Defense Policy', in Men and Masculinities [online]. Available at: https://doi.org/10.1177/1097184X15583906 (Accessed: 6 October 2020).

Krook, Mona Lena and Mackay, Fiona, (eds) (2011). *Gender, Politics and Institutions. Towards a Feminist Institutionalism*. Basingstoke: Palgrave-Macmillan.

Landström, C. (2006) 'A Gendered Economy of Pleasure: Representations of Cars and Humans in Motoring Magazines', *Science Studies*, 19(2), pp. 31–53.

Larsen, G. A. (2015) *Elites in Denmark: Identifying the Elite*. Copenhagen: Department of Sociology, University of Copenhagen, pp. 62–63.

Lautrup-Larsen, P. (2017) Ole Birk Olesen med løkken om egen hals. TV2 Nyheder. 10 Marts [online]. Available at: https://nyheder.tvhttp://2.dk/politik/2017-03-09-ole -birk-olesen-med-loekken-om-egen-hals (Accessed: 29 September 2020).

Lévy, C. (2013) 'Travel choice reframed: "Deep distribution" and gender in urban transport', *Environment and Urbanization*, 25(1), pp. 47–63.

Lisandrello, E. (2015) 'Stories of Performativity in the politics of urban planning in two Scandinavian cities', Paper: International Conference on Public Policy, Milan. 1–4 juy 2015

Lissandrello, E. (2016) 'Three Performativities of Innovation in Public Transport Planning', *International Planning Studies*, 22(2), pp. 1–15.

Mikkelsen, S. (2004) 'Transportforskning og ligestillingspolitik', *Kvinder, Køn og Forskning*, 1, pp. 72–73.

Mikkelsen, S. (2020) Interview with Sonja Mikkelsen by Hilda Rømer Christensen. 28 April 2020.

Nielsen, M. (1999) Ministrenes skræk. Ekstra Bladet. 24 september 1999. Available at: https://apps-infomedia-dk.ep.fjernadgang.kb.dk/mediearkiv/link?articles=Z4840802

Palle, H. (2009) Henrik Palle. Det jeg hører dig sige er: lirum larum kitten gør æg. Politiken. 11 juni 2009.

Phillips, A. (1998) *The Politics of Presence*. Oxford: Oxford University Press.

Politiken (1999a) Oles Zacchi. 60 år. Politiken. 23 september 1999.

Politiken (1999b) Trafik minister Sonja Mikkelsen vil ikke lade sig slås af banen af den politisk betændte sag om Combus. Politiken. 7 November 1999.

Rehling, D. (1998) Motorvejshader. Information. 24 Marts 1998 [online]. Available at: https://www.information.dk/1998/03/motorvejshader (Accessed: 28 September 2020).

Ritzau (1998) Nu bestemmer fru Mikkelsen. Ritzaus Bureau. 23 marts 1998.

Ritzau (1999) Sonja Mikkelsen forholdt folketingets trafikudvalg oplysninger og undlod at invitere udvalget til rejsegilde på Øresundsbroen. Ritzaus Bureau. 4 August 1999.

Salih, S. (2002) 'Why Butler?', in Sarah Salih (ed), *Judith Butler*. London: Routledge Critical Thinkers. p. 63.

Socialdemokratiet. (2019) Danmark skal have en ny langsigtet infrastrukurplan [online]. Available at: https://www.socialdemokratiet.dk/media/7979/danmark-skal-have-en-ny -langsigtet-infrastrukturplan.pdf (Accessed: 28 September 2020).

Sørensen, H. C. and Gudmonsson, H. (2010) 'Målstyret transportpolitik.—hvad kan Danmark lære af Sverige og Norge?', *Oekonomi og Politik*, 83(2), pp. 3–19 [online]. Available at: https://www.djoef-forlag.dk/openaccess/oep/files/2010/2_2010/2_2010 _2.pdf (Accessed: 5 October 2020).

Squires, J. (2005) 'Is Mainstreaming Transformative? Theorizing Mainstreaming in the Context of Diversity and Deliberation' *Social Politics: International Studies in Gender, State & Society*, 12(3), pp. 366–388. https://doi.org/10.1093/sp/jxi020

TiNNGO. [online]. Available at: https://www.tinngo.eu/ (Accessed: 29 September 2020).

Toft, E., Rasmussen, H., and Nielsen, H. (eds) (2000) Hundrede års trafik Trafikministeriet 1900–2000 © Trafikministeriet, 2000. Available at: https://www.trm.dk/publikationer /2001/hundrede-aars-trafik-trafikministeriet-1900-20file:///

Tornbjerg, J. (2020). Danmarks grønneste trafikminister. Politiken. 30 januar 2000. Available at: https://apps-infomedia-dk.ep.fjernadgang.kb.dk/mediearkiv/link?articles =Z5457408 (Accessed: 15 October 2020).

Trafikmafia. Available at: https://da.wikipedia.org/wiki/Den_jyske_trafikmafia

Trafikministeriet. Available at: https://www.trm.dk/ministeriet/ministeriet-artikler/bes tyrelser-raad-mv/ Bestyrelser, Radiat medv.

TRANSGEN. (2007) Gender Mainstreaming European Transport Research and Policies: Building the Knowledge Base and Mapping Good Practices. University of Copenhagen: Coordination for Gender Studies. Available at: https://curis.ku.dk/ws/files/21597551/ EU-rapport-Transgen.pdf (Accessed: 5 October 2020).

Trafikministeriet. Available at: https://www.trm.dk/ministeriet/ministeriet-artikler/bes tyrelser-raad-mv/ (raad-mv/ (Accessed: 29 September 2020).

Venstre. (2016) Venstres Principprogram: Fremtid frihed og fællesskab [online]. Available at: https://www.venstre.dk/_Resources/Persistent/2/0/a/e/20ae5a4934215ebc17db9d 0af1c06ca6e445d1d7/Principprogram%202016.pdf (Accessed: 28 September 2020).

Winther Jørgensen, M. and Phillips, L. (1999) *Diskursanalyse som teori og metode*. København: Samfundslitteratur.

Woodcock, A., Christensen, H.R. and Levin, L. (2020) 'TInnGO: Challenging Gender Inequality in Smart Mobility', *Journal of Road Traffic Engineering*, 129-Article Text-544 -1-10-20200622.pdf (Accessed: 5 October 2020).

8 Wasting resources
Challenges to implementing existing policies and tools for gender equality and sensitivity in climate change–related policy

Susan Buckingham

Introduction

Climate change policy-making has a broad reach into a range of scales and sectors. It is, therefore, too limiting to discuss climate change policy-making in isolation from decision-making in the sectors which produce greenhouse gas emissions. Waste management is one such sector which contributes to greenhouse gas emissions and, consequently, has the potential to make a significant impact on their reduction. Waste management can be considered as including waste prevention actions, waste collection, sorting, material separation, recycling and final disposal. Other than waste prevention, all other waste management activities produce greenhouse gas emissions. The European Union's commitment to the circular economy emphasises the need for discarded materials to be treated in ways that mean they can be reused in products which are both long-lasting and can be dismantled for future recycling. This will eliminate the need for final disposal to landfill or incineration (European Commission, 2020b).

As a profession and an activity, waste management is highly gendered, with unpaid waste work in the household mostly done by women, while paid work – particularly operative and senior decision-making – is mainly undertaken by men. Professionally, waste management is dominated by an engineering tradition, which determines how decisions are made. The resulting path dependency resembles the privileging of the climate debate as 'scientific, elitist, technical and masculinised' – referred to in Chapter 1 of this volume. Recognising what structures the production, management and minimisation of waste is a critical step to achieving the drastic reduction of greenhouse gas emissions necessary to avoid a climate catastrophe; greenhouse gas emissions from the management of waste contribute almost 3% of the EU's total. The EU's dual commitment to both reducing greenhouse gas emissions and increasing gender equality has informed its environmental research programme Horizon 2020, which funded the Innovations project *Urban Waste*. This chapter uses *Urban Waste* as a case study to examine how open waste management organisations are to developing waste reduction innovations in a gender-sensitive way.[1] Amongst other things, *Urban Waste* promoted and investigated the role of gender in developing waste reduction innovations by applying gender mainstreaming throughout the life of the project. Table 8.1 sets out the stages at which this took place. From the

DOI: 10.4324/9781003052821-8

proposal design stage, I was involved as a gender specialist to ensure that sensitivity to gender was embedded throughout the project, which involved 11 pilot cities in eight European countries.[2] The cities all depended on a tourist-based economy, which presents particular waste management challenges that the waste reduction innovations were designed to address. However, they were diverse in terms of geographical distinctiveness, size, and, of particular importance for the gender dimension of the project, social and economic profile, which included differences in the degree of gender equality.

My involvement in the *Urban Waste* project emerged from my work on gender and waste management in 2002, when I led a research project to explore the role of gender mainstreaming in waste management. This small-scale pilot of gender mainstreaming was initiated by the EU's Environment Directorate to explore how the EU's gender mainstreaming commitment could be introduced into environmental policy. The study focused on three member states (Ireland, Portugal, and the United Kingdom), and broadly established that waste management

Table 8.1 Gender considerations at each stage of the project, following EIGE methodology

STAGE OF PROJECT (EIGE STAGE)	GENDER INITIATIVES
Proposal design (DEFINE)	Gender dimensions embedded throughout proposal; gender specialist involved.
Organisation of project (PLAN)	Gender strategy developed
Preparation/benchmarking (PLAN)	Survey of pilot organisations to collect gendered employment data. Stakeholder groups assessed for gender balance.
Surveys of waste managers, tourism employees, and tourists (PLAN)	By gender as well as by age, ethnicity, education, and place of residence to assess socio-economic dimensions of waste attitudes and behaviour.
Focus groups with waste managers, tourism employees, and tourists (PLAN)	Specifically on gender to explore attitudes and behaviour towards waste and waste management.
Training and mutual learning (ACT)	First mutual learning (ML) event was devoted to gender mainstreaming training; subsequent short sessions at various ML events. Three gender webinars on communication; budgeting; reporting.
Measures (ACT)	Training on how measures will have gendered impacts; measure evaluation forms gender proofed.
Final evaluation (CHECK)	Changes to gender balance of stakeholders; gendered changes to organisational structures; gender impact of measures; identifying good practices in and barriers to gender mainstreaming.
Final report and deliverable	Changes to the gender balance of stakeholders; informs future proposals and projects.

teams which were gender (and professionally) diverse were more likely to develop broader strategies for waste management which led to diversions from waste disposal to recycling (Buckingham et al., 2005).

The scale of change needed to address what has now escalated to a climate emergency requires policy transformation rather than incrementalism. A feminist gender analysis can inform and motivate transformative action in ways that a more co-opted form of gender mainstreaming may not. I open the argument which informs this chapter with a critique of gender mainstreaming as it has mostly been applied, and with a consideration of its transformative potential. In reflecting on the role of gender mainstreaming in the *Urban Waste* project, I explain how gender was, or was not, recognised and included in case study programmes and their home organisations to explore ways in which this was variously used to 'add on', 'gender-wash', or, occasionally, to more fundamentally restructure waste reduction. I reflect on the main challenges to including gender in waste reduction policy-making and propose how these challenges can be understood and conceptualised. By considering the synergies between the available gender-mainstreaming tools and waste reduction programmes, I question where the challenges lie – the tools themselves and their transformative potential; their promotion; and/or resistance to their adoption. I propose that these lessons may well be transferable between sectors and institutions responsible for creating and needing to reduce greenhouse gas emissions.

Specifically, my aim in this chapter is to reflect on the conditions under which gender mainstreaming can become a transformative process. The questions this process of reflection asks are:

- What conditions are necessary for gender mainstreaming to become transformative?
- What work is necessary to encourage these conditions to be achieved?
- What are the challenges that gender mainstreaming confronts at different stages of its implementation?
- What can we reasonably expect gender mainstreaming to achieve in environment – specifically waste reduction – policy-making?

Accordingly, this chapter will explain what is distinctive about waste management as a sector, including how it is gendered, and why it should be considered as an integral part of climate change–related policy. It will then review what gender mainstreaming provides as a principle, process, and set of practical tools, and how effective these have been to date. Following these context-setting sections, the chapter will reflect on how gender mainstreaming was applied and accepted (or not) in the *Urban Waste* project.

Managing waste as a contribution to climate change policy

A consumer society, in which most Europeans play a significant part, is a wasteful society. For example, over 80% of discarded textiles are sent to landfill and incineration (Labfresh, 2019) and only 46% of electrical and electronic equipment waste, one of the fastest-growing waste streams in the EU, is recycled (Eurostat, 2020). It is also an unfair consumer society. While the 88 million tonnes of food waste in EU member states represent around 20% of all food produced in Europe, 43 million Europeans (around 8 %) are estimated to be in food poverty (European Commission, n.d.). And while consumption is determined by household income and related factors such as age, it is also important to note that consumption is a gendered activity (Buckingham, 2020).

While reducing overall consumption requires legislation and behavioural changes beyond the scope of waste management policy, managing waste locally and encouraging residents to remove items from the waste stream are important contributors to climate change policy.

In 2017, waste managed within the EU (not all that is produced, because of the global trade in waste) contributed approximately 2.7% of the emissions in EU countries,[3] a drop of 42% in Kt CO_2e since 1990, when waste represented approximately 4% of the EU's greenhouse gas emissions. Waste features in the European Green Deal, which is committed to making a 'just and inclusive transition for all', although the terms 'women' and 'gender' are not found in a search of the document (EU, 2019b). Waste, like all the other sectors recorded as being the main Kt CO_2e emitters, is generally a highly masculinised and male-dominated industry, as this chapter will later demonstrate.

Being the gender advisor on the EU waste reduction innovation project *Urban Waste* has given me access to data on gender in waste management organisations (mostly municipalities) not otherwise generally available, because it is not normally counted. This has enabled me to assess the role, uses, and limitations of gender mainstreaming across the 11 participating case study cities. As the chapter will later discuss, the eight countries in which the pilot cities are located have very different gender profiles including that for paid work, and these different experiences appear to have an impact on their recognition and understanding of gender difference. The project facilitated and evaluated the identification and introduction of locally determined waste reduction innovations in tourist cities across Europe. While this included research with and into tourists and tourism businesses, the reflections in this chapter focus on the research with waste management institutions and staff.

Gender mainstreaming critiqued

Gender mainstreaming was adopted by the European Union (EU) in 1996 as a consequence of the Fourth United Nations World Conference on Women in 1995, which proclaimed the Beijing Declaration and Platform for Equality, Development and Peace as an agenda for women's empowerment. This and the UN Conference on Environment and Development in 1992 argued that to address environmental damage, gender (and other social) inequality must also be tackled. Gender mainstreaming requires that policies take gender equality into account at all stages, from development through to enactment and assessing impact. This becomes a cyclical process when the impact assessment informs the adjustment of future policies, and represents a significant departure from existing environmental decision-making, with the potential, and hope, to be transformative.

However, in practice, the embedded or 'sticky' institutional structures which dominate environmental decision-making in Europe, as elsewhere, have been hard to penetrate and existing path dependencies have dominated (see Chapter 1 in this volume). EIGE has reported that 'gender mainstreaming is strikingly weak within the EU's environmental policies' (EIGE, 2020, p. 18). In EIGE's 25th anniversary review of the platform, it found that:

> Currently, women remain under-represented in environmental policymaking, planning and implementation (European Parliament, 2017d). Women are also substantially under-represented in key sectors such as energy; transport; water and waste; and agriculture, forestry and fishery. The low level of gender diversity in the energy sector is considered to affect innovation and restrict efforts to address climate change.
>
> (EIGE, 2020, p. 128)

Not only are women underrepresented in environmental decision-making, but environmental policy areas are found to lack gender sensitivity or any kind of gender perspective. (EIGE, 2020b).

Accordingly, gender mainstreaming has been variously welcomed and criticised, and often both where it is valued as a tool to achieve gender equality but observed to have been used instrumentally to bolster policies which are inherently unequal. For example, Grosser and Moon have argued that gender mainstreaming benefits the corporate social responsibility agenda, 'simultaneously good for both business and wider society' (in Walby, 2005, p. 457). This sees a dual role for gender mainstreaming which, if effectively implemented, requires that women should have equal rights to men (for example, to participate in careers and civic engagement), and that operational benefits accrue to organisations which strive to achieve gender balance in decision-making. Mieke Verloo was positive about the aims of gender mainstreaming to prioritise the lives and experiences of individuals; in its potential to lead to better government, to involve women as well as men; in acknowledging the diversity amongst men and women; and to make gender equality issues visible 'in the mainstream of society' (1999, p. 8). However, applying the tool that is gender mainstreaming is likely to reflect the values which underpin the organisation implementing it, thereby reinforcing existing path dependencies in a 'logic of appropriateness' as introduced in Chapter 1 of this book. The EU has been observed to strategically frame the concept of gender mainstreaming within a discourse of competitiveness and employment policies (Vida, 2017), and this may have contributed to criticisms of the EU's practical application of gender mainstreaming, including its reliance on 'soft' strategies such as training, for implementing gender sensitivity, although this may change with the Gender Equality Strategy proposed for 2020 to 2025 (EC, 2020). Moreover, those expected to implement and 'police' gender mainstreaming are not always well equipped to do so (Mergaert and Lombardo, 2014). Verloo has also been sceptical that sufficient expertise existed amongst professionals to challenge prevailing discourses, and to align the necessary interests from those 'at the top' with those 'down under' (1999, p. 8).

While Verloo's observation is now 20 years old, the same observations continue to be registered, which is interesting considering that a substantial body of expertise provided by feminists, consultants and trainers informs development policy and practice (Prugl, 2013). For example, while senior policy-makers and politicians have championed gender mainstreaming as a principle, and specialist organisations such as EIGE have developed mechanisms to encourage its application, there has been a failure to do more than tinker with the design of major research policy structures – themselves embedded in institutions through path dependencies, as I discovered when invited to address the (mainly male) first Horizon 2020 policy development team in 2013. Later, in the *Urban Waste* project, our European Commission liaison officer admitted that support and evaluation teams lacked the expertise to assess the role of gender in research or to advise (EC, pers com, 2016), which was certainly evident in our experience of project evaluation. Minto and Mergaert (2018) remain unconvinced about the institutional embeddedness of gender mainstreaming within the EU, observing a lack of resources for comprehensive gender-mainstreaming training and a lack of gender mainstreaming in evaluation at an institutional level. These technical challenges associated with gender mainstreaming can be overcome but are compounded by ideological difficulties.

As global institutions have utilised 'gender mainstreaming' and employed higher numbers of women, a liberal version of feminism – individualised and femocratic – appears to have been co-opted as, in Nancy Fraser's words, capitalism's 'handmaidens' (2013). Penny Griffin examines 'the promotion of co-opted, governance-friendly, "feminist knowledge"' which can be applied to international political, as well as economic, activity. She specifically refers to how a 'crisis governance feminism' is 'a form of feminist strategy friendly to existing neo-liberal governance and supportive of the resuscitation of neo-liberal global finance' (Griffin, 2015, p. 51). In the development context, Bernadette Resurreccion and Rebecca Elmhirst note how gender mainstreaming is 'streamed away' as it is framed 'through a neoliberal discourse with a vocabulary of effectiveness, efficiency, impact assessment and "smart" economics [which] diminishes any power to address feminist agendas' (Resurreccion and Elmhirst, 2020, p. 2). They explore the tensions that gender experts face 'as they seek transformative social change whilst working in technical-environmental contexts that demand simplifications and technical fixes' (Resurreccion and Elmhirst, 2020, p. 4). Nowhere is this more apparent than in the need to address climate change both at the global scale and in local and sectoral organisations, where the transformative potential of feminism offers at least the possibility to make real and lasting change. For this to happen, however, institutional structures need to be receptive to the need for and scope of change, and alive to the necessity to include all women and not just white women who are disproportionately dominating promotions in academia

(Bhopal, 2020) or environmental organisations (Taylor, 2014). As Kalwant Bhopal suggests, it is politically easier to concentrate on what will enable more women to achieve equality in the workplace than to address institutional barriers to Black and minority ethnic women (Bhopal, 2020). For gender mainstreaming to be transformative, both the principles on which it is founded, and the practical ways in which it is applied, need to be intersectional. This means being attuned to the needs of Black and minority ethnic women, disabled women, women of different ages, religions, cultures, care responsibilities, sexualities and sexual orientation. In my own experience of working with *Urban Waste* city partners, acknowledging intersectionality has been particularly difficult, especially when, as is generally the case, decision- and policy-makers are just adjusting to the need to address gender (in)equalities. Exclusively, the women managers who represented the pilot cities were 'white European' (as were the men). And as some of the women were the main carers of young children, this sometimes caused them difficulties in travelling to partner cities for quarterly meetings, and reduced the amount of time they were able to interact.

That gender mainstreaming has not been transformative is found in the persistence of gender inequality, and the disproportionate impact on women of the global financial crisis (Griffin, 2015) and more recently the COVID-19 pandemic (UN, 2020). Griffin quotes Otto as arguing that 'crises everywhere are a particularly dangerous time for feminism, and indeed for all progressive ways of thinking' (2015, p. 58). Beth Bee's analysis of REDD+[4] in Mexico illustrates how the appointment of urban senior professional women to the national programme ENAREDD+ has done little to promote the needs and interests of rural women who do not own land. She also reports that women are appointed as legal representatives in order to score the points now needed for successful forestry applications but that while they can participate as agents of change, they can just as likely be included instrumentally, as signers of forms under the guidance of male figures of authority (Bee, 2017). Bee's work shows clearly how an emphasis on numerical parity does not, in itself, secure social equality.

Griffin characterises a form of influential international decision-making through groupings of individual states (G7, G20), supra-national bureaucracies (European Union), intergovernmental organisations (World Bank and IMF) and UN agencies as 'neo-liberal global governance' (Griffin, 2015, p. 53). She particularly focuses on recent global financial crises, and the international financial sector which, for centuries has been 'dominated not just by men but by dominant models of behaviour... [that have concentrated] historical privilege in the hands of white men' (ibid., p. 55). The persistence of such 'masculine subjectivities' encouraged Virginia Held to cite Carol Gilligan's argument that women who 'advance occupationally learn to think like men' (Held, 2006, p. 27), implying the existence of a dominant masculinity in occupational structures. A similar observation has also been made by Magnusdottir and Kronsell (2015) in their exploration of whether a numerical balance between women and men makes a difference to climate change planning policy; and it informs my own investigation into the gendering of waste management, itself a highly masculinised profession through its most common training routes as well as its professional organisations and workplaces. The EU refers to descriptive and substantive equal representation, which recognises that while equal numbers are important to achieve a critical mass to provide a decision-making environment in which both women and men have the confidence to make effective interventions that are respectfully received, it is not in itself sufficient (EIGE, 2015).

Gender mainstreaming and waste management in the EU

As the gender advisor on *Urban Waste*, my responsibility was to ensure attention to gender and see that gender equality was achieved at all stages of the project, for which an EIGE gender-mainstreaming methodology was used (see Table 8.1).[5] This enabled any correspondence between greater gender sensitivity of waste management departments and higher CO_2 equivalent savings to be observed. This relationship was also noted in an earlier European project which examined the potential for gender mainstreaming waste management. This identified that municipalities where there were women in decision-making positions (and, interestingly, women who were not from the engineering backgrounds which normally dominate), were those that were recycling more and through gender-sensitive activities. Notable amongst these was a project in southern England which was providing reusable nappies and nappy-cleaning services free of charge to low-income families (Buckingham et al., 2005).

That waste management in the European Union is dominated by men is evident from Table 8.2 which provides the gender balance of jobs in 'water supply, sewerage, waste management and remediation' calculated by the OECD. Despite sex-disaggregated data being necessary for effective gender mainstreaming, gendered employment data at the sector level is still not comprehensive (if available at all), and there is no cross-EU comparison of gendered employment in waste management.[6] The OECD data is only available for five of *Urban Waste's* eight case study countries, and it does not breakdown jobs by seniority, although from data elsewhere it is probably safe to assume that senior decision-making and technical jobs are more likely to be held by men (see also Buckingham et al., 2005; EY, 2015), just as they are across science- and research-related jobs (European Commission, 2019a) and in the energy sector (Catalyst, 2018; IRENA, 2019).

Table 8.2 Employment by gender in water supply, sewerage, waste management and remediation activities, 2017. Source: OECD, 2017

COUNTRY	MALE (%)	FEMALE (%)
Denmark	96,000 (72.2)	37,000 (27.8)
Greece	212,000 (77.4)	62,000 (22,6)
Italy	2,015,000 (87.5)	287,000 (12.5)
Portugal	286,000 (78.4)	79,000 (21.6)
Spain	1,128,000 (84.2)	244,000 (17.8)

Table 8.3 Gendered attitudes and behaviour towards avoiding waste. Sources: European Commission, 2014; 2017

GENDER	% reduce or separate waste	% reduce consumption	Avoids food waste	Avoids over-packaged goods	Drinks tap rather than bottled water	Thinks reducing waste is not important
Female	73	59	85	65	70	12
Male	68	52	81	59	63	20

The EU's own Eco-Management and Audit Scheme (EMAS) for the environment (which includes waste management) makes no reference to gender (Dri et al., 2018), although studies and reports, including those from the sector itself, have stated the importance of a gender-sensitive perspective (Charrington, 2017; EY, 2019). This also contrasts strongly with waste management attitudes and behaviour in the household, shown in Table 8.3.

These gendered proportions have remained strikingly consistent across times and places (see, for example, Brough and Wilkie, 2017, for data beyond Europe, as well as earlier Eurobarometer surveys) and indicate the value of a waste management policy which is gender-sensitive and draws on the expertise and concerns of women and men equally. This point is reinforced by the gendered waste behaviour that was noted amongst those *Urban Waste* survey respondents working in waste management. Three of women employees' top-four priorities related to behavioural change (behaviour change, awareness-raising and waste prevention), compared to three of men's top four being to improve operations, but without necessarily reducing the amount of waste produced (improving waste collection, optimising recycling, reducing landfill). This suggests a benefit of gender-balanced teams to have the potential to deliver a broader mix of policies, and proposes that women working in waste management may not be socialised into thinking the same way as men. This contradicts some of the literature reviewed earlier on the institutionalisation of women working in male-dominated fields, but concurs with findings from the earlier gender-mainstreaming waste management project conducted for the EU, and which will be returned to in this chapter's conclusion (Buckingham et al., 2005).

Gender mainstreaming the Urban Waste project

In addition to using the EIGE methodology for gender mainstreaming, the *Urban Waste* project adopted a participatory action research methodology. First, the gender expert was both actively participating in the project as well as researching the application and influence of gender mainstreaming. Secondly, participating representatives from each of the 11 case study cities contributed to identifying the issues to be addressed, the design of online questionnaires, focus group discussions and identification and evaluation of measures. The city partners, who were self-selected participants in the design of the initial project, each established a stakeholder group for their city, with the guideline to achieve gender balance. City partners also recruited for local focus group discussions, adapting the topic discussion guide to local conditions. With their stakeholders, partners identified waste reduction innovation measures to be piloted by stakeholders, which they then evaluated together. The project ran a 'mutual learning' programme in which city partners learned from each other. Bringing these actors together in a participatory research action, the gender expert conducted qualitative surveys of the partners at the beginning and end of the project to assess what had changed in terms of commitment to gender equality, and how this might have affected local practices.

The countries that participated in the *Urban Waste* project (Croatia, Cyprus, Denmark, France, Greece, Italy, Portugal, Spain), had different profiles of women in paid work, and experience of overall gender equality (see Table 8.4). This may have had a material impact on their enthusiasm to acknowledge and address gender inequality. Within the project, the intersection of country of residence and gender appeared to signify differences in waste behaviour.

Each pilot city was required to invite a (gender-balanced) group of stakeholders to jointly identify, develop and implement a number of waste minimisation measures, innovative for its own city. This emphasis on a participative process created scope for gender-balanced stakeholder groups, or where there were male-dominant staff teams in the municipal office, to recruit more women stakeholders to counterbalance male dominance. In this chapter, I focus my reflections on the data generated by surveys and focus groups, and qualitative surveys of city partners conducted at the beginning and end of the three-year project.[7] In addition, observations were made through a continuous programme of training and mutual learning for city partners, in which gender played a key part, particularly to ensure that the context for identifying and implementing measures was gender-aware.

The departments in which the pilot city partner representatives worked displayed a slightly more complex gender profile than that revealed in the survey of waste managers across each of the pilot cities (Buckingham et al., 2020). In the participating waste management departments, men were marginally more likely to

Table 8.4 EIGE Gender Equality Index ranking of participating countries (2019)

Participating Countries (case studies)	EIGE Gender equality ranking in the EU	Participation in paid workforce ranking in the EU
Denmark (Copenhagen)	2	6
France (Nice)	3	17
Spain (Santander, Tenerife)	9	22
Italy (Florence, Syracuse)	14	28
Portugal (Lisbon, Azores)	16	7
Cyprus (Nicosia)	20	9
Croatia (Dubrovnik)	22	24
Greece (Kavala)	28	27

be dominant in the most senior posts, but not necessarily in professional, associate professional and technical jobs. It was quickly apparent that the level of knowledge about gender amongst most project participants was low, and there were many misconceptions (including overestimations) about the extent of gender equality in participants' own countries and organisations. In some cases, there was an initial indifference. For example, at the beginning of the project, only two partners collected gender-disaggregated data, only one partner could (tentatively) respond to the question about providing good practice of involving gender sensitivity in waste management, and only three could provide me with the name of a woman on the waste management team who might be able to talk to me about her experiences. Three partner cities acknowledged that they had, or were about to launch, gender equality training. These attitudes indicated that the pathways and power structures already well embedded in institutions would not be easy to redirect.

Using the outcomes of baseline data, cities could be categorised as (a) already or almost immediately gender-aware, which to a large extent informed processes and outcomes; (b) developed gender-awareness which impacted positively on the process; (c) developed direct gender-awareness amongst partners but had limited or no impact on the process; (d) insistently gender-denying, with no impact on outcomes. How the case study cities performed in terms of gender-awareness can be compared with their carbon emission savings. Although it is not possible to establish a causal relationship between the two, some correspondence can be noted which serves as instrumental encouragement for adopting gender mainstreaming, and this observation will be returned to later in this chapter.

Gender-proofing waste reduction measures

From the outset, gender mainstreaming was signalled as a key dimension of the project which all partners were expected to incorporate. Indeed, it was a feature of the initial proposal and commended by the European Commission representative assigned to the project, aware of a lack of gender expertise in the Innovations programme (EC, pers comm, 2016). Recognising the low level of gender awareness and gender expertise amongst most participants, the first mutual learning event focused on building gender-awareness, with specific training on how to run gender-sensitive focus groups. The Danish partner also provided input into this training as the City of Copenhagen had commissioned a survey of gendered waste behaviour. Copenhagen's degree of gender-awareness was significantly different to other partners, and they were in the process of trying to engage more men in waste management both at citizen level and in their waste management office, which was dominated by women. This session was followed by other mutual learning events, and three gender webinars on gender-sensitive communication, gender budgeting and gender reporting.

Each of the pilot cities was required to identify, through participatory stakeholder decision-making, a number of potential waste minimisation practices which would be innovative to their area. They were asked to monitor how these measures were gendered in decision-making, the extra work involved and in communication. It transpired that women were marginally more likely to incur extra work compared to men. Mindful that ICT use is gendered (EIGE, 2016), much attention was devoted to ensuring that the 'wasteapp', designed to locate recycling facilities and cafes, restaurants and hotels participating in waste reduction, was developed in a gender-sensitive way. Notwithstanding, the 'wasteapp' proved more popular among male tourists (Figure 8.1). However, given the lower recycling and waste minimisation practices of men referred to earlier, a tool which engages more with men could also be seen as a positive achievement. Just as Danish waste managers reported they had found, one US/Chinese study also concluded that men are more resistant to environmentally aware/sensitive behaviour, seeing it as too 'feminine' (Brough and Wilkie, 2017).

Correspondence between gender-awareness and sustainable environmental practices

The extent to which observations about gendered work have led to changes in practices was identified by the end-of-project survey. Eight pilot cities were identified as introducing changes to ensure more gender-sensitive practice or greater gender equality in employment in the longer-term. One pilot region, Tuscany, reported no further changes from an employment practice they already claimed to be 'family-friendly' and supporting women, but reported that they had been prompted by the project to consider introducing a gender strategy in regional waste planning. The remaining two pilot cities did not complete this element of the survey, from which we conclude that no longer-term gender-sensitive measures have been established.

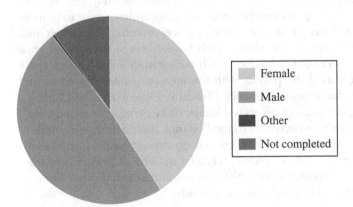

Female
Male
Other
Not completed

Figure 8.1 Gendered use of 'wasteapp' (More details of the evaluation of all innovations can be found in Buckingham et al., 2020.)

Table 8.5 Case studies' environmental performance relative to gender equality and sensitivity

CASE STUDY	Ranked by performance (average/ measure)	T CO_2e prevented/ recycled (number of measures)*	% female stakeholders (= remains same/ + increased)	Gender of senior staff	Gender sensitive changes to practice
Copenhagen	§, 5 (3)	*(1) 2.17 (1)	62 (=)	Retains female majority	√
Dubrovnik	nd	nd	47 (nd, 2016)	nd	None reported
Kavala	7 (7)	0.57 (3)	20 (+)	Retains male majority	√
Lisbon	3 (5)	7.147 (4)	68 (+)	Retains female majority	√
Nice	6 (6)	2.11 (2)	40 (nd, 2016)	Incomplete data	None reported
Nicosia	8 (9)	0.32 (1)	47 (+)	Retains male majority	√
Ponta Delgada	9 (8)	0.051 (1)	62 (+)	Retains female majority	√
Santander	4 (4)	4.228 (2)	44 (+)	New female head	√
Siracusa	10 (10)	0.025 (1)	46 (+)	Retains male majority	√
Tenerife	1 (2)	13.161 (4)	31 (=)	Retains male majority	√
Tuscany	2 (1)	13.12 (3)	53 (+)	Retains female majority	√

*Obersteiner et al. (2019)

To explore the impact of gender balance on waste reduction strategies, an analysis was made of any correspondence between pilot cities which appear to be more gender-aware (through reporting changes to practice, ensuring stakeholder and/ or employee gender balance), and those which have achieved the greatest gross carbon dioxide equivalent (CO_2e) reductions. It is clear from Table 8.5, that those case studies which started the project with the greatest gender-awareness, and where women had the strongest presence (Tuscany, Copenhagen and Lisbon), were amongst the greatest CO_2e savers (ranked by CO_2e savings per measure they rank first, third and fifth, respectively; by total savings, second, fifth and third).

Of the three cities which showed the greatest gender-awareness, Lisbon seems to have learned more from the emphasis on gender in *Urban Waste*, evident from the qualitative end-of-project survey. She also volunteered that other forms of discrimination needed to be tackled simultaneously:

> In my opinion, it's important to keep addressing and raising this subject to ensure equal treatment, opportunities and respect regardless of gender but also race and religion. Also, and regarding the Stakeholders, since the UW

Project takes gender in consideration this might raise awareness in establish-
ments that don't yet provide equal opportunities for all.

(Female Waste Manager, Lisbon)

Those cities which showed the least amount of gender-awareness, judged by the
lowest percentage of female stakeholders, and retaining male-dominated senior
waste management (Kavala, Nicosia, Siracusa), achieved some of the lowest
CO_2e savings (seventh, ninth and tenth gross savings; seventh, ninth and tenth
per measure). The exceptions to this pattern were Tenerife and Ponta Delgada.
Tenerife made the highest gross CO_2e reductions overall and second-highest
average per measure. One possible explanation for this is that Tenerife is located
in the Canary Islands, the government of which led the project; it, therefore, may
have had a particular interest in performing highly. Tenerife also instigated the
highest number of measures. On the other hand, Ponta Delgada showed a strong
commitment to gender equality but ranks low in both gross and average CO_2e
reductions (ninth and eighth, respectively). Ponta Delgada introduced one of the
most radical employment changes during the course of the project that requires
all external waste collection and street cleaning contracts to have a minimum
percentage of jobs to be held by women (the lead in Ponta Delgada was female).
This suggests caution in using arguments of efficiency whereby gender equality
is claimed as synonymous with better environmental performance, although it is
also likely that such a move is likely to manifest environmental improvements
in the longer term. Regardless, such a move towards gender equality in a heavily
male-dominated sector should be seen as a positive achievement in its own right.

The most interesting cities, as far as embedding gender mainstreaming is
concerned, are those which increased their number of female stakeholders and
changed the gender composition of their waste management team during the
project, in the course of which they developed a strong commitment to gender
equality. Santander, in particular, which ranks fourth by both gross savings and
average by measure, found *Urban Waste*'s focus on gender helpful, stating that:
'the simple fact of taking into account gender equality in *Urban Waste* project-
related activities is a way of influencing the organisation and the stakeholders'.
The city is also recommending to:

Follow the general strategy maintained throughout the project, bearing in
mind gender equality in all tasks. For example, in convening stakeholders,
in focus groups, in Communities of Practice, in communication campaigns,
in data collection, statistics, surveys and surveys.

(Female Waste Manager, Santander)

It was also possible to detect positive shifts in cities whose statistics belied their
enthusiasm. For example, while Nicosia is at an early stage of developing a gender-awareness, they also identified the utility of the gender approach for them:

We found it very useful. Men and women must be able and have equal opportunities to participate and successfully fulfil the different tasks assigned to them

in the workplace. Therefore, the manager of each department must remember this issue/term. Even if we consider gender equality in our organisation/department, the *Urban Waste* project through webinars, examples and via the implementation of the measures helped us to become more familiar with this factor.

(Male Waste Manager, Nicosia)

Five of the case study cities (two of which, Copenhagen and Tuscany, had already provided evidence of high gender-awareness) claimed that their city/country was already attending to gender equality. Three of these, however, had not provided evidence of commitment to gender equality requested at the start of the project, nor demonstrated particular enthusiasm for gender equality during its course. These responses could, therefore, be interpreted as 'blocking' mechanisms for change, especially as their employment and stakeholder data reflected gender inequality. An example of this likely overclaiming of attention to gender can be found from the waste manager in Tenerife who, despite not increasing the 31% of women in its stakeholder group nor changing the gender balance of a male dominant waste management team, claimed that (emboldened by the author for emphasis): '*Urban Waste*['s] **gender focus has been consistent with the strategy concerning gender equality** developed by the Cabildo de Tenerife from its cross-cutting approach'.

One pilot city (Nice, with a male lead) failed to respond to the question about how a focus on gender had influenced their project, and another case study (Dubrovnik, with a female lead) did not find the emphasis on gender at all helpful other than instrumentally to help secure grants in the future: 'It is not very useful. It is useful only in terms of future Horizon 2020 project where we can use experiences from this project'. Dubrovnik's persistent resistance to acknowledging gender inequality is despite, and perhaps reflects, Croatia being one of the lowest (24th of 28) performing countries in the EU on gender equality (EIGE, 2017).

Conclusion

The aim of this chapter was to explore the conditions under which gender mainstreaming can become a transformative process within the context of a climate change–related policy area, and asked a number of questions. The first question was what conditions are necessary for gender mainstreaming to become transformative. It was clear from the experience of advising on gender for *Urban Waste* that a prior awareness of and commitment to gender equality made a difference to performance on both gender-awareness and carbon emission reductions. One condition, then, for gender mainstreaming as a transformative process would seem to be a prior and evident commitment to gender equality. It is also clear that the more women who were involved as waste management decision-makers in the individual pilots, the better gender-sensitivity was achieved. This suggests the importance of descriptive equality, and challenges some of the gender mainstreaming analyses reviewed earlier in the chapter which argue that women in occupations dominated by men overall tend to mimic male behaviours. It also challenges, to some extent, the notion that path dependency in waste management institutions and the 'stickiness' of their masculine environments are universally resistant to change. Critical mass, however, can also be seen to be important as the pilot cities with larger numbers of women in decision-making positions were those more likely to support gender equality actions and climate-critical acts. Descriptive equality would, therefore, seem to be a necessary condition for transformative change, but not in itself sufficient.

The second question concerned the work necessary to encourage these conditions to be achieved. One reason proferred for the failure of gender mainstreaming is a lack of resources for training (Minto and Mergaert, 2018). However, given the resources available to the *Urban Waste* project over an extended period of three years, and given the patchy success of achieving gender equality, let alone mainstreaming gender concerns across the pilots, resources alone cannot be seen as a pre-condition for transformation. Even in this three-year project, with many interventions to raise the issue of gender equality, and which was commended in the final evaluation for its work on gender, the success rate was limited. The responses which emerged during the life of the project spanned 'embracing gender equality' to outright resistance. In between lay examples of compliance without enthusiasm, and box-ticking. Certainly, Mergaert and Lombardo's (2014) identification of resistances at the individual and institutional level resonates with some cases in the project, where either individuals resistant to the idea of promoting gender equality, or institutions which are not receptive, are able to block gender mainstreaming. This was certainly the case in Dubrovnik, but experience elsewhere was equivocal. These findings support one of the concepts underpinning this book: that embedded and 'sticky' institutional structures can prevent transformative action regarding gender equality. This, then, is one of the challenges facing gender mainstreaming at different stages of its implementation, the third question posed by this chapter. Perhaps one of the problems is when gender mainstreaming is viewed as something which is added on to an existing workload which people feel ill-equipped and poorly resourced to manage, rather than an opportunity to re-envision how an organisation or part of an organisation might work. For example, Tuscany had provided the example of introducing family-friendly policies, such as flexi-time and working from home, which benefited women and improved the workplace culture for all employees.

The chapter's last question relates to how gender mainstreaming can be expected to achieve change in climate change–related policy-making. The project does provide some evidence that better gender-awareness and gender equality practices are linked with better waste reduction and environmental performance. This reflects research at the global scale, where countries which perform highly on gender equality also do so on environmental measures (Ergas and York, 2012). Although the relationship between the two is not necessarily straightforward, there are indications that greater attention to gender balance and gender equality amongst stakeholders and employees could lead to improved waste reduction, which makes it all the more critical that policies such as the European Union's Circular Economy Action Plan, or EMAS, do actually refer to gender and indicate its importance, which is not yet the case.

Notes

1 The *Urban Waste* project was funded by EU Horizon 2020: Grant Agreement Number: WASTE-6a 2015 690452.
2 The participating cities/countries were: Dubrovnik in Croatia; Nicosia in Cyprus; Copenhagen in Denmark; Nice in France; Kavala in Greece; Florence/Tuscany and Syracuse in Italy; Lisbon and Ponta Delgada (Azores) in Portugal; Santander and Tenerife (Canary Islands) in Spain.
3 The European Union combines sectors to compile annual greenhouse gas emission statistics; there are nine combined sectors, of which waste is currently the lowest emitter of greenhouse gases (EU, 2019a).
4 REDD+ refers to the United Nations 'reducing emissions from deforestation and forest degradation in developing countries, and the role of conservation, sustainable management of forests and enhancement of forest carbon stocks in developing countries'. ENAREDD+ is the related strategic policy of Mexico.
5 In this role, I ensured that the *Urban Waste* project was gender mainstreamed, and supported pilot cities to adopt gender mainstreaming as a principle and practice, at least for the waste innovations they developed.
6 Although the EU considers sex-disaggregated data to be a decisive factor for the effective implementation of gender mainstreaming, the availability of sex-disaggregated data by employment sector is not yet mandatory and varies by country. It will be required by the Gender Equality Strategy 2020–2025, although this had not been ratified at the time of writing (European Commission, 2020).
7 Specifically, this comprises baseline data on staff in the 11 case study waste management offices; a survey of waste management staff in the 11 cities, and follow up focus groups; evaluations of gender, and tonnes of CO_2 equivalent emissions prevented, in the waste reduction innovation measures adopted by each case study city; an end-of-project evaluation which assessed gender impact in each of the case study cities.

References

Bee, B. (2017) 'Safeguarding gender in REDD+: reflecting on Mexico's institutional (in) capacities', in Buckingham, S. and Le Masson, V. (eds.) *Understanding Climate Change through Gender Relations*. London: Routledge, 190–204,

Bhopal, K. (2020) 'UK's white female academics are being privileged above women—and men—of colour', *The Guardian*, 28 July 2020.

Brough, A.R., and Wilkie, J.E.B. (2017) Men resist green behaviour as unmanly. A surprising reason for resistance to environmental goods and habits. *Scientific American* [online]. Available at: https://www.scientificamerican.com/article/men-resist-green-behavior-as-unmanly/# (Accessed: 05 May 2019).

Buckingham, S. (2020) *Gender and Environment*. London: Routledge.

Buckingham, S., Perello, M. and Lopez-Murcia, J. (2020) Gender mainstreaming urban waste reduction in European cities. *Journal of Environmental Planning and Management*. Published online 15 October. doi: 10.1080/09640568.2020.1781601.

Buckingham, S., Reeves, D. and Batchelor, A. (2005) 'Wasting women: The environmental justice of including women in municipal waste management', *Local Environment*, 10(4), pp. 427–444.

Catalyst. (2018) Quick take: Women in male-dominated industries and occupations [online]. Available at: https://www.catalyst.org/research/women-in-male-dominated-industries-and-occupations/ (Accessed: 01 May 2019).

Charrington, J. (2017) Inspiring women: improving gender equality in waste Chartered Institute for Waste Management—Journal [Online]. Available at: https://ciwm-jo

urnal.co.uk/inspiring-woman-improving-gender-equality-in-waste/ (Accessed: 01 May 2019).

Dri, M., Canfora, P., Antonopoulos, I.S. and Gaudillat, P. (2018) Best environmental management practice for the waste management sector, JRC science for policy report, EUR 29136 EN. Publications Office of the European Union, Luxembourg, 2018, ISBN 978-92-79-80361-1, doi:10.2760/50247, JRC111059.

Ergas, C. and York, R. (2012) 'Women's status and carbon dioxide emissions: a quantitative cross-national analysis', *Social Science Research*, 41, pp. 965–76.

European Commission, D-G Environment (2014) *Flash Eurobarometer 288 Attitudes of Europeans Towards Waste Management and Resource Efficiency*. Brussels: European Union.

European Commission, D-G Environment. (2017) *Special Eurobarometer 468 Attitudes of Europeans Towards the Environment*. Brussels: European Union.

European Commission Liaison Officer (2016) Conversation with Susan Buckingham 7th June 2016.

European Commission (2019a) SHE Figures 2018 [online]. Available at: https://ec.europa .eu/info/publications/she-figures-2018_en (Accessed: 30 May 2019).

European Commission (2019b) European Green Deal [online]. Available at: https://eur-lex .europa.eu/legal-content/EN/TXT/?qid=1588580774040&uri=CELEX:52019DC0640 (Accessed: 01 May 2020).

European Commission (2020a) A union of equality: gender equality strategy 2020– 2025 COM(2020) 152 final. Brussels [online]. Available at: https://eur-lex.europa .eu/legal-content/EN/TXT/?uri=CELEX%3A52020DC0152 (Accessed: 16 October 2020).

European Commission (2020b) A new circular economy action plan for a cleaner and more competitive Europe COM(2020) 98 final. Brussels [online]. Available at: https:// ec.europa.eu/environment/circular-economy/ (Accessed: 16 October 2020).

European Commission (n.d.) Food waste [online]. Available at: https://ec.europa.eu/food/ safety/food_waste_en (Accessed: 15 August 2020).

European Institute for Gender Equality (EIGE) (2015) *Gender Equality in Power and Decision Making*. Luxembourg: European Union.

European Institute for Gender Equality (EIGE) (2016) *Gender and Digital Agenda*. Luxembourg: European Union.

European Institute for Gender Equality (EIGE) (2017) Gender equality index 2017: measuring gender equality in the European Union: 2005–2015 report [online]. Available at: https://eige.europa.eu/publications/gender-equality-index-2017-measuri ng-gender-equality-european-union-2005-2015-report (Accessed: 28 May 2019).

European Institute for Gender Equality (EIGE) (2019) Gender Equality Index Vilnius, Lithuania: EIGE [online]. Available at: https://eige.europa.eu/gender-equality-index/ 2019/compare-countries/work/1/bar (Accessed: 27 April 2020).

European Institute for Gender Equality (EIGE) (2020) Beijing 25+: the fifth review of the implementation of the Beijing platform for action in the EU member states. Area K— Women and the environment: climate change is gendered. Lithuania: EIGE.

European Parliament (2019)[online]. Available at: http://www.europarl.europa.eu/ RegData/etudes/ATAG/2019/630359/EPRS_ATA(2019)630359_EN.pdf (Accessed: 26 October 2020).

EY (2015) Women in power and utilities index 2015 [online]. Available at: https://ww w.ey.com/Publication/vwLUAssets/EY-women-in-power-and-utilities-index-2015/ $FILE/EY-women-in-power-and-utilities-index-2015.pdf (Accessed: 28 May 2018).

EY (2019) Could gender equality be the innovation boost utilities need? EY, 8 March 2019. https://www.ey.com/engl/women-power-utilities/could-gender-equality-be-the-innovation-boost-utilities-need (Accessed 20.2.2021).

Eurostat (2020) Waste statistics. Electrical and electronic equipment [online]. Available at: https://ec.europa.eu/eurostat/statistics-explained/index.php/Waste_statistics_-_electrical_and_electronic_equipment (Accessed 15 August 2020).

Fraser, N. (2013) *Fortunes of Feminism. From State Managed Capitalism to Neoliberal Crisis.* London: Verso.

Griffin, P. (2015) 'Crisis, austerity and gendered governance', *Feminist Review*, 109, pp. 50–72.

Held, V. (2006) *The Ethics of Care. Personal, Political and Global.* Oxford: Oxford University Press.

IRENA (International Renewable Energy Agency) (2019) *Renewable Energy: A Gender Perspective.* IRENA: Abu Dhabi.

Labfresh. (2019) Fashion Waste Index [online]. Available at: https://labfresh.eu/pages/fashion-waste-index?lang=en&locale=en (Accessed 15 August 2020).

Magnusdottir, G. and Kronsell, A. (2015) 'The (In)Visibility of Gender in Scandinavian Climate Policy-Making', *International Feminist Journal of Politics*, 17(2), p. 308–26.

Mergaert, L. and Lombardo, E. (2014) 'Resistance to implementing gender mainstreaming in EU research policy', in Weiner, Elaine and Heather MacRae (eds.) *The Persistent Invisibility of Gender in EU Policy' European Integration Online Papers (EIoP) Special Issue 1*, Vol. 18, Article 5, pp. 1–21. http://eiop.or.at/eiop/texte/2014-005a.htm.

Minto, R. and Mergaert, L. (2018) 'Gender mainstreaming and evaluation in the EU: comparative perspectives from feminist institutionalism', *International Feminist Journal of Politics*. doi.org/10.1080/14616742.2018.1440181

Obersteiner, G., Gollnow, S., Erikksen, S., Lopex-Murcia, J., Gonzalex Martin, G., Wojtowicz, M., Bjorn Olsen, T., (2019) *D7.1 – Environmental economic and social impacts of tested measures, including gender issues* Urban Strategies for Waste Management in Tourist Cities: Government of Canary Islands

OECD. (2017) Employment by gender in water supply, sewerage, waste management & remediation activities, 2017 [online]. Available at: https://stats.oecd.org/viewhtml.asp x?datasetcode=ALFS_EMP&lang=en# (Accessed: 17 April 2019).

Prügl, E. (2013) 'Gender expertise as feminist strategy', in C. Gülay, E. Prügl, and S. Zwingel (eds.) *Feminist Strategies in International Governance*, London: Routledge pp. 57–73.

Resurreccion, B. and Elmhirst, R. (2020) *Negotiating Gender Expertise in Environment and Development: Voices from Feminist Political Ecology.* London: Routledge.

Taylor, D. (2014) The State of Diversity in Environmental Organizations. Green 2.0 Working Group.

Verloo, M. (1999) 'On the conceptual and theoretical roots of gender mainstreaming', The Interface between Public Policy and Gender Equality, Seminar 1. Sheffield Hallam University, Centre for Regional Economic and Social Research.

Vida, B. (2017) *Failure of Gender Mainstreaming—Renegotiating Gender Equality Norms in Horizon 2020.* CEU MA-thesis. Budapest.

Walby, S. (2005) 'Introduction: Comparative Gender Mainstreaming in a Global Era', *International Feminist Journal of Politics*, 7(4), pp. 453–470.

UN Women. (2020) *The Impact of COVID-19 on Women.* New York: United Nations.

9 Gender analysis of policy-making in construction and transportation

Denial and disruption in the Canadian green economy

Bipasha Baruah and Sandra Biskupski-Mujanovic

Introduction

Countries around the world are figuring out ways to make their economies less carbon-intensive by creating new green jobs, developing less-polluting technologies and by retrofitting existing sectors such as manufacturing, construction, transportation, energy production, water and waste management (ILO, 2011). A gender-focused analysis of such green growth and development strategies reveals two major blind spots. First, women are known to have weaker access to new technologies almost everywhere in the world (Rosser, 2005; Hafkin and Huyer, 2006) so there are likely to be unequal access issues inherent in the transition to low-carbon economies. Second, it is well-established that women are already very poorly represented globally in some of the sectors that are deemed critical to the green economy. For example, women account for 9% of the global workforce in construction, 12% in engineering and 24% in manufacturing (UN Women, 2012). In the absence of appropriately targeted training, education, apprenticeships, employment placement, financial tools and supportive social policies, transitioning to a green economy may exacerbate existing gender inequities and reproduce existing norms and path-dependent knowledge in governmental strategies in the green economy.

In this chapter, we present findings from our synthesis of existing scientific and practitioner literature on promising programmes and policies for promoting employment equity in two sectors, construction and transportation, that are deemed critical for the Canadian green economy and in which women are currently severely underrepresented. In Canada, women currently comprise only 23% of those employed in transportation and 12% in construction (Statistics Canada, 2019). This could indicate a lack of commitment to equality in the green economy sector by state authorities and their policy-makers.

We choose to focus on employment in construction and transportation[1] in this chapter because these sectors tend to produce well-paid stable jobs, but they are extremely male-dominated in Canada and in other Organisation for Economic Co-operation and Development (OECD) countries. We analyse employment in these two sectors while fully cognizant that in addition to increasing women's participation in male-dominated occupations that are well-compensated, we

DOI: 10.4324/9781003052821-9

should be very concerned about how undervalued and underpaid women are in sectors like early childhood education, primary education, social work and health and library services that are already green. The care industry, for example, is 30% less polluting (in terms of GHG emissions) than the construction industry and the education industry is 62% less polluting than the construction industry (Women's Budget Group, 2020). As Hirshman (2008) notes:

> Maybe it would be a better world if more women became engineers and construction workers, but programs encouraging women to pursue engineering have existed for decades without having much success. At the moment, teachers and childcare workers still need to support themselves. Many are their families' sole support.

Improving the status, wages and working conditions for people working in feminised green sectors is as important, if not more, as increasing women's participation in well-compensated male-dominated occupations (Baruah, 2017).

The remainder of this chapter includes the methodology and findings of our research. We provide an overview of the major barriers and opportunities that women encounter in seeking, retaining and advancing within employment in the construction and transportation sectors in Canada. This is followed by an overview of the blind spots and oversights about employment equity in Canada's green transition, and a description of promising practices and policies being pursued in the Canadian context to improve women's participation in construction and transportation. We conclude with a summary of the trends we observed in employment in these sectors in Canada and a reflection on the policies, practices and social awareness necessary to close the gender gaps.

This chapter is not intended as a comprehensive survey of women's employment or gender equality in the green economy in Canada. We were unable to undertake an intersectional analysis of gender along cross-cutting identities such as class, ethnicity and sexuality because the empirical data does not currently exist to enable such an assessment. Thus, when we refer to gender equity in this paper, we are limited to discussing those who identify biologically and socially as either women or men. In fact, the need to collect more gender-informed and gender-diverse data to inform evidence-based policy-making is one of the clearest findings from this study and those of others who have attempted to study the green transition in Canada (see, for example, Lieu Et al., 2020; MacArthur Et al., 2020). We hope that the issues identified by our research will provide the grounding and detail against which other related issues and research, perhaps using very different methodologies as well as broader conceptualisations of gender equality (including those of gender-diverse LGBTQ persons) and intersectional gender analysis, can be tested, verified and advanced.

Methodology

This knowledge synthesis is based on both scientific and practitioner literature. Peer-reviewed scholarship on women's employment in transportation and construction is limited. Most of the information for this study was gathered from practitioner organisations. We drew upon information from websites, annual reports, policy reviews, position papers and survey results of Canadian governmental and non-governmental organisations (NGOs), labour associations, professional networks, public policy institutes, think-tanks, financial institutions and social enterprises.

We asked ourselves several questions while reading the assembled academic literature and policy documents: Does this piece discuss employment equity? Is there mention of gender equity specifically? Does it make a reference to green or sustainable employment in transportation or construction? Although we intended to specifically collect data on women's employment in green or low-carbon transportation and construction, we found that most of the available employment information is from the 'brown' or traditional (fossil fuel–based) forms of these sectors. We found very few specific examples of women's inclusion in employment in green transportation and construction, which supports a major finding from this study that gender equality (or employment equity, more broadly) has not been meaningfully addressed or even considered in mainstream green transition policies and practices in Canada.

Findings

We have organised findings from our knowledge synthesis under three themes. First, we identify and discuss the specific barriers and opportunities faced by women in seeking and retaining employment in these two sectors in Canada. Next, we outline blind spots and oversights in Canada's green transition that may contribute to women's underrepresentation in construction and transportation employment. Finally, we selectively present promising programmes and practices undertaken by municipalities, provincial governments, corporations and NGOs in Canada in order to reduce the gender gap in construction and transportation.

Barriers and opportunities for women's employment in construction and transportation in Canada

The challenges and opportunities women face in the transportation and construction industries are similar given that they are both non-traditional occupations for women in Canada. A non-traditional occupation (NTO) is defined as any occupation in which women or men comprise less than 25% of the workforce. Of course, it is important to emphasise that although certain occupations become male- or female-dominated over time in specific contexts due to historical and current processes of socialisation and the gendered 'logic of appropriateness' (Chappell, 2006), such categorisations are rarely static or stable. As evidenced by the parity or near-parity reached by women in fields such as medicine and law that were previously male-dominated, different sectors of employment may change over time due to shifts in societal attitudes as well as economic and political necessity. Also, some occupations within male-dominated sectors (for example, human resources, administrative and clerical services, public relations and communication, financial services) may have more than 25% (or even 50%) female workers, but women tend to be a minority in the most well-compensated and stable positions, namely in trades and operations, in technical positions that require science, technology, engineering and math (STEM) training, and in management, senior leadership and boards of directors of companies. Women's underrepresentation in executive and managerial positions is, indeed, often because experience in engineering and operations is considered a prerequisite for leadership roles. In other words, the presence of women in some roles in these sectors does not necessarily translate into, the representation of their ideas, needs and priorities at the institutional level (Phillips, 1998). The global shift to low-carbon economies demands a growing array of skills – technical, business, administrative, economic and legal, among others (Baruah, 2016). Widening the talent pool in these industries is, thus, an instrumental reason for boosting the participation of women, in addition to intrinsic reasons of gender equity and fairness (IRENA, 2019).

Findings from our knowledge synthesis point to four major barriers for women in these sectors, which need to be addressed both by state and local institutions as well as other stakeholders. The barriers are a lack of adequate information and awareness about these fields of employment, gender bias and gender stereotyping, male-biased work culture and working conditions, and concerns about sexual harassment and violence against women.

Barrier: Lack of adequate information and awareness

We found that women often lack information about the opportunities and prerequisites for employment in the transport industry because careers in transportation are rarely pitched to women and girls through avenues such as recruitment sessions and career fairs (APGST, 2017). Much more needs to be done to introduce girls at an early age to the potential in these careers. Encouraging girls to study STEM fields is critical since employment in the most well-paid jobs in these two sectors (for example, installation, engineering and architecture) tend to require STEM training (Antoni Et al., 2015). Because these sectors were almost exclusively dominated by men for so long, employment information continues to travel disproportionately through male professional networks. Young men tend to find job opportunities in these fields through references from friends and relatives while women usually do not have similar social connections. Therefore, there is an urgent need to mainstream – i.e. improve equity in access to job information (ILO, 2011).

Male candidates tend to apply for jobs in these two sectors even when they meet *some* of the requirements, but most female candidates tend to not apply for jobs unless they meet *all* requirements (APGST, 2017). Women are also less likely to negotiate salaries and benefits. Women must often outperform men in male-dominated industries just to fit in and certainly to progress. The preference for male recruits in most transport jobs is very much a 'chicken-and-egg' problem

– women often lack the necessary educational qualifications for many jobs in the transport sector, but these jobs have usually not been designed with women in mind and are, therefore, not particularly attractive (Turnbull, 2013). Thus, when it comes to selection, (male) managers are less likely to regard women as suitable candidates. Women appear to encounter both 'sticky floors and glass ceilings' in transportation (Turnbull, 2013). Careers may never get off the ground because of persistent and confining stereotypes of feminised roles. And the absence of role models and gender-balanced initiatives make moving up the ranks more challenging for women than men.

Barrier: Gender bias and gender stereotyping

Since transportation is presently such a gender-unbalanced industry, Turnbull (2013) emphasises that it is neither attractive nor welcoming to women. For this to change, the masculinist culture of the industry must be dismantled and career opportunities within the transportation industry must be publicised more directly to women. Women who do work in the transportation industry are routinely underestimated or discounted because of gender-based stereotypes. Of course, women are keenly aware of this and will often forgo applying for certain types of jobs, for example those involving fieldwork, because they know gender stereotypes will work against them. This explains, at least partially, why more women in these industries end up in administrative and clerical jobs rather than technical ones (APGST, 2017). Turnbull (2013) emphasises that social attitudes play a large part in discouraging women from pursuing careers in these fields when he writes:

> the fact that sex discrimination has not disappeared from the world of work may be due, in part, to a lack of political commitment and – in some contexts – legal laxity, but the underlying cause remains embedded in social attitudes.

Thus, women in construction are either ridiculed as masculine or they are caught between expected roles and stereotypes – such as being treated as sexual objects or labelled as lesbian for choosing a more masculine field to work in (Moir Et al., 2011). Clearly, this issue highlights much broader gender inequities and problematic social attitudes based on misunderstandings and misperceptions about gender.

The transportation industry is frequently referred to in the literature as having a work culture and environment that include risks of sexual harassment and isolation. Turnbull (2013) traces the primary causes of women's low participation in the transportation sector to working conditions and argues that working time, shift work (24/7) and location of employment (i.e. at sea, driving long distances from home) are often unappealing to women (especially those with childcare or eldercare responsibilities) and the transportation sector may not only be unwelcoming, but hostile towards women. However, it is important to remember that although such factors may, to some extent, explain women's underrepresentation in transportation or construction employment, many women may already work in less-than-optimal environments for much less pay than they would make in the construction or transportation industry. Given the option, some women would probably prefer to work in construction or transportation simply because of the potential to earn higher wages. Because of persistent and entrenched male-biased norms in these industries, even women who are able and willing to work may not be given the option to choose between difficult or dangerous working conditions with low pay and similar conditions with higher pay (McKee, 2014; Carpenter Et al., 2017). Instead, women are assumed not to want to work in such jobs and tracked into feminised occupations in administrative and support services within these sectors. The fact that women's careers can be adversely affected by 'benevolent sexism' has also been documented in other sectors of employment (Dizik, 2016).

Barrier: Male-biased work culture and working conditions

It is impossible to neatly demarcate the barriers and opportunities women face in entering the construction and transportation sectors, from those that influence their decision to leave or remain. Turnbull (2013) analysed the barriers women face in the transportation industry using the career-cycle model. He identified issues women faced at every stage of career: attraction, selection, retention, interruption, re-entry and realisation. Attraction to the transport sector comes from exposure at school, home and community which he argues is 'heavily influenced by the HR policies of transport organisations (e.g. corporate image, commitment to equal opportunities) and societal values (e.g. prevailing views on what constitutes "men's work" and "women's work"'. Most departures (resignations and dismissals) in the transportation industry appear to occur within the first five years of employment. Therefore, women's initial experiences – how they are welcomed and treated, and whether they are supported and promoted – are critical.

Women in these sectors tend to leave work more frequently after maternity leave than at any other point in their careers (Baruah, 2018). Success in the retention of female employees, especially in the aftermath of interruptions for childbearing or other caring work, appears to be dependent both on the support of the organisation and the support of co-workers. Attitudes of male co-workers were deemed particularly important in retaining women after career interruptions. While studying the transportation sector, Peter Turnbull distinguished between gender-specific barriers, such as stereotypes about men's and women's work, and gender-intensified barriers, such as the absence of working arrangements to accommodate childcare and other reproductive responsibilities. Gender-specific barriers tend to interact with gender-intensified barriers in influencing women's decisions to leave these sectors. An example that was presented frequently in the literature was the double standard women faced when they returned to work after parental leave. While men who took parental leave were welcomed back to work and often valourised for their commitment to parenting, women were more likely to find their commitment to work being implicitly or explicitly questioned, to be taken less seriously by colleagues and superiors and to feel that they were no longer competitive or competent in their positions (Baruah, 2018). Such findings shed light upon why ostensibly gender-neutral institutions have differential effects upon women and men. Feminist institutionalist approaches are used to answer questions about power inequalities, mechanisms of continuity, and the gendered limits of change (see e.g. the introduction to this volume). Feminist institutionalist literature emphasises that it is crucial to identify how informal norms and behaviours interact with formal institutions and policies in order to understand wider processes of continuity and change as well as variable outcomes for women and men (Krook and Mackay, 2010).

Arcand (2016) arrives at similar findings about women's experiences in the construction industry:

> Women tend to enter the industry with fewer skills than their male counter-parts, not having access to informal networks commonly used to find work, being let go once diversity quotas have been met, and encountering hostile work environments and sexual harassment on the job.

The fact that informal networking is still the norm rather than the exception for seeking apprenticeships and employment in the trades is identified repeatedly as a barrier to women's entry into and advancement in these fields (National Women's Law Centre, 2014). Thus, female construction workers refer (not facetiously) to their inability to access FBI (friends, brothers and in-laws) networks as a major impediment in gaining full-time employment in the trades (National Women's Law Centre, 2014). Gender-disaggregated data on the number of new registrations in apprenticeship programmes in Canada paint a stark picture. In 2015, for example, there were just 66 women enrolled as machinists, compared to 1,545 men. There were 54 women registered as sheet metal workers, compared to 1,476 men (Smith, 2017). The need for policies aimed at enabling fair and equitable access to apprenticeships is urgent and critical for promoting equity in these sectors.

Barrier: Concerns about sexual harassment and violence against women

Violence against women transport workers is frequently reported as a factor limiting women's ability or willingness to seek and keep transportation jobs (Turnbull, 2013). Moir Et al. (2011) emphasise the same about the construction industry. They suggest that because construction work is carried out autonomously or with minimal supervision and often in dark corners, trenches, and small or secluded spaces, there are concerns about safety, especially for female workers. The Health and Safety of Women in Construction (HASWIC) Workgroup reported that 88% of female construction workers surveyed had experienced sexual harassment at work (Moir Et al., 2011). The immediate safety risks for women can certainly be mitigated by certain actions and practices (better lighting and working in pairs, for example) but eliminating violence against women will require much deeper and more proactive engagement with the social structures and power relations that sustain and reproduce it. We found very little engagement in the literature on construction and transportation with the structural causes of violence against women.

As more jobs are created in these sectors due to the transition to low-carbon economies, equitable representation of women in transportation and construction is becoming even more important. In Canada, women's employment in the transportation industry ranges between 22% and 25% while women's employment in the overall labour force is 48% (APGST, 2017). The lowest employment rate for women is in rail and truck transportation, 13% and 14% respectively, while the postal service and air transportation have the highest representation of women, at 50% and 39%, respectively. Canada will create 200,000 new jobs in the construction industry over the next decade but women only account for 4% of new registrants in the construction trades (Status of Women Canada, 2017). Retrofitting and upgrading residential buildings will make up a third of projected global green jobs (Sustainlabour, 2009). Additionally, the transportation sector is expected to contribute substantially to green growth through fuel-efficient vehicles, new infrastructure and public transport.

Opportunities

We found wages in construction and transportation to be more gender-equitable than wages in other sectors of the economy. For example, women in construction earn 98 cents for every dollar earned by a man compared to earning just 75 cents for every dollar earned by a man in the personal care and services field (Arcand, 2016). The gender wage gap can be narrowed further by encouraging women's participation in construction and other non-traditional occupations. Hegewisch and O'Farrell (2015) argue that women may be especially valuable in green construction because of their higher interest in environmental issues compared to men. They write that in the US and Canada women represent more than one in five environmental engineers, but only one in eight civil engineers. The same study reports that people who are already employed in the traditional construction industry can access training in green construction skills quite easily. Since women are a minority in the traditional construction sector, their ability to access green construction training is significantly weaker than their male counterparts. Most women who are in the traditional construction sector in North America indicated that green training and work is part of their construction trade and not a separate route into the trades. This highlights the need to influence equity practices in the traditional construction sector and not just in green construction. Thus, our findings from Canada echo those of Clarke and Sahin-Dikmen (in this volume) from the European Union, who emphasise that there is scant evidence to suggest that gender has even entered mainstream green transition approaches in construction.

In a report about good practices for the construction sector, Peters and Katalytik (2011) emphasise that companies with more women on their boards fare better than their rivals, with a 42% higher return on sales, 66% higher return on invested capital and 53% higher return on equity. This data should encourage employers not just to recruit women for entry-level positions but also to make the effort to retain and promote them to leadership positions. Better gender balance in male-dominated professions has been demonstrated to contribute to the improvement of working conditions for both men and women, with positive effects on wellbeing, work culture and productivity. A respectful workplace is a workers' issue, not just a women's issue (Kong, 2020).

There is tremendous potential in Canada, and globally, to create and optimise livelihoods for women in transportation and construction. However, women can gain optimal traction from employment in these fields only within the context of wider socially progressive pro-women policies, as well as more transformative shifts in societal attitudes about gender roles. This is as true for developing countries and emerging economies as it is for industrialised nations (Baruah, 2016). The green transition in these two sectors should benefit both women and men, but policy-makers must be proactive about enabling women to establish a stronger equity stake to compensate for historical and contemporary economic injustices and unequal outcomes. This will require more concrete and proactive actions and policies. As evidenced by our findings from Canada, simply creating opportunities for training and employment in new fields and suggesting that women are not unwelcome in them is not enough (Baruah and Biskupski-Mujanovic, 2018).

Blind spots and oversights about employment equity in the green transition in Canada

Some of the challenges faced by women in transportation and construction may be exacerbated by blind spots and oversights in the green transition in Canada. Planning and implementing measures for employment equity in the green economy is critical for Canada. Most green initiatives in Canada have been driven by the private sector, municipalities and provincial governments. The federal government has, at least until very recently, not played an active role in framing and implementing effective policies to enable the transition to a low-carbon economy. Despite growing evidence of the potential for the green economy to generate a larger volume of employment than the 'brown' (fossil fuel–based) economy, even organisations committed to advocating for equity and social justice in environmental sustainability in Canada have never specifically mentioned or addressed gender inequity. As an example, the One Million Climate Jobs Campaign – an alliance of labour, social movements and other civil society organisations in Canada – is mobilising for practical solutions to the threat of climate change. Although the campaign claims to place the interests of workers and the poor at the forefront of strategies to combat climate change and demands that governments work to create 'one million climate jobs', it makes no mention whatsoever of the importance of addressing gender inequity. Reports that do highlight opportunities to employ underrepresented groups, including women, in the green economy stop short of calling for specific policies and concrete actions required to ensure gender equity (see, for example, Blue Green Canada, 2012; Katz, 2012). We found more evidence in such reports for calls for equity measures to improve the representation of other underrepresented groups in Canada, namely, workers with disabilities and Indigenous peoples, than for improved representation of women.

It is difficult to identify straightforward explanations for why gender inequality has not permeated the public or policy consciousness in the transition to a green economy in Canada. The conversation about gender issues in the green economy has been much more prominent in the context of developing countries (see, for example, ILO, 2011; IUCN, 2011; UN Women, 2012), where the gender gaps in education and employment may be more prominent and visible. The lack of attention paid to gender equity in the green economy in Canada and other industrialised countries may be a result of the assumption that because women and girls increasingly enjoy parity with men and boys in educational opportunities and employment, they cannot possibly be underrepresented or marginalised in some sectors. The fact that women represent more than 50% of university students and almost half – 48% – of the labour force in Canada (Statistics Canada, 2019) may have obscured their extreme underrepresentation and marginalisation in traditionally masculinist trades and occupations. Several authors have also noted that popular mythologies about Canada being a 'post-gender' and 'post-racial' society often end up encouraging complacency among policy-makers and deflecting attention away from persistent social, economic, and political inequalities between Canadians based on gender and race (Subramaniam, 2020). Recent developments such as the COVID-19 pandemic in 2020 – which is being described as a 'she-cession' (Trichur, 2020) because of the disproportionate job losses in feminised sectors such as retail and hospitality services compared to male-dominated industries such as construction, transportation, and manufacturing (that are more easily adaptable for social distancing requirements) – may lead to a reckoning that compels more attention in the future to gender inequity in the green economy.

Some of the most well-paid future green jobs in Canada will be in occupations in which women are currently underrepresented, such as engineering and the skilled trades (Katz, 2012). A Statistics Canada study found that in 2007 women only accounted for 1–2% of completions in apprenticeship training in major trade groups (McMullen Et al., 2010). In 2018, women made up only 18% of newly licensed engineering graduates in Canada (Engineers Canada, 2020). According to a 2016 report by the Canadian Apprenticeship Forum (CAF-FCA), in 2012 women accounted for only 14.2% of all registrations in the trades. However, most women in the trades are employed in the feminised service sectors, including some green trades (as bakers, hairstylists, nail technicians, and pet groomers, for example), not in the male-dominated manufacturing and construction trades – especially those governed by the Red Seal Interprovincial Standards Program. Red Seal tradespeople have been trained to the same standards across the country, allowing them to move more easily between provinces. In fact, women's representation in many trades – including as automotive service technicians, electricians, and carpenters – is at less than 5%. The numbers are even smaller when it comes to women of colour. The same CAF-FCA report says, 'nearly half of Red Seal trades have no visible minority women, and the other half only between one and 12%' (quoted in Kong, 2020). Since tradespeople have been known to transition from jobs in the 'brown' economy (which is heavily male-dominated) to the 'green' (Baruah and Gaudet, 2016), it seems inevitable that women will also be underrepresented in green jobs unless gender equity in employment is planned and implemented proactively. Recent media reports confirm this trend, indicating that 25% of students studying to be wind turbine technicians in Alberta, Canada, were once oil and gas workers (Canadian Press, 2017), and that laid-off oil and gas workers in Alberta are using their unemployment benefits to retrain and find employment in the clean energy sector (Bickis, 2016).

The conversation about gender equity or social justice (more broadly) in Canada's green economy is at best incipient and tokenistic. Raising awareness is, therefore, urgent and critical. In 2016, the OECD reported that although Canada performs better than the OECD average for the gender employment gap (6.1% compared to 11.7%), the gender wage gap is above the OECD average (19% compared to 15.5% overall). Not only does Canada have a large gender wage gap, but women are severely underrepresented in all green growth sectors (Thirgood Et al., 2017). The types of policies and programmes put in place by governments and corporations will promote or impede opportunities for women. Although employment in the green economy in Canada is a growing topic of research and policy engagement, the need to promote gender equity in green jobs is rarely a topic of conversation. For example, the One Million Climate Jobs campaign proposes a 5% investment of the annual federal budget in renewable energy, energy efficiency and public transportation in order to create one million new jobs while reducing greenhouse gas emissions by 25–35%. Although this campaign discusses the types of jobs that will be created through such an effort quite extensively, there is no explicit or even subtle mention of employment equity.

A 2015 report prepared by the Steering Committee of the Green Economy Network in Canada identifies priority areas to stimulate the transition to a green economy, including a renewable energy development strategy, improved energy efficiency of homes and buildings, and expanded public transit. The report emphasises that millions of new decent jobs would be created by transitioning to a green economy. In order to ensure social equity, this report suggests creating a transition fund to assist workers displaced by the phasing out of fossil fuel production, to promote affirmative action for members of marginalised communities, and to enable consultation with Indigenous peoples. Among the suggested equity measures, there is no mention of gender and no consideration of the underrepresentation of women in the industries that will help Canada reduce greenhouse gas emissions.

Promising policies and programmes in Canada

In 2019, the Canadian Apprenticeship Forum (CAF-FCA) published a labour market report that showed that from 2019 to 2023 an estimated 67,000 new certified workers will be required to sustain workforce certification levels across the ten largest Red Seal trades in Canada. Keeping pace with the demand for skills and workforce certification will require attracting 167,739 new apprentices over the next five years (Kong, 2020). This will not be possible or sustainable without concrete efforts to bring in populations that are currently severely underrepresented: namely women, Indigenous peoples, and people with disabilities.

Despite evidence of severe skill shortages in construction and transportation, and adequate knowledge about the barriers to entry and retention for women (described in a previous section of this chapter), we did not find much evidence to suggest that concerted effort was being made in Canada, especially at the federal level, to design policies or to establish targets to attract and recruit more women to the trades. As with the invisibility of gender issues during the green transition, it appears that the tremendous gains made by Canadian women in post-secondary education and the labour market in recent decades may have obscured the fact that they remain severely underrepresented in the trades. There is some evidence of trade schools and educational institutions trying to alleviate financial barriers women may face. Conestoga College's School of Engineering and Technology, for example, provides free training for women in the skilled trades. These are called pre-apprenticeship programmes since they provide basic training that prepares women for entering a trade and securing an apprenticeship (Kong, 2020).

The Centre for Skills Development and Training, which has five locations in the province of Ontario offers resources for women looking to enter the residential construction industry with its Women in Skilled Trades (WIST) programme. Women learn about tools, energy star building practices, the building code, and marketing and communications. Pre-apprenticeship programmes that provide women with flexible childcare, financial aid, and wraparound supports, from counselling to subsidies for equipment and boots, also help. Something as simple as providing women with safety clothing can make a difference since there are safety hazards for women working in men's clothing.

There is also some evidence of efforts aimed at addressing other challenges women face in transportation and construction. In most instances, the efforts are driven by **provincial and municipal governments, non-profit organisations, and private sector companies**. For example, violence against women has been identified as a barrier for women's equitable access to and employment in public transit. The city of Toronto has initiated the Metropolitan Toronto Action Committee on Violence against Women and Children (METRAC), a collaborative relationship formed by community-based women's organisations, the Toronto Transit Commission, and the Toronto Police Department, to conduct safety audits of the city's transport system. The Toronto Transit Commission (TTC) has a mandate to staff its workforce with 'employees as diverse as the riders it serves' (McIntyre, 2015). The City of Toronto's benchmark for women's employment in transportation in 48.7%. While women account for 57% of ridership, they only accounted for 15% of its workforce in 2014. To respond to this unimpressive number, the TTC has introduced the Diversity and Inclusion Lens and Toolkit to guide 'managers and employees to consider the potential impacts of projects, actions and initiatives on diverse employees and customers' (TTC, 2017). The toolkit includes a guide to inclusive language on disability, race, ethnicity, Indigenous peoples, gender, gender identity, and gender expression, sexual orientation, and LGBTQ+. One guideline from the toolkit states: 'Avoid making assumptions based on stereotypes/unconscious biases, such as assuming clerks are female, or assuming mechanics are male'.

Some **provincial governments** in Canada have also made efforts to improve employment equity in construction and transportation. For example, the government of Alberta began an outreach initiative called 'Women in Non-traditional Occupations: Stories to Inspire' to make role models visible for women. This initiative reaches out to women who are interested in, but wary of, careers in non-traditional occupations by sharing stories of how successful women addressed issues of feminised stereotypes, attitudes of men, physical requirements, and assumed incompetence for non-traditional work (Alberta Human Services, 2013). Also, in the same province, the Alberta Council of Turnaround Industry Maintenance Stakeholders (ACTIMS), which includes oil sands owners, heavy industrial maintenance contractors and labour providers run the Women in Trades Awards/ Bursaries Program to recognise female members of the Canadian Building Trades and to create awareness about opportunities for women in the skilled trades. Each year, this organisation provides three recipients with $1,500 (one for any woman Journeyman,[2] and two for apprentices). Although the intent of the project is laudable, ACTIMS should take note of the power of language in their quest to combat women's underrepresentation in the field; 'woman Journeyman' is not as inclusive as 'journeywoman' or 'journeyperson'.

In British Columbia, Cohen and Braid (2000) conducted research on the Vancouver Island Highway Project (VIHP) and its efforts to integrate women and Indigenous communities as part of its commitment to equity initiatives. They found that at its peak production periods, the equity hires constituted more than 20% of the workforce, ten times higher than usual. Women's participation in the project reached 6.5% and Indigenous workers reached 7.5% (9% and 11% at peak times, respectively). Further, 93% of the workforce consisted of local hires. The equity provisions in this programme had support at the highest levels, including from the Premier of British Columbia, and they were continually monitored and supported by women's groups and Indigenous peoples' organisations on Vancouver Island. The local community strongly supported the initiative since local hires dominated the workforce of the project and all labour on the VIHP was unionised. Initially, few members of the targeted equity groups applied but recruitment practices were changed as a result. Active recruitment on Indigenous reserves and in women's centres was a remedy to this barrier and the interview process was even adjusted to consider volunteer or other non-paid labour as work experience. Cohen and Baird note that the workplace culture on the VIHP project changed over time. Female workers were frequently reported to have better attitudes and work ethic. The project also had an Employment Equity Coordinator, whose role was to advocate for employment equity, and an Equity Integration Committee, which ensured constant monitoring of the equity initiative.

Another example of a Canadian provincial programme that encourages women's employment in transportation is in Newfoundland and Labrador, where the Office to Advance Women Apprentices offers a wage subsidy of up to 75% for two contracts as an incentive for employers to hire female apprentices. However, we were unable to find any evidence to understand what happens to these women's careers once the subsidy runs out. Speaking more broadly about findings from this research project, we found very little information about the long-term sustainability or outcomes of gender equity initiatives in construction and transportation. Future research should address the long-term impact of these programmes, identifying where there is room for improvement, providing detailed documentation of what has worked well and what has not, and evaluating the potential to scale up successful initiatives at the provincial or national level.

The **non-profit sector** in Canada has also driven some equity initiatives, sometimes in partnership with provincial and municipal governments. For example, Building Up and BUILD Inc. are two non-profit social enterprises that offer employment and training in the construction trade to those with barriers to employment (i.e. visible minorities, those with criminal records, high school dropouts). Building Up is based in Toronto. It provides energy and water retrofits in affordable housing while employing those living in poverty, often from within the public housing units they live in. Building Up calls itself a 'social contractor' and specifically trains and employs those with the most barriers to employment. They offer 16-week paid apprenticeships and 90% of their graduates go on to full-time employment in the skilled trades. BUILD Inc. partners with Manitoba Hydro (the provincial utility company) to improve insulation in low-income housing to reduce carbon emissions and to lower utility bills. They offer a six-month paid training programme for those who 'would often be deemed unemployable'. BUILD Inc. also offers life skills, vocational training, and remedial education on topics such as math and money management. Although there is no explicit focus on gender, there is a very clear mandate to work with those who have the most significant barriers to employment, often including women and Indigenous workers.

The record of **trade unions and industry associations** in Canada has been somewhat erratic in promoting employment equity. For example, Canada's Building Trades Union (CBTU) released a report in July 2017, the first original study on the impacts of Canada's transition to low-carbon development upon the construction industry. The report identifies new employment opportunities that will emerge if Canada meets its commitment to the 2015 Paris Agreement. The report contained no sex-disaggregated data and no mention of women, gender, or equity. It proposes that moving to an electrical supply grid would result in over 1.75 million jobs but fails to consider who might get these jobs and who might be left out. Despite such glaring oversights, CBTU does run Women in the Building Trades, an initiative that promotes creating 'the space for conversation on how to engage women' in construction. The initiative focuses on outreach activities including tradeshows, career fairs, mentorship events and networking functions.

'Build Together', a Canadian national programme focused on the recruitment and retention of workers from underrepresented portions of the population, has compiled a list of suggestions for what a union or employer can do in order to attract women to the construction sector. Their recommendations to unions and other representative organisations for building a more inclusive and diverse construction sector include training for diversity and respectful workplaces, putting pressure on industry leaders to promote inclusivity, mentorship programmes, networking and conferences to build community and ease isolation, ensuring proper safety equipment and gear for women, flexible hours to navigate childcare and family needs, and starting a women's committee. Unions and other organisations that represent workers do have the opportunity and responsibility to play an important role in promoting women's employment in transportation and construction. With 80% of female apprentices participating in union-sponsored programmes, union apprenticeship programmes have had greater success at recruiting women and ensuring completion of training than non-unionised programmes (Arcand, 2016). However, as several researchers writing about unions have pointed out (see, for example, Baker and Robeson, 1981) unions are often themselves masculinist organisations that have historically ignored women and other socially marginalised groups. Depending on the setting, unions may be in as much need of institutional reform as other organisations.

Conclusion

In Canada, women make up only 23% and 12%, respectively, of those employed in transportation and construction. Concerns about climate change and fossil fuel insecurity have ensured that there is significant interest in the technologies and financing for transitioning to a low-carbon economy, but far too little attention is being paid in Canada to the employment equity implications of such a transition. Although we intended to specifically collect data on women's employment in green or low-carbon transportation and construction, we found that most of the available data and information on employment is from the 'brown' or traditional (fossil fuel–based) forms of these sectors. We found very few specific examples of women's inclusion in employment in green transportation and construction, which supports a major finding from this study that gender equality (or employment equity, more broadly) has not been meaningfully addressed or even considered in mainstream green transition policies and practices in Canada.

In the absence of appropriately targeted training, education, apprenticeships, employment placement, financial tools, and supportive social policies, transitioning to a green economy may exacerbate existing gender inequities and social hierarchies. This chapter synthesised and assembled existing scientific and practitioner literature on promising programmes and policies for promoting and optimising women's employment in two sectors that are critical for the Canadian green economy and in which women are currently severely underrepresented, namely, construction and transportation. Our findings reveal that there are currently significant gender inequities in employment trends in these sectors in Canada.

Our findings suggest that people who are already employed in the traditional (fossil fuel–based) versions of these industries can access green skills and training more easily than people who have never worked in these sectors. Since women are severely underrepresented in the traditional construction and transport sectors, their ability to access green skills and training is significantly weaker than men. Further, our findings confirm that although green training is part of the existing trades in these sectors, it does not presently constitute a separate route into them. This highlights the need to influence and implement equity practices in the traditional construction and transport sectors in order to ensure equitable representation of women in the low-carbon iterations of these industries.

Most existing equity initiatives in Canada in these two sectors have been driven by the private sector, municipalities, and provincial governments. The federal government has, at least until very recently, not played an active role in framing and implementing effective policies to enable the transition to a green economy. Canada has been hindered in this regard by a political structure that divides responsibility for policies related to climate and energy between the federal government and the provinces. This has led to a patchwork of policies at the provincial level exacerbated by an absence of federal leadership on the issue. The current Liberal federal government has made more explicit commitments to a green economy and climate change mitigation. But there has been virtually no effort from the federal government to initiate policies on employment equity in the transition to a green economy. The federal government must play a stronger leadership role in implementing employment equity policies in order to motivate and optimise the efforts of other actors in the green economy.

Our findings highlight the need for a collective and cohesive national strategy that has measurable outcomes and actual numerical targets monitoring to improve women's access to, and retention, in the growing green economy. Since women are already underrepresented and marginalised in key sectors of the green economy, growth in green industries may further exclude women if proactive measures are not adopted. McFarland (2013) argues that given women's current pattern of participation in jobs and training in the trades, almost none of the green jobs we expect to be created would go to women. She stresses the dire need for equity programmes and quotas that are monitored and enforced. In addition to federal and provincial leadership on policy issues, Canada needs all stakeholders, including federal, provincial, and municipal governments, civil society organisations, corporations, labour associations, public policy institutes, and think-tanks to work together to ensure that women are not further marginalised as more green jobs are created in transportation and construction.

While emphasising the need for effective policy interventions to ensure employment equity in the transition to low-carbon economies, our findings suggest that even well-planned and well-designed policy interventions may not adequately subvert the broader social structures that create the inequities in the first place. Most policies designed to address women's underrepresentation in these fields tend to be reactive responses that do not engage adequately with structural social, economic, and political inequality. Improving the lighting in construction sites in order to prevent sexual assaults against women and requiring women to work in pairs instead of alone are classic examples of reactive policies that have already been adopted by some employers but that may, ironically, end up reinforcing and institutionalising gendered social hierarchies rather than challenging them (Connell, 1990; Krook and MacKay, 2010). We found that the most common response to women's underrepresentation in these fields has been to 'add women and stir', simply adding women to industries with unchanged masculinist values and work cultures (Harding, 1995). Policy responses also often reinforce affirmative gender essentialisms. Women tend to be valourised in essentialist ways that reinforce existing social hierarchies. For example, assumptions that women are gentler with machinery than men and therefore maintain machinery better or that women bring specific valuable qualities and skills (empathy, patience) to the job simply by being women reinforce social hierarchies since most women acquire these skills because of historical and current social oppression and not because they are biologically female.

Our findings suggest that although the formal realm of law and policy can and does play a significant role in optimising women's employment and retention in the labour force, legislation, and policy cannot be the sole vehicles for social change because they are not enough to alter entrenched gender roles, social hierarchies and entitlements. Consciousness-raising initiatives that raise awareness among women, as well as men, about the benefits of greater equity and women's equal entitlements to employment in all fields are as crucial as policy reforms and state or corporate actions that protect women's interests and facilitate their agency.

Notes

1 The energy sector, which is both critical for the green economy and in which women are also extremely marginalised, is not included in this chapter because we have already published extensively on this topic elsewhere (see, for example, Baruah, 2016; Baruah and Gaudet, 2016).
2 'Journeyman' is a term used in Canada to describe a skilled tradesperson who has successfully completed an official apprenticeship qualification.

References

Alberta Human Services (2013) *Women in Non-Traditional Occupations: Stories to Inspire* [online]. Available at: https://open.alberta.ca/publications/9780778569374. (Accessed: 13 October 2020).

Antoni, M., Janser, M. and F. Lehmer. (2015) 'The Hidden Winners of Renewable Energy Promotion: Insights into Sector-Specific Wage Differentials', *Energy Policy*, 86, pp. 595–613.

Arcand, C. (2016) 'Women in Construction and the Workforce Investment Act: Evidence from Boston and the Big Dig', *Labor Studies Journal*, 41(4), pp. 333–354.

Asia Pacific Gateway Skills Table (2017) Women in Transportation Careers: Moving Beyond the Status Quo [online]. Available at: https://www.westac.com/application/files/8015/0161/4819/WIT-Report-2017-Final-Web.pdf (Accessed:13 October 2020).

Baker, M. and M., Robeson. (1981) 'Trade Union Reactions to Women Workers and Their Concerns', *Canadian Journal of Sociology*, 6(1), pp. 19–31.

Baruah, B. (2016) 'Renewable Inequity? Women's Employment in Clean Energy in Industrialised, Emerging and Developing Economies', *Natural Resources Forum*, 41(1), pp. 18–29.

Baruah, B. (2017) 'Renewable Inequity? Women's Employment in Clean Energy in Industrialised, Emerging and Developing Economies', in Cohen, M.G. (ed.), *Gender and Climate Change in Rich Countries: Work, Public Policy and Action*. London, UK: Routledge, pp. 70–86.

Baruah, B. (2018) 'Barriers and Opportunities for Women's Employment in Natural Resources Industries in Canada', Report presented at Natural Resources Canada, Ottawa, November 27.

Baruah, B. and C., Gaudet. (2016) 'Confronting the Gender Gap in Canada's Green Transition', *Western News*, September 21 [online]. Available at: https://news.westernu.ca/2016/09/baruah-gaudet-confronting-gender-gap-canadas-green-transition/ (Accessed 13 October 2020).

Baruah, B. and S. Biskupski-Mujanovic. (2018) Identifying Promising Policies and Practices for Promoting Gender Equity in Global Green Employment [online]. Available at: https://institute.smartprosperity.ca/library/research/identifying-promising-policies-and-practices-promoting-gender-equity-global-green (Accessed 1 October 2020).

Bickis, I. (2016) 'Renewable Energy, Other Industries Draw laid-off Oil and Gas Workers', *Huffington Post*, 27 January[online]. Available at: http://www.huffingtonpost.ca/2016/01/27/oil-and-gas-career-change_n_9089360.html (Accessed 1 October 2020).

Blue Green Canada (2012) Building Ontario's Green Economy: A Road Map [online]. Available at: http://environmentaldefence.ca/report/report-building-ontarios-green-economy-road-map/ (Accessed 13 October 2020).

Canada's Building Trades Unions (CBTU) (2017) Jobs for Tomorrow: Canada's Building Trades and Net Zero Emissions [online]. Available at : https://adaptingcanadianwork.ca/jobs-for-tomorrow-canadas-building-trades-and-net-zero-emissions/ (Accessed 13 October 2020).

Canadian Apprenticeship Forum (CFA-FCA) (2016) Apprenticeship Analysis Women and Apprenticeship in Canada [online]. Available at: https://caf-fca.org/wp-content/uploads/2017/07/Member9_Women-and-Apprenticeship.pdf (Accessed 13 October 2020).

Canadian Press (2017) Laid Off Oil and Gas Workers Train for Alternative Energy Jobs as Wind Blows Alberta in New Direction [online]. Available at: https://www.cbc.ca/news/canada/calgary/alternative-energy-training-laid-off-oil-and-gas-1.4463217 (Accessed 1 October 2020).

Carpenter, J., Matthews, P. and A. Robbett. (2017) 'Compensating Differentials in Experimental Labor Market', *Journal of Behavioral and Experimental Economics*, 69, pp. 50–60.

Chappell, L. (2006) 'Comparing Political Institutions: Revealing the Gendered "Logic of Appropriateness"', *Politics & Gender* 2 (2), pp. 223–235.

Cohen, M. and K. Braid. (2000) 'Training and Equity Initiatives on the British Columbia Vancouver Island Highway Project: A Model for Large-Scale Construction Projects', *Labour Studies Journal*, 25 (3), pp. 70–103.

Connell, R.W. (1990) 'The State, Gender, and Sexual Politics: Theory and Appraisal', *Theory and Society*, 19 (5), pp. 507–544.

Dizik, A. (2016) 'Where are All the Expat Women?', *BBC Worklife* [online]. Available at: https://www.bbc.com/worklife/article/20160929-where-are-all-the-expat-women (Accessed 1 October 2020).

Engineers Canada (2020) Women in Engineering [online]. Available at: https://enginee rscanada.ca/diversity/women-in-engineering/30-by-30 (Accessed 13 October 2020).

Hafkin, N. and S., Huyer. (2006) *Cinderella or Cyberella? Empowering Women in the Knowledge Society*. Boulder: Kumarian Press.

Harding, S. (1995) *'Just Add Women and Stir?' in United Nations Commission on Science and Technology for Development (ed.) Missing Links: Gender Equity in Science and Technology for Development*. Ottawa: International Development Research Centre, pp. 295–308.

Hegewisch, A. and B., O'Farrell. (2015) Women in the Construction Trades: Earnings, Workplace Discrimination, and the Promise of Green Jobs [online]. Available at: https ://iwpr.org/wp-content/uploads/wpallimport/files/iwpr-export/publications/C428-W omen%20in%20Construction%20Trades.pdf (Accessed 13 October 2020).

Hirshman, L. (2008) 'Where Are the New Jobs for Women?', *New York Times*, December 9 [online]. Available at: http://www.nytimes.com/2008/12/09/opinion/09hirshman.ht ml?_r=0 (Accessed 13 October 2020).

ILO (2011) Promoting Decent Work in a Green Economy [online]. Available at: https ://www.ilo.org/employment/Whatwedo/Publications/WCMS_152065/lang--en/index. htm (Accessed on 13 October 2020).

IRENA (2019) Renewable Energy: A Gender Perspective [online]. Available at: https:// irena.org/-/media/Files/IRENA/Agency/Publication/2019/Jan/IRENA_Gender_perspe ctive_2019.pdf (Accessed 13 October 2020).

IUCN (2011) A Gender Perspective on the Green Economy [online]. Available at: https ://www.iucn.org/content/a-gender-perspective-green-economy (Accessed 13 October 2020).

Katz, J. (2012) Emerging Green Jobs in Canada: Insights for Employment Counsellors into the Changing Labour Market and its Potential for Entry-Level Employment [online]. Available at: http://www3.cec.org/islandora-gb/en/islandora/object/islandora%3A 1193 (Accessed 13 October 2020).

Kong, S.L. (2020) 'Why We Need More Women in These Particular Careers Than Ever Before', *Macleans*, February 3 [online]. Available at: (https://www.macleans.ca/work/w omen-in-skilled-trades/(Accessed on April 26, 2020).

Krook, M. and F., Mackay. (eds.) (2010) *Gender, Politics and Institutions: Towards a Feminist Institutionalism*. London: Palgrave Macmillan.

Lieu, J., Sorman, A., Johnson, O., Virla, L. and B. Resurrección. (2020) 'Three Sides to Every Story: Gender Perspectives in Energy Transition Pathways in Canada, Kenya and Spain', *Energy Research and Social Science*. https://doi.org/10.1016/j.erss.2020 .101550

MacArthur, J., Hoicka, C., Castleden, C., Das, R. and J. Lieu. (2020) 'Canada's Green New Deal: Forging the Socio-Political Foundations of Climate Resilient Infrastructure?', *Energy Research & Social Science*. https://doi.org/10.1016/j.erss.2020.101442

McFarland, J. (2013) 'The Gender Impact of Green Job Creation', Paper presented at Work in a Warming World International Conference, Toronto, November 29–December 1.

McIntyre, C. (2015) 'TTC Falls Short on Gender Equality Targets', *Torontoist*, 23 December [online].Available at: http://torontoist.com/2015/12/ttc-falls-short-on-g ender-equity-targets/ (Accessed 13 October).

McKee, L. (2014) 'Women in American Energy: De-feminizing Poverty in the Oil and Gas Industries', *Journal of International Women's Studies*, 15(1), pp. 167–178.

McMullen, K., Gilmore, J. and C. Le Petit. (2010) Women in Non-traditional Occupations and Fields of Study [online]. Available at: https://www150.statcan.gc.ca/n1/pub/81-0 04-x/2010001/article/11151-eng.htm (Accessed 13 October 2020).

Moir, S., Thomson, M. and C. Kelleher. (2011) Unfinished Business: Building Equality for Women in the Construction Trades. Labor Resource Center Publication [online]. Available at: https://scholarworks.umb.edu/cgi/viewcontent.cgi?article=1004 &context=lrc_pubs (Accessed 13 October 2020).

National Women's Law Center (2014) Women in Construction: Still Breaking Ground [online]. Available at: https://nwlc.org/wp-content/uploads/2015/08/final_nwlc_ womeninconstruction_report.pdf (Accessed 13 October 2020).

OECD (2016) Promoting Green and Inclusive Growth in Canada [online]. Available at: http://www.oecd.org/canada/promoting-green-and-inclusive-growth-in-canada.pdf (Accessed 13 October 2020).

Peters, J. and M., Katalytik. (2011) Equality and Diversity: Good Practice for the Construction Sector. Equality and Human Rights Commission [online]. Available at https://www.equalityhumanrights.com/sites/default/files/ed_report_construction_s ector.pdf (Accessed 13 October 2020).

Phillips, A. (1998) *The Politics of Presence*. London and New York: Oxford University Press.

Rosser, S. (2005) 'Women and Technology Through the Lens of Feminist Theories', *Frontiers: A Journal of Women's Studies*, 26(1), pp. 1–23.

Smith, J. (2017) 'Women Account for Just 4.5% of Skilled Trade Workers in Canada', *Canadian Press*, 25 October [online]. Available at: https://globalnews.ca/news/382 3621/women-skilled-trades-manufacturing/ (Accessed 13 October 2020).

Statistics Canada (2019) Employment by Industry and Sex [online]. Available at: http://www.statcan.gc.ca/tables-tableaux/sum-som/l01/cst01/labor10a-eng.htm (Accessed 13 October 2020).

Status of Women Canada (2017) The Competitive Advantage: A Business Case for Hiring Women in the Skilled Trades and Technical Professions [online]. Available at: https://cfc-swc.gc.ca/abu-ans/wwad-cqnf/bc-cb/business-case-en.pdf (Accessed 13 October 2020).

Steering Committee of the Green Economy Network (2015) Making the Shift to a Green Economy: A Common Platform of the Green Economy Network [online]. Available at: https://canadianlabour.ca/research/issues-research-making-shift-green-economy/ (Accessed 13 October 2020).

Subramaniam, V. (2020) 'Before You Declare Canada is Not a Racist Country, Do Your Homework', *National Post*, June 2 [online]. Available at: https://nationalpost.com/ opinion/vanmala-subramaniam-before-you-declare-canada-is-not-a-racist-country-do -your-homework (Accessed 13 October 2020).

Sustainlabour (2009) Green Jobs and Women Workers: Employment, Equity and Equality [online]. Available at : http://www.greengrowthknowledge.org/sites/default/files/down loads/resource/Green_jobs_and_women_workers_employment_equity_equality_Sust ainlabour.pdf (Accessed 13 October 2020).

Thirgood, J., McFatridge, S., Marcano, M. and J. Van Ymeren. (2017) Decent Work in the Green Economy [online]. Available at: https://munkschool.utoronto.ca/mowatcentre/ decent-work-in-the-green-economy/ (Accessed 13 October 2020).

Toronto Transit Commission (2017) Annual Report on Diversity and Human Rights Achievements [online]. Available at: http://www.toronto.ca/legdocs/mmis/2015/ex/bg rd/backgroundfile-85432.pdf (Accessed 13 October 2020).

Trichur, R. (2020) 'It's a 'She-Cession'. Governments Must Put Women First During the Recovery', *Globe & Mail*, May 1 [online]. Available at : https://www.theglobeandma il.com/business/commentary/article-legislators-must-prioritize-women-combat-work place-gender/ (Accessed 13 October 2020).

Turnbull, P. (2013) Promoting the Employment of Women in the Transport Sector: Obstacles and Policy Options [online]. Available at: https://www.ilo.org/sector/Res ources/publications/WCMS_234880/lang--en/index.htm (Accessed 13 October 2020).

UN Women (2012) 'Fast-forwarding Women's Leadership in the Green Economy', *Women*, June 19 [online]. Available at: https://www.unwomen.org/en/news/stories/2012/6/fast -forwarding-women-s-leadership-in-the-green-economy (Accessed 1 October 2020).

Women in Skilled Trades [WIST] (2017) [online]. Available at: https://www.westofwindsor .com/ (Accessed 13 October 2020).

Women's Budget Group (2020) What would a Feminist Green New Deal Look Like? [online]. Available at: https://wbg.org.uk/wp-content/uploads/2020/05/A-Feminist-Gr een-New-Deal.pdf (Accessed 13 October 2020).

10 Why radical transformation is necessary for gender equality and a zero carbon European construction sector

Linda Clarke and Melahat Sahin-Dikmen

Introduction

Under capitalism, the domination and exploitation of labour are inseparable from those of nature (Parsons, 1977), and there is perhaps no sector where this is more apparent than in construction, an industry with often poor working and employment conditions and one responsible for a high proportion of carbon emissions, including through extensive use of cement (Bataille, 2019). Construction is also a male-dominated industry where it has proved very difficult to improve women's historically low levels of participation, averaging around 10% in Europe although varying between countries and occupations. This chapter identifies both the constraints involved in meeting zero energy targets and those that have served to exclude women. It seeks to show why and how incorporating women's interests and experiences will help to ensure equitable outcomes and contribute significantly to the successful transformation of the construction sector into an eco-industry.

The climate emergency has highlighted many of the weaknesses of the industry, in particular the priorities of healthy and safe employment and working conditions, and together with the 2020 arrival of a coronavirus crisis, these have prompted an urgent, long-overdue drive for change. Construction is a sector set to gain more employment than any other from the transition to a green economy through policies and programmes for nearly zero energy building (NZEB), renewable energy installations and retrofit across Europe (ILO, 2018). This opens up opportunities for women and, at the same time, for overcoming the structural, organisational and cultural obstacles to their inclusion. However, the pursuit of the European Union's (EU), green transition policy for the built environment is focused on ecologically modernising an industry without addressing the social relations and structures that characterise the sector and hinder its ability to reduce carbon emissions (Foster, 2002; Hampton, 2015; Lundström, 2018). At the same time, the mainstream approach to achieving gender equality in the sector is narrowly focused on recruiting more women and similarly fails to address the employment and working conditions that give rise to their exclusion in the first place.

DOI: 10.4324/9781003052821-10

The chapter begins by outlining women's participation in construction and approaches to addressing gender inequalities. Also drawing on the experiences of women themselves, the reasons for the failure of policies and practices to improve the participation of women are pinpointed, both in the vocational education and training (VET) system and the labour market. There follows a review of the EU's policy to reduce carbon emissions associated with the built environment, revealing its premises and consequent gender blindness and lack of attention to the social aspects of building production. The crucial importance of addressing these for effectively implementing the European Performance of Buildings Directive (EPBD) is then addressed, both in relation to VET and the labour process requirements for low energy construction (LEC). Finally, examples of good practice are presented, highlighting the role of different stakeholders and pointing to what is required for gender-sensitive, equitable, and socially sustainable employment in a green construction sector. The chapter concludes by suggesting that a comprehensive retrofit and new-build NZEB programme can be the means of transforming the industry to become socially useful and carbon-neutral, driven by social concerns and involving women and their representative organisations in shaping institutional strategies at all levels.

Why are so few women found in the construction sector?

Women's participation in the construction sector is low across Europe, averaging 10% in 2018 and ranging from 13% in Germany and Austria, 11% in the UK and 10% in France down to only 7% in Denmark and Greece and 6% in Italy, Poland, Romania and Ireland (ECSO, 2020). These figures include, however, both those in the professions and in the operative workforce and participation is higher for the professions, as apparent from Table 10.1, particularly in Eastern European countries. Women represented 43% of graduates from engineering, manufacturing and construction in Poland and 37% in Romania, in contrast to central

Table 10.1 Graduates in engineering, manufacturing and construction 2018

Country	Number of women graduating	Number of men graduating	Total	% of women graduating	% women employed in narrow construction sector
Denmark	2,607	7,188	9,795	27	7.2
France	2,512	84,421	110,933	24	9.6
Germany	26,379	94,449	120,828	22	12.8
Italy	19,415	42,014	61,429	32	5.9
Norway	1,846	5,458	7,304	25	
Poland	29,049	39,321	68,371	43	6.3
Romania	7,997	13,622	21,619	37	6.2
Switzerland	2,276	11,801	14,077	16	
UK	18,136	55,953	74,090	25	10.7

Sources: Eurostat (2018), ECSO (2020) *Improving the human capital basis,* European Construction Sector Observatory: 28

European countries, Scandinavian countries and the UK at between 22% and 27%, with only Italy faring better at 32% and Switzerland worse at 16%. Across Europe, too, Eurostat figures on female engineering workers employed as a share of the total engineering workforce show relatively high proportions, including in Bulgaria (30%); Slovenia, Poland and Italy (20%); Belgium and Hungary (19%), Spain (17%); and Germany, Ireland and Finland (15%) (Clarke et al., 2015). And in the UK, a significant proportion of women is found in technical positions in construction (24%), such as quality assurance technicians (39%) and quality control and planning engineers (19.1%) (Clarke et al., 2019).

Of those employed in the energy sector, 22% are women and these are generally in lower paid and not managerial jobs, though there is an increasing requirement for those with scientific knowledge and specialist expertise (EIGE, 2016a). In the renewables sector, too, though women are estimated to represent 35% of those employed across Europe, in Germany, Italy and Spain they hold less than 30% of jobs, most of which are similarly low paid, non-technical, or administrative rather than technical, managerial, or involving policy-making (IRENA, 2019). Occupations belonging to the operative workforce in the renewables sector, many of which are construction, tend to be heavily male-dominated, including metal workers, insulation specialists, plumbers, pipefitters, electricians and heating and cooling experts (Clancy and Feenstra, 2019).

Though there has been more progress in the professions, what is surprising concerning the participation of women in the construction operative workforce is the shockingly slow or non-existent pace of change. In 2016, only 3% of those employed across the EU were women (Eurostat, 2016), a figure that has remained relatively stable, apart from during the world wars, for well over a century. The reasons for this poor representation have been well-researched and are generally attributed to: structural, organisational and cultural obstacles, including the lack of formalised recruitment practices and procedures and inappropriate selection criteria; employment conditions, in particular the fragmented nature of employment and lack of support and work–life balance possibilities; inflexible and hazardous working conditions, such as long working hours and inappropriate equipment; and less tangible obstacles, such as the lack of knowledge and poor image of the sector, traditional stereotypes, sexist attitudes and a male-dominated culture, network, and environment (Fielden et al., 2000; Clarke et al., 2004, 2015; Worrall et al., 2010; Sang and Powell, 2013; Baruah, 2018; Clancy and Feenstra, 2019). In terms of ability, however, equal competence has been shown (e.g. Arditi et al., 2013).

The most pervasive reason for exclusion is the working conditions, above all the long working hours, both on sites and in the professions (Watts, 2009; Styhre, 2011; Caren and Astor, 2013). Indeed, in a 'virtual' meeting with women working in construction from the USA and different European countries held during the coronavirus pandemic, working conditions, above all related to health and safety, were the dominant source of complaint, including shift hours, lack of washing, toilet and changing facilities and of personal protective equipment suitable for women, social distancing, site cleansing and risk assessments. The increasing fragmentation of the construction process through subcontracting and self-employment have also aggravated the exclusionary nature of the industry, but its stubborn persistence, despite all the efforts to increase the number of women, suggests deeply embedded structural obstacles and a gendered division of labour that is unlikely to change without a social transformation.

The solutions to greater female participation also vary, with much emphasis placed on corporate social responsibility and human resource management (HRM), including mentoring and networking (Clancy and Feenstra, 2019; ECSO, 2020). In addition to these, Worrall et al. (2020, p. 280), in a perfect illustration of 'neo-liberal' onus placed on women themselves to accommodate to male domination (Fraser, 2013; Rottenberg, 2018; Ferguson, 2020), advocate support systems and continuing professional development:

> providing women with the necessary 'soft skills' in communication, people management and confidence building that equips them to negotiate difficult working environments and male-dominated organisational cultures.

The remit of the unfortunately named EC-funded initiative, *High heels: building opportunities for women in the construction sector* (ECSO, 2020, p. 81), covering Bulgaria, Greece, Romania and Cyprus, follows this advice, seeking to train women to 'strengthen their soft skills in order to improve the performance of the construction sector'. Such strategies, in their acceptance of gender discrimination, devalue women and fail to challenge – and even succeed in maintaining – the gendered nature of construction and engineering (Powell et al., 2009; Watts, 2009). Other HRM-inspired solutions involve awareness-raising, including through training trainers in 'gender sensitivity' (ECSO, 2020, p. 81). As Clarke et al. (2018) argue, such employer-led, top–down, 'business case' approaches to achieving diversity in STEM occupations lack effectiveness through the absence of involvement from other stakeholders, especially employees and target groups, allowing only a fraction of diversity-related issues to be 'visible' in the organisation, while others remain 'suppressed'.

Stronger and more systematic solutions have been more successful, if only temporarily. These include targeted measures and monitoring and enforcement to improve access to and retention in green jobs, though too often employers meet procurement diversity requirements only to let women go as soon as the contract is secured (Baruah, 2018). Incremental measures are also proposed that seek to directly alleviate and improve working conditions for those women working in construction, including working in pairs and improved lighting. Overall, however, despite all the efforts over many decades, little has changed. As Baruah (2018, p. 3) complains, most policies are 'reactive responses that do not engage adequately with broader societal structures and institutions that produce and maintain inequality'. Indeed, over two decades ago, Dainty et al. (1999) questioned efforts to increase the number of women in construction given that workplace practices were geared to men's needs, including long working hours, geographical instability and the subordination of personal lives. They showed how men progress in the industry vertically through networking and informal mechanisms, whilst women focus on coping with a male environment, and concluded that the only way for women to further their careers was to leave. In raising doubts as to whether women should be attracted to an industry 'ill-equipped for employing them', therefore, Dainty et al. (1999, pp. 356–357) challenged existing measures and insisted that:

> It is only through a genuine commitment to the development of a more equitable industry from the highest level, that women are likely to be able to develop their careers in parity with men.

But how does gender inequality in construction fare in policies formulated at the highest level in the EU today? Does gender feature in visions of a future sustainable industry? More specifically for the purposes of this chapter, to what extent does the EU green transition policy for the built environment address gender inequalities?

The EU energy-efficiency policies and gender

Being responsible for 36% of CO2 emissions and 40% of energy consumption, the built environment is targeted for a major transformation as part of EU climate change action plans, (EC, 2019a). The energy strategy up to 2020 is set out in EU2020 development plans and designed in accordance with Paris Agreement targets, aiming to reduce energy use, improve energy efficiency and the use of renewable energy, each by 20%, compared to 1990 levels (EC, 2010). For 2021–2030, adaptation measures for the built environment are set out in *Clean Energy for All Europeans* (EC, 2019a). In the medium term, these include a package of initiatives with the aim of improving energy efficiency by 32.5%, increasing the share of renewable energy by 32% and reducing CO2 emissions by 40% by 2030. It is planned to be progressed in tandem with the European Green Deal (EC, 2019a), the EU's strategic programme for implementing the UN's 2030 Agenda. Improving the energy efficiency of buildings is fundamental to achieving these targets and this is driven by the EPBD (2010, 2018), which requires that all new buildings are NZEB by the end of 2020. The EPBD provides the overarching legislative framework, setting out the technical definition of NZEB and guidance for its implementation, whilst member states are responsible for its transposition into national law. Thus, although the exact technical specifications vary, NZEB means higher energy performance standards for all EU countries.

On closer review, it is apparent that these key policies concerning the green transition in the built environment are underpinned by the logic of consensus of the technologically driven ecological modernisation approach, which is oriented to innovation and argues for the economic benefits of environmentalism (Mol et al., 2009; Machin, 2019). This approach is evident from the stated objectives of the energy policy over the last two decades of achieving an energy-efficient and carbon-neutral European economy (including in the built environment and transport) and a fully integrated pan-European energy market by 2050 by ensuring energy security through cooperation between EU countries, increasing renewable energy, supporting research and innovation into clean energy technologies, empowering consumers and strengthening the EU's external energy relations (EC, 2015, 2019a). The Green Deal (EC, 2019b) similarly sets out decarbonisation strategies and emphasises the role of technology in increasing the share of renewable energy and the potential of renovating Europe's building stock, giving further considerations to the legislation, public investment and private financing and education and training. Both the energy policy and the Green Deal highlight job creation as an anticipated outcome, particularly through energy efficiency improvement measures in buildings and growth in the renewable energy sector, which alone are expected to employ two million people by the end of 2020, mostly in the construction industry.

EPBD implementation measures and plans required of the member states, such as setting minimum energy performance standards, adaptation of technical building systems, energy performance certificates and inspection regimes, further illustrate the technocratic framing of the transition to sustainable construction. For example, between 2007 and 2020 implementation was facilitated by National Energy Efficiency Action Plans (NEEAPs), setting out the adapted definition of NZEB, the energy efficiency measures to be pursued, energy performance certification, inspection schemes and financial incentives, as well as developing renovation strategies and other complementary measures to achieve the now superseded EU 2020 targets (EPBD, 2010, 2018; EC, 2016a). The ten-year National Energy and Climate Plans (NECPs) for the next decade are required to outline, in a similar way, what legislative, financial and regulatory measures will be put in place to meet the 2030 targets and indicate a long-term strategy towards 2050 (EC, 2019a).

In none of these policy documents on energy, the EPBD, or the NECPs is there any reference to women or gender. The energy efficiency strategy of the last two decades has made no provision to address women's participation in this transition, neither catering to their education and training needs nor taking measures to ensure that they gain advantage from emerging employment opportunities. The only nod to the social aspects of the transition is the Build Up Skills initiative (BUS), launched to increase the number of building workers trained in the competencies needed for low energy construction (LEC) and subsequent Horizon 2020 funded training programmes (EC, 2014, 2016b, 2018). However, these made no provisions to address gender inequalities in the sector, despite EU attempts to 'add' gender (Allwood, 2014) to existing policies, focused on increasing the number of women particularly in STEM (science, technology, engineering and mathematics) education and the renewable energy sector (EIGE, 2016a, b).

This does not mean that the EU does not legislate for gender equality. On the contrary, gender equality remains an explicit objective and gender mainstreaming has been a commitment since 1996, following the formation of the Beijing Platform of Action (EIGE, 2012). This commitment is articulated, for example, in Article 23 of the Charter of Fundamental Rights (European Parliament, 2012c) and in calls by the European Parliament for gender equality in the green economy and for the inclusion of women at all levels of decision-making (European Parliament, 2012a, b). Further, gender equality is comprehensively covered by the European Pillar of Social Rights (European Parliament, 2017) in terms of employment (access, progression, pay), equal opportunities (non-discrimination to cover education and all other social services) and work–life balance. The recently updated Gender Equality Strategy (EC, 2020) renews the commitment to gender mainstreaming to ensure women's participation in decision-making, including on climate change, and acknowledges intersectionality as critical to understanding the complexity of disadvantage and inequality, away from essentialising male–female binaries.

There is, therefore, a seeming contradiction between formal and loud commitment to gender equality on the one hand and the 'silence' (Bakker, 2015) of green construction policies on gender inequalities on the other. For explanation, feminist institutional analyses turns the lens to institutions and the structures, power relations and formal and informal processes that work to prevent a gender perspective from being integrated into specific public policies (Weiner and MacRae, 2014). Whilst women's presence is no guarantee that a gender perspective will have a fair hearing, what difference are women participating in climate policy actually making? Few ministers of energy, within which the green construction strategy is subsumed, are women, and the construction sector too is particularly male-dominated (Clancy and Feenstra, 2019). Even when women are present, power imbalances within institutions, established and taken for granted prioritisation of sector-specific issues, and the interpretation of gender mainstreaming by the different actors involved, all contribute to the marginalisation of gender issues in the policy-making process (Weiner and MacRae, 2014). Organisations are 'sticky' and path dependent and there is a limit to how far institutional agendas can be challenged. The result is that gender equality, rather than being a fundamentally cross-cutting issue, continues to be tackled as an add-on to discreet areas of policy and gender mainstreaming becomes a box-ticking exercise that does not lead to change on the ground (Arora-Jonsson, 2017; MacGregor, 2017).

This discretion is exemplified in documents produced by the European Institute for Gender Equality (EIGE), in particular on *Gender and Energy* (EIGE, 2016a) and *Gender in environment and climate change* (EIGE, 2016b). These recognise the serious underrepresentation of women in higher technical and scientific education and employed in the energy sector, including in renewable energy and climate change decision institutions. They call for: a 'more balanced representation of women' at all levels in the field of climate change mitigation; support for women in science and technology; the elimination of gender stereotypes and promotion of gender equality in education, training, and working life; and the integration of the principle of gender mainstreaming into relevant legislation, policy measures and instruments related to climate change mitigation (EIGE, 2016b). However, there is no consideration given to social and economic relations, such as the gendered division of labour, which give rise to the low participation of women in the first place or to structural problems in the sectors themselves that perpetuate gender inequality and need to be transformed. Policy is, instead, directed to inserting women and increasing participation in sectors structurally aligned to reproduce gender segregation and falls far short of the social transformation needed.

Whilst the democratisation of and equal representation in environmental governance are critical, therefore, a deeper challenge to the dominant policy paradigm is necessary to put gender equality at the centre of visions for a sustainable Europe. The current EU approach to gender mainstreaming is, on one level, a manifestation of the underlying neo-liberal political and economic rationale that externalises social reproduction and separates gender inequality, which is dealt with as a question of equal rights and opportunities and discrimination, from issues of the 'real economy', including industrial, economic, or climate policy-making (Mellor, 2017; McGregor, 2010). Relegating gender inequality and the gendered division of labour in society to a separate platform is, thus, predicated on giving individual women the equal opportunity and 'support' they need to compete for jobs, while the patriarchal structures and practices and existing exploitative and unsustainable employment practices that shape their participation and progress in paid employment, including in construction, are ignored (Fraser, 2013; Rottenberg, 2018; Ferguson, 2020). This lack of attention to the gender implications and context of construction parallels and is not separate from the neglect of entrenched problems of construction labour markets and VET, which also hinders the pursuance of green transition policies.

Putting the 'social' back into green construction policies: Labour and VET

EU green construction policy is thus gender blind and generally lacking a social perspective; climate change, energy transition, and energy efficiency are formulated as neutral, scientific systems and processes with little social context or implications (Machin, 2019). The EU's ecological modernisation approach and climate change policies, driven by technical targets and market priorities, are all the more disquieting given that the transition to NZEB has significant consequences for the employment, education, and training of the construction workforce. LEC is fundamentally different from the traditional construction process as it introduces the concept of energy performance into a production system driven by building to time and budget. Buildings must meet specific energy performance targets through such measures as airtight building envelopes, thermal-bridge free construction, and on-site renewable energy sources. LEC, thus, needs a greater degree of precision and careful co-ordination so that building components are put together to constitute a system and function to restrict energy use to pre-determined limits. For the construction labour process, these requirements imply overcoming interfaces between different occupations, integrated team working, and improved communication given the complex work processes involved (Clarke et al., 2017a). Meeting these depends on an adequately trained workforce and expertise distinct from that developed in traditional construction VET, calling for broader qualification profiles, deeper theoretical knowledge, and higher technical competences to acquire thermal literacy, interdisciplinary understanding, and a broad range of transversal abilities such as project management and problem-solving. Incorrect and poor-quality installation results in a performance gap, the difference between the energy standards intended and those actually achieved, jeopardising the EPBD stipulated emission savings (Zero Carbon Hub, 2013; Johnston, 2016). Evidence suggests that failure to build to standards required indicates structural problems in terms of work organisation, employment, and the quality of VET (EC, 2014; Gleeson, 2016; Clarke et al, 2017a, 2019).

The BUS EU initiative (2010–2017), launched to develop NZEB competencies in the workforce, illustrates the complexity and the sheer scale of the task facing the industry (EC, 2014, 2016b, 2018). The BUS national status quo analyses, completed by member states highlight weaknesses in national VET systems and reveal that the number of construction workers in need of training runs into millions across the EU, with many having low general education levels and lacking formal training or qualifications. The overview report shows that most EU countries have a long way to go in upgrading occupational competences and learning resources, developing new courses and qualifications, and training the trainers, entailing a major programme of work. These challenges are compounded by under-resourced VET systems in many countries, with several also undergoing major reforms. Moreover, the scale of what is needed varies substantially between countries, and evaluation of VET for LEC reveals distinctive approaches, not all adequate for providing the expertise needed (Clarke et al., 2019).

Other imperatives also drive the requirement for more and improved VET, including an ageing workforce, the 'skills drain', increasing digitalisation and automation, changes in work organisation and the division of work, and the development of a circular economy (EFBWW and FIEC, 2020). In criticising the EU skills agenda for its 'homogeneous vision' and neglect of 'responsible partnership', the European construction social partners – the European Federation of Building and Woodworkers (EFBWW) and the European Construction Industry Federation (FIEC) – stress that the skills and VET agenda are interconnected with collective bargaining discussions on wages and working conditions. Indeed, the structure of the industry and its labour market characteristics – dominated by micro firms and casual and self-employment, facing a severe recruitment crisis, reliant on migrant labour, with many employers neither valuing nor seeing the need for qualifications – present a momentous challenge to VET and to retraining the construction workforce (Clarke et al., 2019). The challenge, therefore, is not simply a technical one of adapting to the demands of construction at a time of climate change, but of transforming the structure and organisation of the industry and the agency required of workers and reforming VET systems to take account of increased requirements for worker autonomy, integrated teamwork, project management awareness, and applied knowledge, as well as specific skill gaps.

Some of these problems are recognised, as recently highlighted in a report of the European Construction Sector Observatory (ECSO, 2020), *Improving the human capital basis*, which estimates that between three and four million workers need to develop energy efficiency–related skills in the industry. It identifies structural obstacles to this, including the fragmentation of the market and of construction value chains, cyclical factors, and the fact that 'many companies adopt a temporary employment model … limiting incentives for long-term investment in the workforce' (ECSO, 2020, p. 53). Moreover, 75% of companies struggle to follow occupational safety and health (OSH) requirements and 40% do not work safely, so that OSH-related training must increase by 60% (ECSO, 2020). This is detrimental to the 'attractiveness' of the industry, whose poor image is associated with low job security, tough working conditions, and health and safety concerns. However, these obstacles and detrimental factors are not then the direct target of policy, and increasing women's participation in construction is even seen as one solution to improving the attractiveness of the sector (ECSO, 2020, p. 9). Just as policies to achieve a zero carbon built environment fail to address employment and working conditions, similarly, so too do those seeking to increase female participation and improve the 'image' of the industry, even though these conditions are underlying obstacles to achieving both gender equity and NZEB.

The very use of the term 'human capital' in the title of the ECSO (2020) report helps to explain why the social structures forming and constraining the quality and quantity of labour and the complexities of 'skill formation' at different levels are neglected. Human capital theory (Becker, 1993) has long been challenged for regarding the 'skills' or 'human capital' of the workforce as the property of individual workers and associated only with the work processes of particular firms (Maurice et al., 1986). This narrow conception of 'skills' inevitably ignores the socialisation of labour, including women, into production through structures of employment, wage relations, and training (Campinos Dubernet and Grando, 1991). Just as the ecological modernisation approach to a green transition of the European Commission (EC) is without concern for the quality of labour and employment involved or worker agency, so too is the human capital approach taken by ECSO, under the EC's Competitiveness of Enterprises and Small and Medium-sized Enterprises (COSME) programme on market conditions and policy developments in the European construction sector.

Learning from examples of good practice

Embedding gender equality in policy and enabling women to shape and participate in the emerging green transition in construction require a paradigm shift, implying a re-definition of terms. It has also been suggested that greater female participation could make for more sustainable practices, enhancing women's opportunities, and accelerating social and technical change (Pearl-Martinez and Stephens, 2016). Examples of the extensive involvement of women in the green transition are, however, rare and more common in the professions than for the operative construction workforce. There are, though, local and organisational cases where women have been employed on a sustained basis in construction which, on examination, serve to identify some of the ingredients necessary for their successful participation. Where these ingredients are also conducive to effective LEC, and above all where organisations are also actively engaged in NZEB, a model is given for developing an eco-friendly and inclusive sector, in particular in terms of the coalition of actors involved.

One of the first ingredients needed for women's successful involvement in LEC is their ability to acquire the necessary qualifications. Given the requirement for high-level qualifications with LEC, good communication, and coordination skills, and the ability to manage the project, the way is potentially opened up for greater involvement of women, especially considering their generally higher educational achievements and greater presence on environmentally oriented courses and in technical, professional, administrative, and clerical functions (Clarke et al., 2017b). However, as indicated, any involvement is predicated on good and inclusive employment and working conditions. In this regard, large infrastructure projects in which women have been significantly involved, especially in the more professional areas, are useful for drawing out the measures that can facilitate their employment. These have included UK projects such as Heathrow Terminal 5, the 2012 Olympics, Crossrail, Thames Tideway Tunnel, and Hinkley Point Nuclear Power Station (Clarke et al., 2015; Baruah, 2018). The longevity, size, complexity, and nature of such major infrastructure projects mean that they are often highly regulated and subject to scrutiny, opening up the possibility for a more inclusive employment policy, setting ambitious targets, and new ways of working. For instance, the all-embracing Common Framework Agreement between EDF (Électricité de France) and unions at Hinkley Point places great emphasis on establishing integrated teams and new working practices and is structured to optimise opportunities to bring new people into the workforce (EDF, 2013). For the Olympics, contract compliance, continuous monitoring, and the guarantee of direct employment were critical to meeting equality targets (Wright, 2014). All in all, from examining these different mega projects, particular factors critical to the achievement of greater inclusivity are indicated – the roles of public procurement; the significance of regulated agreements secured with key stakeholders, including local authorities and unions; the involvement of the workforce and women's groups; systematic, targeted, and controlled recruitment, proactively applying equal opportunities policies; guarantees of direct employment; close monitoring; well-conceived training programmes, facilitating broader occupational profiles, formal links with colleges and universities, work placements and work experience; and good working conditions, including structured working hours, childcare provision, flexible working arrangements, and mentoring (Clarke et al., 2015).

Whilst these examples point to some of the measures required for the successful involvement of women, they are at the same time short-lived and without significant environmental credentials, especially concerning the use of concrete and the building of nuclear power stations and airports. More sustained participation of women in construction in the UK is found in the building departments of local authorities, known as Direct Labour Organisations (DLOs), which have always prided themselves on good employment and working conditions. These have made concerted attempts to include women since the 1970s, particularly following the introduction of the Sex Discrimination Act (1975), making it illegal to discriminate on the grounds of sex in employment or education, and grassroots campaigning by women to set up women-only training workshops. Women, consequently, accessed training in construction occupations, which was consolidated by joining DLOs. By the mid-1980s, Hackney DLO in London, for example, was running one of the largest training schemes for building workers in Britain, in which over 50% of the adult trainees were women, many going on to permanent jobs in construction (Clarke et al., 2015). Local authorities committed to changing their male-dominated construction workforce created a framework of support for women through the provision of designated women's officers; regular meetings; placing more than one woman on any site; flexible hours of work; and a clear and transparent set of equal opportunities guidelines backed up by internal procedures to address grievances. The success of these measures is evident in the presence of 266 women in construction operative occupations in seven Inner London DLOs in 1989. This legacy of the 1980s survives, despite political challenges, and DLOs have continued to address the low numbers of young women seeking construction training. For example, of the 283 apprentices at Leicester DLO between 1985 and 2002, 84 (30%) were women and, by 2012, 123 of its 431 strong workforce were women, as were 18 of the 75 apprentices employed in all occupations – as carpenters, electricians, plasterers, painters and decorators, bricklayers, heating and ventilating engineers, gasfitters, and metal workers (Clarke et al., 2015). This, therefore, provides us with further aspects necessary for the sustained employment of women: the involvement and commitment of public authorities, stable and direct employment, and a framework of support.

The different ingredients implied for the successful inclusion of women in construction can, therefore, be summarised as:

- Public sector involvement, especially, as in the aforementioned cases, the municipal authorities
- Good organisational employment and working conditions, targeted recruitment and retention, family-friendly policies
- Stakeholder involvement, as recommended in the Beijing Platform for Action, including women's organisations, professional bodies and unions
- Comprehensive training linked to employment

Whilst less impressive in terms of the participation of women, one organisation discovered in Scotland that fulfils most of these different criteria and has, at the same time, a clear commitment to NZEB is City Building Glasgow, which represents a sustained alternative employment model to the private sector. City Building Glasgow is a not-for-profit organisation, jointly owned by Glasgow City Council and the Wheatley Group Housing Association, and formed in 2006 from the original DLO of Glasgow City Council. Most of the 2,200 permanent construction employees of City Building Glasgow are unionised and the Joint Trade

Union Council is actively engaged in the organisation and underpins its strong social ethos (Clarke et al., 2018a). The organisation is unique in directly employing, under decent standards, a large construction workforce, regulating and monitoring subcontracting through a framework agreement, and running an in-house training centre providing a comprehensive and acclaimed four-year programme for a diverse intake, including women, with most trainees staying on as employees. The organisation's LEC schemes built to varying energy efficiency standards include social housing, care homes, schools, hostels, and retrofitting social housing estates, including through the installation of district heating using air source heat pumps, as part of efforts to tackle fuel poverty. Environmental measures are, thus, intertwined with employment and training practices that prioritise workers in a model shaped by the enhancement of labour capacity and opportunities for direct engagement in the green transition and underpinned by the traditions of municipal socialism.

These case studies suggest that the transformation of construction into an inclusive, eco-industry needs the input of a range of other stakeholders such as public authorities, VET institutions, unions, women in construction and their representative organisations, environmental organisations, and employers and their associations. Thus, even though gender is given scant attention and women rarely sit around the table in European and national energy policy-making, stakeholder involvement and institutional interventions at more local, sectoral, and organisational levels can be exemplary and indicate potential and effective drivers of change.

Conclusion

The lack of gender diversity in construction is a critical issue, one that relates to barriers in terms of the nature of VET and employment, human resource policies and practices for the industry, and the lack of employee engagement. Many of these are also barriers to achieving effective NZEB, including the need for a comprehensive and high-standard VET system and a stable, safe, and healthy system of employment. The suggestion is that meeting the challenge of a green transition in construction opens up the possibility to include women. Raising standards in construction VET and in employment and working conditions will also help address the current Europe-wide recruitment crisis in construction. Technologically up-to-date, well-resourced and high-level VET leading to qualifications valued in the sector could make a career in construction an attractive option for women. And interesting, eco-friendly, socially useful, not-for-profit and challenging construction work, including large-scale retrofit programmes, can turn that option into a reality over which women can have a decisive say and impact. What is needed to strike a blow at carbon emissions in construction is a major retrofit initiative in which women and men, trained and employed on an equal basis, play a decisive role – one that serves as a demonstration of how the industry can be transformed to become inclusive and socially useful through producing a carbon neutral building stock.

This chapter has sought to show how the EU's ecological modernisation approach and gender-blind policies in relation to energy mean that implementation of the EPBD is regarded as a technological fix. Yet, without addressing the social and structural problems besetting the construction industry, it will not only be impossible to achieve gender equality but also difficult to realise NZEB. Transforming the employment and working conditions that serve to exclude women and impede LEC is necessary both in order to meet energy-efficient and low carbon emission targets and to become an inclusive eco sector. It is also essential to develop VET systems capable of qualifying the many thousands of construction workers required for NZEB and retrofit programmes across Europe. While some countries, including Germany and Belgium, are well on the way to upgrading their VET systems to incorporate LEC elements, their construction industries continue to exclude women. The examples given show where women have been successfully included and where NZEB has been achieved, and suggest the combination of factors and the coalition of stakeholders that will be needed to transform construction into an inclusive, socially useful eco-sector. These factors include direct employment, good working conditions, comprehensive and targeted VET for LEC programmes, NZEB and measures to support the inclusion of women, whilst the actors include local authorities, not-for-profit building organisations, unions, colleges and training centres and, above all, women. In this way, the chapter reveals how a gendered analysis helps to identify the structural problems in the sector that need to be overcome if a green transition and gender equality in construction are to be accomplished.

References

Allwood, G. (2014) 'Gender mainstreaming in EU climate change policy', in Weiner, E. and MacRae, R. (eds) The Persistent Invisibility of Gender in EU Policy. *European Integration online Papers* (EIoP), Special Issue 18 (1), Article 6, http://eiop.or.at/eiop/texte/2014-006a.htm, pp 1–26.

Arditi, D., Gluch, P. and Holmdahl, M. (2013) 'Managerial competencies of female and male managers in the Swedish construction industry', *Construction Management and Economics*, 31(9), pp. 979–990.

Arora-Jonsson, S. (2017) 'Gender and environmental policy', in MacGregor, S. (ed) *Routledge Handbook of Gender and Environment*. Abingdon and New York: Earthscan, pp. 289–303.

Bakker, I. (ed) (1995) *Strategic Silence*. London: Zed Books.

Baruah, B. (2018) *Identifying Promising Policies and Practices for Promoting Gender Equity in Global Green Employment*. Clean Economy Working Paper Series. Canada: Western University.

Bataille, C. (2019) *Low and Zero Emissions in the Steel and Cement Industries—Barriers, Technologies and Policies*. Paper prepared for OECD Green Growth and Sustainable Development Forum. 26–27 November. Paris: OECD.

Becker, G. (1993) *Human Capital: A Theoretical and Empirical Analysis, with Special Reference to Education*. 3rd edn. Chicago: University of Chicago Press.

Campinos-Dubernet, M. and Grando, J.-M. (1991) 'Construction, constructions: a cross-national comparison', in *The Production of the Built Environment, Proceedings of the 11 Bartlett International Summer School. Paris 1989*, pp 17–34. London: University College London, Bartlett School.

Caren, V. and Astor, E.N. (2013) 'The potential for gender equality in architecture: an Anglo-Spanish comparison', *Construction Management and Economics*, 31(8), pp. 874–882.

Clancy, J. and Feenstra, M. (2019) *Women, Gender Equality and the Energy Transition in the EU*. Report requested by the Committee on Women's Rights and Gender Equality of the EU Parliament. European Union.

Clancy, J. and Feenstra, M. (2019) *Women, Gender Equality and the Energy Transition in the EU*. Policy Department for Citizens' Rights and Constitutional Affairs, European Parliament.

Clarke, L., Gleeson, C., Sahin-Dikmen, M. and Winch, C. (2019) *Vocational Education and Training for Low Energy Construction (VET4LEC)*. i) Final Report for the European Commission and ii) Country Summaries, Brussels.

Clarke, L. and Sahin-Dikmen, M. (2018a) *City Building (Glasgow): An Inspirational Model of Low Energy Social Housing and Public Building Production*. York University, ON: ACW programme: https://adaptingcanadianwork.ca.

Clarke, L., Gleeson, C. and Winch, C. (2017a) 'What kind of expertise is needed for low energy construction?', *Construction Management and Economics*, 35(3), pp. 78–89.

Clarke, L., Gleeson, C. and Wall, C. (2017b), 'Women and low energy construction in Europe: a new opportunity?' in Cohen, M.G. (ed.) *Climate Change and Gender in Rich Countries*, pp 55-69. New York: Earthscan/Routledge.

Clarke, L., Michielsens E., Snijders, S., and Wall, C. (2015) *No More Softly, Softly: Review of Women in the Construction Workforce*. ProBE: University of Westminster https://core.ac.uk/download/pdf/161107132.pdf.

Clarke, L., Frydendal Pedersen, E., Michielsens, E., Susman, B. and Wall, C. (2004) *Women in Construction*. Brussels: CLR/Reed.

Clarke L., Michielsens E. and Snijders S. (2018b) 'Misplaced Gender diversity policies and practices in the British construction industry: developing and inclusive and transforming strategy', in *Valuing People in Construction* edited by Fidelis Emuze and John Smallwood, Taylor and Francis/Routledge.

Dainty, A., Neale, R.H. and Bagilhole, B. (1999) 'Women's careers in large construction companies: expectations unfulfilled?', *Career Development International*, 4(7), pp. 353–357.

ECSO (2020) *Improving the Human Capital Basis. Analytical Reporter March*. Brussels: European Construction Sector Observatory.

EDF (2013) *Hinkley Point C Construction Project: Industrial Relations Common Framework Agreement*. Electricité de France r.

EFBWW and FIEC (2020) *Joint Reaction of European Social Partners of the Construction Industry on the Consultation on the Update of the Skills Agenda for Europe*. January 29. Brussels: European Federation of Building and Woodworkers and the European Construction Industry Federation.

EPBD (2018) *Directive 2018/844/EU of the European Parliament and of the Council of 30 May 2018 Amending Directive 2010/31/EU on the Energy Performance of Buildings and Directive 2012/27/EU on Energy Efficiency*. Brussels: European Commission.

EPBD (2010) *Directive 2010/31/EU of the European Parliament and of the Council of 18 May 2010 on the Energy Performance of Buildings*. Brussels: European Commission.

European Commission (EC) (2019a) *Clean Energy for All Europeans*. Directorate General for Energy. Brussels: European Commission.

European Commission (EC) (2019b) *The European Green Deal*. Brussels: European Commission.

European Commission (EC) (2018) *Final Report on the Assessment of the Build UP Skills Pillar II*. EASME. Brussels: European Commission.

European Commission (EC) (2016a) *Synthesis Report on the National Plans for Nearly Zero Energy Buildings*. JRC Science for Policy Report 97408. Brussels: European Commission.

European Commission (EC) (2016b) *Evaluation of the Build Up Skills Initiative under the Intelligent Energy Europa Programme 2011–2015*. EASME. Brussels: European Commission.

European Commission (EC) (2015) *A Framework Strategy for a Resilient Energy Union with a Forward-Looking Climate Change Policy*. Brussels: European Commission.

European Commission (EC) (2014) *Build-up Skills: EU Overview Report, Staff Working Document. Intelligent Energy Europe*. Brussels: European Commission.

European Commission (EC) (2010) *Energy 2020—A Strategy for Competitive, Sustainable and Secure Energy*. Brussels: European Commission.

European Construction Sector Observatory (ESCO) (2020) *Improving the Human Capital Basis*. Updated March. Brussels: European Commission.

European Institute for Gender Equality (EIGE) (2016a) *Gender and Energy*. Luxembourg: European Union.

European Institute for Gender Equality (EIGE) (2016b) *Gender in Environment and Climate Change*. Luxembourg: European Union.

European Institute for Gender Equality (EIGE) (2012) *Review of the Implementation in the EU of Area K of the Beijing Platform for Action: Women and the Environment—Gender Equality and Climate Change*. Luxembourg: European Union

European Parliament (2012a) *Resolution of 20 April 2012 on Women and Climate Change (2011/2197(INI))*. Luxembourg: European Parliament.

European Parliament (2012b) *(Report on the Role of Women in the Green Economy,(2012/2035(INI))*. Luxembourg: European Parliament.

European Commission (EC) (2020) A Union of Equality – Gender Equality Strategy 2020-2025. Brussels: European Commission.

European Parliament (2017) *European Pillar of Social Rights*.

European Parliament (2012c) *Charter of Fundamental Rights of the European Union*.

Eurostat (2016) *Share of Men and Women in 20 Most Common Occupations*. European Labour Force Survey.

Ferguson, S. (2020) *Women and Work: Feminism, Labour and Social Reproduction*. London: Pluto Press.

Fielden, S.L., Davidson, M.J., Gale, A.W., and Davey, C.L. (2000) 'Women in construction: the untapped resource', *Construction Management and Economics*, 18(1), pp.113–121.

Foster, J.B. (2002) *Ecology Against Capitalism*. New York: NYU Press.

Fraser, N. (2013) *The Fortunes of Feminism: From State-Managed Capitalism to Neoliberal Crisis*. London/New York: Verso.

Gleeson, C. (2016) 'Residential heat pump installations: the role of vocational education and training', *Building Research and Information*, 44(4), pp. 394–406.

Hampton, P. (2015) *Workers and Trade Unions for Climate Solidarity – Tackling Climate Change in a Neoliberal World*. London/ New York: Routledge.

ILO (2018) *Greening with Jobs: World Employment Social Outlook*. Geneva: International Labour Office.

IRENA (2019) *Renewable Energy: A Gender Perspective.* Abu-Dhabi: International Renewable Energy Agency.

Johnston, D. (2016) 'Bridging the domestic building fabric thermal performance gap', *Building Research and Information*, 44(2), pp. 147–159.

Lundström, R. (2018) 'Greening transport in Sweden: the role of the organic intellectual in changing union climate change policy', *Globalisations*, 15(4), pp. 536–549.

MacGregor, S. (2017) *Routledge Handbook of Gender and Environment.* Abingdon and New York: Earthscan.

MacGregor, S. (2010) 'A stranger silence still: the need for feminist social research on climate change', *Sociological Review*, 57, pp. 124–140.

Machin, A. (2019) 'Changing the story? The discourse of ecological modernisation in the European Union', *Environmental Politics*, 28(2), pp. 208–227.

Maurice, M., Sellier, F. and Silvestre, J.J. (1986) *The Social Foundations of Industrial Power—A Comparison of France and Germany.* Cambridge, MA: MIT Press.

Mellor, M. (2017) 'Ecofeminist political economy: a green and feminist agenda', in MacGregor, S. (ed) *Routledge Handbook of Gender and Environment.* Abingdon and New York: Earthscan, pp. 86–100.

Mol, A.P.J., Sonnenfeld, D.A. and Spaargaren, G. (eds.) (2009) *The Ecological Modernisation Reader: Environmental Reform in Theory and Practice.* London/ New York: Routledge.

Parsons, H.L. (1977) *Marx and Engels on Ecology.* CT/London: Greenwood Publishing Group.

Pearl-Martinez, R. and Stephens J.C. (2016) 'Toward a gender diverse workforce in the renewable energy transition', *Sustainability, Practice and Policy*, 12(1), pp. 8–15.

Powell, A., Bagilhole, B. and Dainty, A. (2009) 'How women engineers do and 'undo' gender: consequences for gender equality', *Gender Work and Organizations*, 16(4), pp. 411–428.

Rottenberg, C. (2018) *The Rise of Neoliberal Feminism.* Oxford: Oxford University Press.

Sang, K. and Powell, A. (2013) 'Equality, diversity, inclusion and work–life balance in construction', in Dainty and Loosemore (eds) *Human Resource Management in Construction Projects: Critical Perspectives.* pp. 163-196. London: Routledge.

Styhre, A. (2011) 'The overworked site manager: gendered ideologies in the construction industry', *Construction Management and Economics*, 29(9), pp. 943–955.

Watts, J. (2009) 'Allowed into a man's world' meaning of work-life balance': perspectives of women civil engineers as 'minority workers' in construction', *Gender Work and Organizations*, 16(1), pp. 37–57.

Weiner, E. and MacRae, R. (2014) 'The persistent invisibility of gender in EU policy: Introduction', *European Integration—Online Papers Special Issue*, 18(1), Article 3.

Worrall, L., Harris, K., Stewart, R., Thomas, A. and McDermott, P. (2010) 'Barriers to women in the United Kingdom construction industry', *Engineering, Construction and Architectural Management*, 1(3), pp. 268–281.

Wright, T. (2014) *The Women into Construction Project: An Assessment of a Model for Increasing Women's Participation in Construction.* Centre for Research in Equality and Diversity, School of Business and Management, Queen Mary: University of London.

Zero Carbon Hub (2013) *Closing the Gap Between Design and As-Built Performance—Interim Progress Report.* London: Zero Carbon Hub.

Part 3

Local, community institutions, and climate practices

11 Addressing climate policy-making and gender in transport plans and strategies

The case of Oslo, Norway

Tanu Priya Uteng, Marianne Knapskog, André Uteng, and Jomar Sæterøy Maridal

Introduction

Land use and travel behaviour remain closely conjunct and, since the 1990s, European nations have developed planning guidelines for integrated land use and transport development. These planning guidelines have also addressed overarching themes like universal design, improving conditions for children and youth and a more sustainable society. Lately, efforts have been directed towards examining the overlaps between sustainable planning and the climate policy-making agenda, but fine-grained analyses are largely lacking. To this end, gendered travel behaviour and its impacts on climate goals present a glaring research gap.

We begin with analysing the Norwegian planning and climate policy agenda by focusing on the case of Oslo. The scope of investigation is limited to presenting a quantifiable base for analysing gendered travel behaviour, both the existing patterns and in light of upcoming solutions like electrification and shared mobility, which are being posited as sustainable solutions. The topic is analysed under two thematic domains – administrative and demographic – which are sewed together in the end.

Through analysis of planning documents and information available to the public and private sector planners, the intention is to uncover the presence of (climate) strategies in plans, tools and other essential factors on the national, regional and local scale and then to analyse the results in light of gendered patterns of daily mobilities.

Three types of plans are especially important to climate policy-making – land use plans, climate action plans and mobility plans. These three different types of plans address crucial aspects of climate policies. In Norway, strategies for modifying travel behaviour saturate the plans and form a core component of strategic land use and mobility plans. Electrification comes across as part of the solutions to abate emissions and become fossil-free. At the European level, other types of sustainable fuels like hydrogen and biogas are more prominent, but Norway has concentrated on electrification. The voluminous growth of electric cars in Oslo is often used as a benchmark for electrification of vehicles across the world.

DOI: 10.4324/9781003052821-11

Distribution impact and acceptability of land use and transport solutions are not explicit in the planning documents. On the surface, the plans treat the entire population as a homogenous unit. The most important organising principle in the plans is the national zero-growth objective (ZGO)[1] for private car use, prioritising those who drive less and use more sustainable modes of transport. Gendered and segregated analyses, however, do not form part of the current approach and discussions. There is much to gain from disaggregated analyses. For example, travel patterns for the city of Malmö in Sweden highlighted that if women were to adopt the travel patterns of men, the modal share by car would increase by 17% and CO_2 emissions from car traffic would increase by 31%. Further, an additional demand for driving and parking space would add up to 190 standard town squares (Svanfelt, 2018). Building on these arguments, this chapter explores the nexus between gendered travel behaviour and climate policy-making by concentrating on the following question: How can transport and climate policy-making be better understood and conceptualised in light of gendered travel behaviour?

The chapter is structured as follows: the second section briefly introduces gendered travel behaviour and the third, the theoretical framing informing this study. The fourth section outlines the method, while the contextual background of the case city Oslo is presented in the fifth section. A brief content analysis of the major policy documents guiding Oslo's climate policy-making in the field of urban-transport planning is outlined in the sixth section, and a quantitative assessment of current and future travel behaviour, based on a macro-based prognosis-model called DEMOTRIPS, is summarised in the seventh section. The final section presents the concluding discussion and argues for more nuanced, disaggregated analyses to be made integral to climate policy-making exercises undertaken in future. The overall framing of the chapter is guided by an acknowledgement that current climate policy-making exercises will fail to deliver the desired results if policies and programmes do not recognize the contextual realities of current and future travel behaviour. Men and women show a consistent difference in their needs, preferences and practices of daily mobility which produces different bundles of climate effects at an aggregate level and, thus, differential marketing and targeting of future interventions are required to align travel behaviours with the goals of climate policies.

Gendered patterns of daily mobilities

In order to build a case for linking gendered travel behaviour and climate goals, this section presents the main findings emerging from a literature review of research and projects with direct impacts for climate policy-making. It is notable that there is a consistent pattern of women's daily mobility across the world with evident differences on a range of topics characterising the travel behaviour of women versus men (Priya Uteng and Creswell 2008).

Though car license-holding for women in the Global North has risen sharply in recent decades (Hjorthol, 2008), this development has not automatically led to an equal distribution of car-driving between men and women. Women are still, to a greater degree, car passengers than men (Polk, 2003). Women have a strong preference for sustainable modes of transport as they both use more public transport and have higher walking frequency compared to men. Given their propensity to use more public transport, to walk more and their high incidences of trip-chaining, women use multiple modes rather than a single mode (Heinen and Chatterjee, 2015). Women travel shorter distances, and their trip duration is also limited compared to men. This is evident in their commuting or work-related trip patterns but holds true for other trip purposes as well (Hjorthol, 2008; Sandow and Westin, 2010; Scheiner et al., 2011; Kim et al., 2012). In comparison to men, long-distance commuting remains much more restricted for women (Sandow, 2008; Scheiner et al., 2011; Hjorthol and Vågane, 2014).

Given the unabated development of car-based societies, it is not surprising that certain travel trends exhibit converging tendencies over time for the respective genders. As Konrad (2015) and Scheiner (2018) comment, elements of license-holding, car availability, mode choice (car driving), trip distances and trip duration seem to be converging over time. However, we need to analyse these converging tendencies in light of what Scheiner (2018) calls attention to – women are perhaps less habitual, more responsible, more sustainable and more sociable (Matthies et al., 2002; Polk, 2003; Hjorthol, 2008). This effectively means that the converging and shifting of women's sustainable travel behaviour towards a car-based, unsustainable one could be in response to the structural conditions that have underpinned city planning in the past decades. These trends also indicate that there has been a gradual evolution of an unsustainable path dependency (overlapping with the feminist institutional approach discussed in the introduction chapter) toward car-use in city planning which has pushed women to choose unsustainable transport mode. This shift has some serious implications, both from the perspectives of climate change (in line with the zero-growth objective) and the inclusivity agenda which insists the cities should be designed to be more inclusive and cater to the needs of its various demographic groups – women, children, elderly, disabled, immigrants etc. However, research simultaneously indicates that this (unsustainable) shift can be steered towards a more sustainable direction through the creation of conditions conducive to walking, cycling, and public transport (Priya Uteng et al., 2020). The ongoing global discussions on creating 15-minutes post-COVID-19 cities also rest on the modus operandi of increasing accessibility by walking and cycling to core facilities like grocery, health and other basic amenities (C40 cities climate leadership group 2020).

Gender and climate policy-making

The literature review highlights that a substantial amount of discussion so far has veered towards the gendered nature of the politics and management of climate change agenda. Findings suggest that females have greater awareness and concern about climate change than males (McCright, 2010; McCright and Dunlap, 2011) and the differences in attitude are often attributed to qualities such as cooperation and care, frequently associated with women (Gilligan, 1982; Beutel and Marini, 1995). Mavisakalyan and Tarverdi (2019) note that there is a positive correlation between a country's female representation and the 'stringency of its climate change policies'. For example, in Denmark, with a female representation of 37%, the stringency of climate change policies is eight times more than Bahrain with a female representation of 2%. An attempt to quantify this relationship highlights that a ten-unit increase in female representation results in a decrease of 0.24 metric tonnes of carbon dioxide per capita (ibid). This correlation provides substantial support to the claim that increased female representation in national parliaments results in more stringent climate change policies, thereby resulting in lower carbon dioxide emissions and campaigns targeting climate policy change are more successful in nations with higher female representation.

Other findings from Scandinavia problematise a simple correlation between female representation and a more sustainable climate agenda (Magnusdottir and Kronsell, 2014; Winslott Hiselius et al., 2019). Kronsell (2013, p. 12) issues an advisory remark:

> Gender sensitivity will not come about simply through the inclusion of women in policy-making. The practice of regimes, and the norms that constitute the social order of the landscape are all important to climate governance, and hence, they too, have to be actively scrutinised for gender and other injustices.

Kronsell (2013) argues for the adoption of the following three processes to bolster this interlocking: to strengthen female participation, to deal with oppressive power relations, and to challenge institutionalised norms. The core message is that there is a categorical lack of knowledge on how to deal with the topic of 'gender', and following Kronsell's claims (2013), for equal representation to have a substantial effect, we need actors who are knowledgeable on gender relevance of climate issues and who are interested in change and transformation, but were previously excluded from climate governance.

A second strand of focus in research builds on *gender mainstreaming*, which ECOSOC (1997) describes as following:

> The process of assessing the implications for women and men of any planned action, including legislation, policies or programmes, in all areas and at all levels. It is a strategy for making gender an integral dimension of the design, implementation, monitoring, and evaluation of policies and programmes in all political, economic, and societal spheres so that women and men benefit equally, and inequality is not perpetuated. The ultimate goal is gender equality.

Alston (2014) argues that while gender mainstreaming has been widely accepted, there exists a lack of understanding of its goals. This argument resonates with the transport field as well. Bacchi and Eveline's (2010) claim, that there is a lack of understanding of the actual problem and therefore the need to perform a rigid gender analysis of policies and outcomes and identify the actual problem being solved, hits the bullseye for the problem at hand in the transport domain. For example, the framing and understanding of the actual problem – *how to decrease car use and thus address the zero-growth objective in Norwegian urban areas* – is currently not being sieved through a gendered lens and consequently gender analyses do not inform the current climate policy-making exercises in Oslo.

Thus, though the first argument about the politics and management of climate policy-making in light of gendered positioning remains important, it needs further bolstering. This argument is repeated in the gender-mainstreaming agenda as well. Alston (2014) argues that whilst gender mainstreaming strategies include providing gender-disaggregated data, gender auditing, gender budgeting and greater transparency on factors such as the number of women in leadership positions, this is not enough to promote substantial change. Gender-mainstreaming goals often conflict with conservative bureaucratic processes and norms in terms of path dependencies (Chapter 1, this volume) which reduces the ability to make substantial changes (Wittman, 2010). Alston (2014) stresses the need to review and reassess policies and actions to ensure that they do not reinforce or create new gender inequalities.

A third line of enquiry is rooted in the gender role congruity theory which proposes that men and women engage in behaviours that are congruent with traditional gender roles (Eagly, 1987; Eagly et al., 2000). Swim et al. (2018) propose that climate policy arguments that focus on science and business are associated with men, whilst climate policy arguments that focus on ethical and environmental justice systems are associated with women.

Relying on any one particular strand might render the arguments unbalanced and, thus, it becomes important to attend to the question of how to bring these strands together. The field of transport provides a unique opportunity to sew together the science–business and ethics–justice frames when discussing climate change policies (Sovacool et al., 2018). Analyses presented in this chapter confirm that, both in current times and in the future, women exhibit sustainable travel behaviour and a higher usage of walking, cycling and public transport, while being less dependent on cars compared to men. These findings demonstrate that rather than being mere proponents of the ethics–justice frame being applied to discussions on climate change, women's travel behaviour makes them active agents of the science and business of designing transport policies. And it is utmost essential that these findings are taken forward in the practices and the norms guiding the designing of transport policies to cater to climate goals.

Methods

The chapter combines the qualitative method of content analysis with a quantitative prognosis model. In order to provide an overview of the connections between gendered travel behaviour and climate policy-making strategies, we employed content analysis of planning documents. The intention was to highlight the climate strategies in plans, tools and other essential factors on the national, regional and local scale, and check *if* and *how* the gender component is being taken forward. Though the primary focus was on strategic planning documents of the municipalities, due attention has also been paid to the regional and national plans as they are an integral part of the framework for climate policies. The content analysis investigated the plans and not the planning processes or political discussions per se. Plans and strategies are crucial as the starting point for further discussions and are, therefore, an essential part of the framing, formulating and providing a vocabulary for possible solutions in designing climate policies.

The second method focuses on a macro-based prognosis-model called DEMOTRIPS, a model used to assess the travel trends and demographic changes of men and women in the city of Oslo. The model uses current travel data, elasticities reflecting the relationship between transport modes, logistical growth modelling and the expected demographic development within specified geographical areas to create a prognosis for future travel behaviour. Additionally, DEMOTRIPS extends the analysis beyond women and men to include the intersectional category of age.

The case of Oslo

Oslo, Norway's capital, is the largest city and urban region of the country. The current platform for city council, cooperation between the Labour Party, the Green Party and the Socialist Left Party in Oslo (2019–2023), has employment for everyone, abatement of climate change and social equity as the most critical priorities. An important goal is to become the world's first emission-free city by 2030. The red–green coalition aims to govern the city according to a climate budget to reach 95% cut in greenhouse gas emissions by 2030, along with reducing Oslo's contribution to greenhouse gas emissions outside the municipality and an approximately one-third reduction in Oslo's car traffic by 2030 compared to the 2015 level. The city council aims to make explicit climate change effects and consequences in all relevant plans that are submitted to the city government.

Oslo is both a county and a municipality. The municipality is responsible for implementing the national, regional and local policies and translating them into local solutions along with attending to budgetary details. Most of the strategic planning takes place at the municipal level. In the first half of the four-year local election period, each council adopts a political strategy for municipal planning and decides ways to revise the municipal master plan. The adopted municipal master plan comprises the framework for preparing local plans and processing cases and is legally binding.

Further, though the regional plans could have been legally binding after the Planning and Building Act was modified in 2008, the county chose not to make the regional plan legally binding. So, in the case of Oslo, the regional plan remains advisory and provides the foundation for the municipality plans guiding the ZGO. For transport-related tasks that cover larger areas or overlap between municipal levels, there are two relevant items to be discussed – Ruter and the urban growth agreement for Oslo and Akershus.

Ruter is the public transport agency in charge of the administration of the tram, metro and buses in Oslo. Oslo and parts of the neighbouring county of Viken own Ruter. As a rule, Ruter oversees both tendering of public transport services and drafting the public transport strategy. In 2019, the national government, the county of Akershus (now part of Viken), the transport agencies, the municipality of Oslo as well as neighbouring municipalities of Skedsmo, Bærum and Ullensaker agreed to sign an urban growth agreement. The aim is to reach ZGO by implementing the regional plan and extending one part of the public transport network.

Table 11.1 shows the relationship between the main plans for climate policy-making on different levels that are valid for Oslo. As a rule, the laws, rules and guidelines come from the national government. At the regional level, the regional plan for land use and transport and the public transport plan is valid both for the city of Oslo and the Akershus region. The municipality oversees the land use, thematic (action) plans, and climate budget.

Content analysis of policy documents

This section presents the results of the content analysis of the policy documents outlined in Table 11.1.

Regional plan for land use and transport

The regional plan for land use and transport in Oslo and Akershus was approved in 2015 by both the Municipality of Oslo and the neighbouring county of Akershus (now part of Viken) (Plansamarbeidet, 2015). The plan, a joint strategic platform outlining the strategies to achieve efficient coordination between land use and transport in the region, is not legally binding but provides the foundation for further strategic planning in Oslo and the municipalities of Akershus County. The plan, gaining further momentum by an urban growth agreement that is accepted at all government levels, covers the following issues: densification,

Table 11.1 Strategic climate policy framework for Oslo 2020

Government level	Area	Plans and policies
State/national level	National	Laws including rules and regulations, National transport plan, thematic initiatives, zero-growth goal (private car use), urban growth agreements
County/regional level	Oslo and Viken (and the wider Oslo region)	Regional plan for integrated land use and transport, public transport plans (Ruter), Policy package for transport (Oslopakke 3)
Municipality	Oslo	Municipal master plan, climate budget, thematic plans and strategies including climate budget, climate action plan, bicycle plan

multicore development and zero-growth in transport demand, supported through better accessibility for walking, cycling and public transport.

The ZGO is the overarching goal implying that private car travel demand should be gradually diminished, in addition to halving greenhouse gas emissions by 2030. The plan seeks to modify travel behaviour through denser land use, prioritising some urban centres or nodes and local communities for development over others, leading to better integration between land use and public transport. The plan envisages increasing public transport supply as a vital measure to lower emissions. Cycling and walking are also posited as necessary tools to get a sustainable transport system and emphasis is put on improving the system, maintenance of the bicycle infrastructure and snow clearing in winter. Besides, the plan instructs municipalities to introduce more restrictive parking regulation that is adapted to the local context.

In summary, the regional plan has a focus on local accessibility in some nodes and towns aiming at access to services, jobs, housing and public transport within walking distance as an organising principle. However, it is not discussed how different demographic groups will be affected by the proposed plan. Further, there is no mention of gendered analyses or upcoming shared modes of travel.

The municipal master plan of Oslo

Oslo City Council approved the municipal master plan for Oslo in 2015 along with the regional plan. The master plan for Oslo is divided into social and land use elements where the latter is legally binding and consists of thematic maps, regulations and guidelines. The social element of the municipal master plan was revised in 2018 and renewed the urban strategy to guide revisions of the land use plan (Municipality of Oslo, 2015 and 2019). The plan, finalised by the Oslo City Council in 2019 and valid until 2040 (Municipality of Oslo, 2019), covers the following main issues: changing the pace of adoption of climate and environmental policies, active and sustainable municipality and equity (or social sustainability).

The plan aims at making Oslo a zero-emission city by 2030, and the strategies include building the city from the inside out and locating businesses in climate-friendly locations with good public transport access. According to the plan, the city has a significant potential to reduce pollution if the vehicle fleet becomes emission-free and there is a modal shift from cars to public transport, cycling and walking.

The plan seeks to modify travel behaviour by making car travel more difficult and at the same time making travel by public transport more accessible and improving the urban structure around the public transport nodes. Neighbourhoods are to become self-sufficient so that the need for daily travel (apart from commuting to work) will be reduced or based on sustainable forms of transport. Additionally, the plan talks about the electrification of the car fleet. The municipality will change its car fleet to electricity, hydrogen, or biogas and facilitate private (primarily electric) car use by supplying charging stations. The electric car facilitation is also a part of Oslo's Smart City project. Further, the plan integrates shared mobility (car sharing, bike sharing) and new modalities (for example, MaaS – mobility-as-a-service) in its outlook to mobility.

In summary, the municipal plan repeats the focus on local accessibility and embraces electrification and upcoming smart mobility solutions. But even though this plan makes explicit mention of equity and social sustainability, it does not delve into how different demographic groups and their travel behaviour will be affected by the proposed plan. There is no mention of gendered analyses even though the focus is at the municipal level and detail, disaggregate, gender data is readily available.

Climate and energy strategy and climate budget

Another important plan for Oslo is the climate and energy strategy (valid from 2016 to 2020) (Municipality of Oslo, 2016). The main themes deal with reducing car traffic, densification along the railway network and public transport hubs, increased bicycle share and reduced vehicle kilometres. The climate management part aims at integrating climate budgets into the municipal budget processes, practising environmentally efficient procurement, working closely with citizens and assisting organisations and other public authorities to develop and implement the right climate solutions.

Implementation is undertaken through the climate budget. In addition to the normal budget focusing on the municipal economy, Oslo has a climate budget highlighting the importance of sustainable development. The climate budget of 2019 was the third budget for Oslo and was essentially a list of measures the city council plans to implement for one year. The plans calculate the impact of emissions for each measure. The overall aim of the climate budget is twofold. Firstly, the climate budget highlights if the set targets will lead to an estimated reduction in emissions. Secondly, the budget stipulates the different municipal units responsible for reporting the results during and after the implementation.

In summary, the principles of gender budgeting are yet to intersect with Oslo's climate budgeting processes. It is a relatively simple exercise to include a gendered analysis of the proposed measures and ways to prioritise and tackle the outcomes if a certain measure favours one demographic group over another. But such calculations haven't been undertaken in Oslo's climate budgeting exercises. For example, Oslo has spent a substantial amount of public money in creating dedicated bike lanes but initial analyses reveal that the relative location of these dedicated bike lanes might be favouring men who undertake high-speed cycling over women who don't (de Jong et al., 2018).

Public transport strategy

The public transport strategy, conceived by Akershus County and Oslo Municipality in 2016 and valid until 2020 (Ruter, 2016), is also rooted in the regional plan with a focus on densification. The strategy suggests combining mass transport with individual solutions. Ruter has calculated how the areas covered in the regional plan can reach the ZGO (see Table 11.2). Recent technological

Table 11.2 The public transport company Ruter's estimate of reaching the zero-growth goal by 2030

	Inner-city		Outer-city		Regional cities and central areas		Sprawling areas		Total	
New trips	63 million		91 million		88 million		44 million		286 million	
Growth	25%		21%		19%		26%		22%	
Market shares	2014	2030	2014	2030	2014	2030	2014	2030	2014	2030
Public transport	33%	37%	27%	35%	19%	26%	10%	10%	23%	29%
Bicycle	8%	11%	5%	7%	4%	5%	4%	4%	5%	6%
Walking	41%	42%	23%	24%	18%	19%	17%	17%	24%	25%
Sum	82%	90%	55%	66%	41%	50%	31%	31%	52%	60%
Car	18%	10%	45%	34%	59%	50%	69%	69%	48%	40%

Source: Ruter (2016)

innovations have led to new shared solutions that Ruter plans to integrate with traditional public transport. The plan shows that car traffic has stagnated while the bicycle and walking share remains unchanged in Oslo. Ruter believes that the substantial growth in public transport needs to be sustained through further investments in infrastructure, capacity enhancement and better connection with both existing and upcoming shared modes of transport.

Bicycle strategy

Oslo does not have a sustainable urban mobility plan or a walking strategy at present but is in the process of making a walking strategy. However, Oslo has a bicycle strategy (Municipality of Oslo, 2014) that was adopted by Oslo City Council in 2014 and is valid for ten years. It aims to increase the bicycle share to at least 16%, which effectively means tripling the current bicycle share. The plan provides the foundation for a systematic working method to continuously improve conditions for cyclists in Oslo. Controlling emissions is a part of the overall strategy, and it is calculated that a bicycle trip that is five kilometres long emits 105 grams of CO_2 (lifecycle analysis for the bicycle and bicyclist's calorie needs) while the same emission for a private car trip is 1,355 grams. The plan addresses shared mobility as well, and the municipality continues to play an active role in the city's bike-sharing scheme. It is the only plan or strategy which recognises gendered behaviour, and highlights that men travel more with bicycles than women. And though the plan further acknowledges that increasing cycling shares will require different approaches for men and women, it is not able to engage with the topic in greater detail than a simple acknowledgement (Municipality of Oslo, 2014).

Summing up

A consistent finding throughout these different hierarchies of plans, guiding future transport solutions, is a lack of disaggregated analyses. Although the plans make perfunctory mentions of different areas and contexts, they do not engage with the gendered nature of travel behaviour. The framing of solutions relies heavily relies on a shift in travel behaviour without understanding these behaviours. Secondly, the technological domains of electrification, shared mobility, digitalisation and mobility-as-a-service (MaaS) are gaining traction in realising the overarching ZGO, and the domain of social sustainability in realising ZGO is lagging behind.

The point of contention is that none of the plans and strategies guiding future transport growth in Oslo undertakes a gendered approach to support the ongoing initiatives directed at increased usage of sustainable modes. The only plan or strategy that acknowledges gender at all is the bicycle strategy. Outside Oslo, the gender perspective is found in the walking strategy (that is still under construction in Oslo) in the small Norwegian town of Haugesund (Municipality of Haugesund, 2014). The walking strategy in Haugesund addresses disaggregated travel patterns, and the issue of safety, suggesting a more gendered walkability planning. But overall, the gendered perspective is almost non-existent in the current plans. Gendered analyses should be made an integral part of the planning strategies and planning documents ranging from the regional to the local levels. Not only will it ensure public support, but also give a grounded basis for infrastructure planning and provision.

DEMOTRIPS analyses

In light of the above conclusion, this section presents gendered analyses of Oslo's current and projected travel behaviour by employing the DEMOTRIPS model. DEMOTRIPS utilises the current travel data, elasticities between transport modes, logistical growth modelling and the expected demographic growth to create a prognosis for future travel behaviour. Through combining different elements, the model estimates how trends in mode usage within specified age groups and gender affect the usage of competing modes and the overall mode distribution in the area (model details available in Saeteroy Maridal 2018).

In order to discuss the gendered implications of current and future travel behaviour, we present two distinct travel scenarios for Oslo – the first is the business-as-usual (BAU) scenario which is built on the assumption that each age group and gender carries forward its present travel behaviour and is not affected by policies and programmes directed at ZGO; the second, or the zero-growth, scenario is built on the perception that car usage will reduce significantly but will be disproportionately divided among the different demographic groups given their ability to adopt other modes. The analyses are based on the following datasets:

- Population projections (extracted from Statistics Norway)
- Distribution of daily trips on different transport modes split by gender and trip length (extracted from The National Travel Survey [NTS] 2013/14) (Hjorthol et al., 2014)
- Mode substitution factors split by trip length

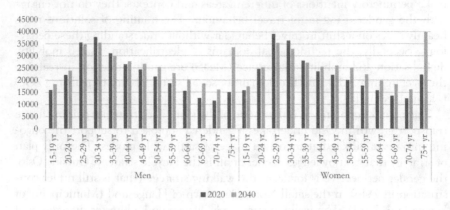

Figure 11.1 Oslo population by gender and age groups, 2020 versus 2040

Both scenarios contain short, medium and long trips, disaggregated by gender and age, as analyses based on travel behaviour surveys have consistently revealed that transport mode usage is highly dependent on the interaction between gender, age and trip length. Similar graphical representations and analyses are presented for the long, medium and short trips, providing a level of convenience to read and compare the results. Population projections indicate that by 2040, the disparity between youth and elders is larger in women than in men, although both genders are projected to have a spike in their elderly population. Women are also projected to see a stronger growth of established workers (40–64 age group) than men. These disparities feed into the scenario results and amplify the deviations in the existing gendered mobility patterns.

Scenario Set A: Business as usual scenario (BAU)

We present the BAU projections for short, medium and long trips disaggregated by gender and age. The first scenario explains and lays out the context of the data in detail to avoid being repetitive.

Long trips

Figure 11.2 illustrates an excerpt of the mobility data, denoting the mean number of daily long trips (> 7.5 km) each age group currently undertakes on different modes of transport. What immediately stands out is that men are much more car-dependent than women on long trips. The disparity in the number of long-distance car trips, both in 2020 and 2040, is significant between the genders while the reverse relationship for public transport is not as strong, meaning that men travel at a higher rate on long daily trips than women. The implications of this, in a BAU framework, is that men's travel habits and population projections will influence overall mobility on long trips to a greater extent than women. In

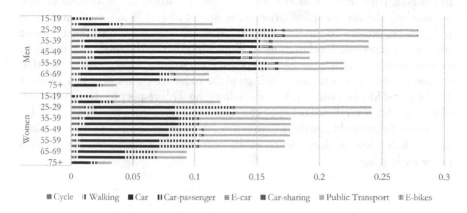

Figure 11.2 Oslo mean number of daily long trips per capita, BAU scenario

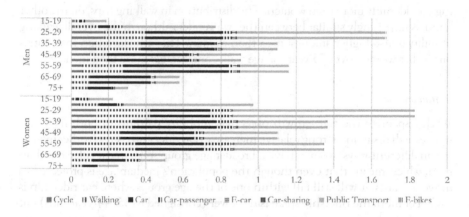

Figure 11.3 Oslo mean number of daily medium trips per capita, BAU scenario

terms of car travel, this is negative for the zero-growth objective because men are more car-dependent than women according to the current modal distribution. However, as the population's median age shifts towards the older demographics, overall mobility falls on account of a reduction in the mean number of trips. Still, car traffic growth is higher than public transport towards 2040.

Medium trips

Deviations between the genders are present in medium trips (2.5–7.5 km) as well (Figure 11.3). Women have a much higher public transport modal share than men and the difference in car dependence between men and women is also furthered by women being car passengers at a much higher rate than men. It becomes clear that though men and women have more similar car mobilities in medium trips than in short trips in terms of distribution across age groups, women lack the car-use spike between the ages of 25 and 34 as seen in men. This is, however, present in public transport, where women have considerably more trips at the aggregate level than men, illustrating that in the medium trip length as well, men are much more car-dependent than women. However, the fact that the number of car trips is relatively high in the older age groups as well as amongst 25–34-year-olds, for both genders, is challenging with respect to the zero-growth objective. An even distribution of car ridership combined with the skewed distribution of sustainable mode ridership towards younger age groups may translate into an increased car dependence in the years to come due to the population's rising median age.

Short trips

Deviations between men and women's mobility patterns exist for short trips (< 2.5 km) as well (Figure 11.4). It is evident that for short-distance trips, men are driving considerably more than women. The distribution in walking trips, on the other hand, is surprisingly similar between the two genders although the overall walking mobility is also higher amongst men than women. The only deviation that stands out is that women over 75 years of age walk considerably more than men.

Summary

With respect to the zero-growth objective, mobility patterns of both men and women will result in a strong relative increase in car ridership *vis-à-vis* other modes, but in different ways: women have a broader age group spike in car ridership than men, which means that even though the population's median age is projected to move upwards, it will still fall within one of the age groups where car ridership is high. For men, the ridership spike is narrower (consigned to 35–44-year-olds) but

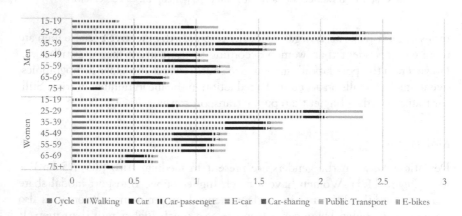

Figure 11.4 Oslo mean number of daily short trips per capita, BAU scenario

the elder demographics in men that today constitute a smaller share of the overall population are even more car-dependent than 35–44-year-olds, according to current mobility distributions. The influx of men aged over 45 will, thus, increase the overall car share for men relative to other modes.

According to the BAU trajectories presented, total car travel is expected to rise by 11% from 2020 to 2040 which means that 39,824 daily long, medium and short car trips need to be substituted by either walking, cycling, or public transport by 2040 if the zero-growth objective is to be met.

There are, however, both positive and negative aspects of the BAU scenarios with respect to the zero-growth objective. On a positive note, overall mobility per capita is projected to fall which means that at the aggregate level, slightly fewer daily trips will be undertaken in 2040 than in 2020. This is derived from population projections where the median age is expected to rise. Across the three scenarios, 25–34-year-olds conduct the highest number of daily trips and an ageing population means that these demographics will soon constitute a smaller share of the overall population, resulting in lower rates of mobility. This is, of course, assuming that the age groups' rates of mobility remain constant towards 2040. As for negative implications, the projected car traffic growth is disproportionally high relative to the projected growth in sustainable modes of transport across the three categories of trip lengths. This is because cars, in general, make up a larger share of overall mobility amongst the elder age groups (cf. men). A rising median population age combined with a general population influx thus translates into elders' travel habits increasingly affecting overall mobility, making Oslo's future mobility more car-oriented.

Scenario Set B: Achieving a zero-growth objective (ZGO)

In presenting the zero-growth scenarios, we plotted the trajectories where the sustainable modes identified by the Nation Transport Plan (NTP) have been set to grow and where mode substitution factors have been introduced to the model framework. The goal is to analyse how a zero-growth development in car traffic could unfold with an emphasis on the gendered responses to the growth rates and constraints set in terms of their projected mobility outputs.

Long trips

The main 'problem' for the elder groups in general and those above 75 years of age, in particular, is not that the trips are not transferred away from cars, but rather that the population increase is so massive in this demographic group that the overall number of trips conducted for *all* modes will grow despite the transfers to more sustainable modes. Analyses clearly demonstrate that public transport is absorbing prospective car trips for all demographics. The differing rates of car ridership transfers between the age groups are due to two factors: population increase and its associated rising median age, and the mode substitution factors' effect on altering mobility habits. Deviations indicate that there is a lower level of probability for elders to shift onto public transport in long trips than what is the case for the younger age groups. The implication of this is that the habitual change needed to substitute public transport for car travel becomes even higher with increasing age.

Medium trips

On medium trips, there is a decreasing rate of relative transfers from car travel in older age. The total number of car trips is up for men over 75 years of age in 2040 compared to 2020. Women also demonstrate the same tendency, but the overall number of car trips still falls across all age groups for women. The same mechanisms are at play here as in long trips – the projected population growth, skewed towards the older age groups, results in these demographics having more trips than before. Note that amongst 25–34-year-olds, the relative decline in car trips is much stronger for women than for men. The same transfers happen to a lesser extent for men, while women in this age group are seemingly easier to shift from cars to the other sustainable transport modes in this scenario. Women, in almost all age groups, will have gone from being relatively car-dependent to relying on public transport, cycling and walking for their trips by 2040, but the same cannot be said for men. Middle-aged and older men are projected to remain car-dependent on account of their mode substitution factors,[2] which explains the lower rate of transfers from cars to other modes like public transport, bicycling and walking.

Short trips

The decreasing rate of negative transfers from cars to sustainable modes is also present for short trips, but only for men. Here, 75+-year-olds end up having slightly more car trips in total in 2040 than in 2020, partially due to this age groups' relatively high population growth. This development is, however, not present in the same way amongst women, where car ridership falls considerably for all age groups. Both in men and women, the relative fall in car ridership is the strongest amongst 20–29-year-olds who seem to be particularly ready to substitute cars for sustainable modes in short trips, and 40–44-year-old women also demonstrate this tendency. It is worth noting that women, in general, conduct much fewer daily short-distance car trips than men. The total number of cycling trips is also projected to fall for 25–39-year-olds. The driving factor behind this projection is not that these age groups do not transfer car trips onto cycling trips, but because the overall population within these groups is projected to decline. There are also strong rates of substitution between walking and cycling on short trips.

Summary

The zero-growth scenarios found a clear indication that growth rates for sustainable modes need to be stronger for older age groups than for the younger ones to

Table 11.3 ZGO Scenario growth rates and aggregate mode shares, long trips

	Cycling	Walking	Car	Car-passenger	El-car	Car-sharing	Public transport	El-cycle
Yearly growth rate	4%	3%	0%	0%	0%	0%	7%	4%
Mode shares 2020	3.5%	2.9%	39.1%	9.5%	1.2%	0.9%	42.8%	0.1%
Mode shares 2040	3.5%	2.8%	22.0%	7.5%	1.2%	0.9%	61.8%	0.1%

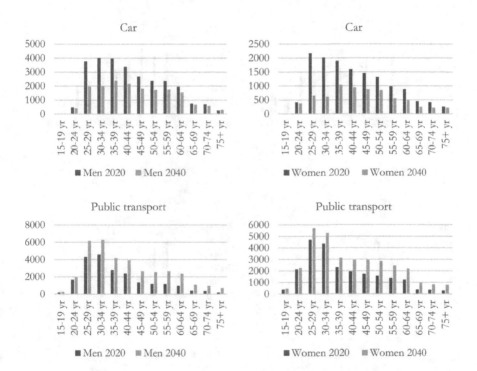

Figure 11.5 ZGO Scenario number of trips on select modes by age group and gender, long trips

achieve zero-growth in car traffic. This applies across all trip lengths. Three factors cause the above effect: older people, in general, being more car-dependent than younger people, the demographic shift in terms of an ageing population, and elders having weaker mode substitution factors from cars onto sustainable modes on some trip lengths. The population effect in particular will make it hard for elders to maintain overall car travel at today's levels. The age groups corresponding to established workers generally have a high car dependency per capita both now and in 2040. However, this demographic can still be projected

Table 11.4 ZGO Scenario growth rates and aggregate mode shares, medium trips

	Cycling	Walking	Car	Car-passenger	El-car	Car-sharing	Public transport	El-cycle
Yearly growth rate	7%	4%	0%	0%	0%	0%	6%	7%
Mode shares 2020	8.3%	16.3%	25.6%	4.6%	0.7%	0.6%	43.5%	0.3%
Mode shares 2040	9.7%	17.7%	12.6%	3.0%	0.7%	0.6%	55.3%	0.4%

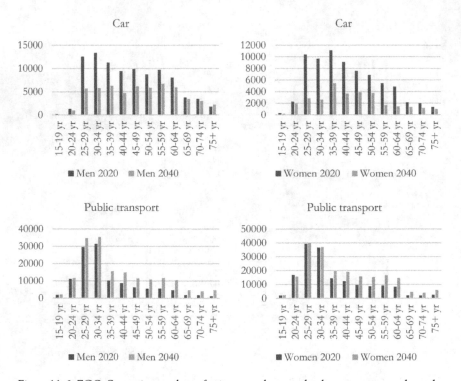

Figure 11.6 ZGO Scenario number of trips on select modes by age group and gender, medium trips

to achieve zero-growth in car traffic because its population growth is expected to be somewhat weaker than the oldest age groups. In general, men prove harder to shift away from cars and to sustainable modes than women, partly due to men being more car-dependent than women. This is particularly true for medium trips (2.5–7.5 km). 25–34-year olds, while having the highest rates of mobility of all age groups, are relatively easily transferred to sustainable modes on account of their mode substitution factors and low car dependence. Successfully shifting

Table 11.5 ZGO Scenario growth rates and aggregate mode shares, short trips

	Cycling	Walking	Car	Car-passenger	El-car	Car-sharing	Public transport	El-cycle
Yearly growth rate	7%	5%	0%	0%	0%	0%	3%	7%
Mode shares 2020	5.6%	72.2%	11.4%	1.8%	0.4%	0.3%	8.1%	0.2%
Mode shares 2040	5.3%	81.9%	2.8%	0.8%	0.5%	0.3%	8.0%	0.3%

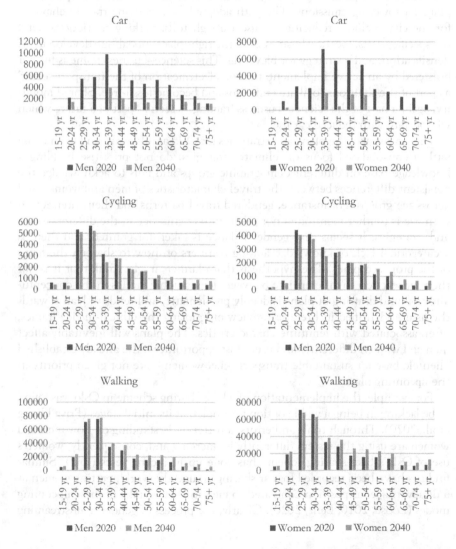

Figure 11.7 ZGO Scenario number of trips on select modes by age group and gender, short trips

this demographic goes a long way in reaching the zero-growth objective at the aggregate level because they constitute a large part of overall daily mobility. The rising median age, however, counters some of the effects in ease of shifting 25–34-year-olds onto sustainable modes because, by 2040, this group will constitute a smaller share of Oslo's overall population and number of daily trips than today's scenario.

Concluding discussion

Our analyses reveal that Oslo has a vision, but it has not successfully linked mobility plans with climate action plans. Strategies for modifying travel behaviour saturate the plans targeting climate goals and are the central component of plans for lowering emissions. The path adopted for modifying travel behaviour for the city includes reducing car use through tolls, parking restrictions, land use specifications, locating businesses in the right place, nodal/hub development, densification, and a strategy for bicycling. This science–business frame is further bolstered by an active implementation of 'electrification of the transport sector' as part of the solutions to abate emissions and become fossil-free. Shared mobility, again a part of the science–business framing, is seen as part of a future which will operate with reduced car use.

On the surface, the plans and strategies for Oslo at all levels, and even those with a consolidated focus on climate strategies, do not prioritise building a knowledge base on different demographic groups and fails to acknowledge the persistent differences between the travel characteristics of men and women and across age groups. For instance, gendered travel patterns and their intersection with age is neither mentioned nor taken into cognizance in the different plans and strategies. It seems that gender balance is taken for granted, and there is a categorical lack of knowledge among planners on how to sharpen the focus of the proposed plans. The hypothesis that planners might think that it is not their jurisdiction to differentiate between demographic groups in the conceptualisation and drafting of plans is deeply problematic, and tends to lean towards the fissure identified in literature review of ignoring the ethical-justice frames, often associated with feminine characteristics. The plans will inevitably affect men and women differently and it is a lost opportunity if the already established clientele base for sustainable transport – i.e. women – are not given priority in the upcoming plans.

For example, the implementation of a bike-sharing scheme in Oslo was found to be lacking in being a viable option for women at its initial stages (Priya Uteng et al., 2020). Through trials and errors, a better understanding on how men and women are using the system differently is emerging and, consequently, women's use of Oslo's bike-sharing scheme has gone up in recent years (ibid). Similar findings have been recorded for car sharing around the world, putting women at a disadvantageous position compared to men in their uptake of these upcoming modes (Loose, 2010; Lenz, 2020). Clearly, the process of gender mainstreaming is lacking in the transport sector, which if applied can certainly strengthen the climate goals.

In short, the case of interlocking between climate goals and transport plans in Oslo has, to date, not been analysed from a gendered perspective. Building on Kronsell's (2013) words, the practice of regimes, and the norms that constitute the social order of the urban transport landscape, remain important for climate governance and, hence, need to be actively scrutinised for gendered outcomes. If the core principles of the plans addressing climate change and ZGO are taken in alliance with the discussions on gender and made more explicit, climate change mitigation and adaptation strategies can be better debated and addressed. Hiding these underlying trends is not only irresponsible but creates serious chasms between public policies and public acceptance of proposed plans.

Further, distribution impacts are not explicit in most of the planning documents. The need for a shift has been seen as necessary at all levels of government, but a disturbing aspect is that the combined plans for Oslo indicate that regional development would continue as before, regional car dependency would continue to increase and the targets for greenhouse gas reduction as well as the regional competitiveness would most likely not be reached (Plansamarbeidet, 2015).

From a methodological perspective, the usage of results from models like DEMOTRIPS should be routinised for municipalities and regional authorities based on regular feeds from different sources of big data, travel surveys and projections from Statistics Norway. In order to plot future travel behaviour scenarios for different regions and cities, disaggregated analyses of passenger transport in light of ageing, gendered travel patterns and preferences, societal megatrends (like digitalisation, etc.) and transport sector trends should inform both transport and climate plans and strategies.

To conclude, the existing knowledge gap on both the gendered profile of daily trips, portfolio of age-groups and their interaction with the different policy efforts aiming at modal shifts means that the current policies run a risk of being much less effective. Seeing that the entire zero-growth objective rests on the rationale that travellers can be simply steered towards alternative modes constitutes a real problem. For creating effective climate policies in the urban-transport planning sector, further empirical research is needed to design targeted measures and interventions addressing both gender and age, based on the principles of gender mainstreaming, in climate policy-making.

Acknowledgement

This study is a part of PLATON – a PLATform for Open and Nationally accessible climate policy knowledge – funded by The Norwegian Research Council (NRC) (grant number 295789).

Notes

1 The zero-growth objective was introduced by the National Public Roads Administration in the National Transport Plan in 2014 for Norway's largest urban regions. The goal entails that the future growth in transport must be absorbed by public transport, cycling and walking and not by private car. In short, vehicle kilometres driven by private cars should either remain constant or go down in future.

2 The mode substitution factors are based on the likelihood of one mode replacing another depending on the prevalence of various trip purposes on the different modes. This was based on the Norwegian Travel Survey (NTS) 2013/14 which specifies the age, gender, trip purposes, trip length and mode used for each trip recorded in the NTS. The rationale for calculating mode substitution factors is that if, for example, cycling trips according to NTS constitute a high number of school trips and a low number of medical trips for a particular age group, one can assume that any new substituted cycling trip for that age group is more likely to be a school trip than a medical trip. These likelihoods are calculated across modes, age groups and trip purposes in the

following way: $LH\left(\text{purpose} = x \middle| y\right) = \dfrac{\text{purpose}_x^y}{\sum_{i=1}^{n} \text{purpose}_i^y}$

The equation denotes the likelihood (LH) f a . trip on mode y. aving trip purpose x. . or each age group. In short, it reflects the probability of a trip on transport mode y. aving travel purpose x.

References

Alston, M. (2014) 'Gender mainstreaming and climate change', *Women's Studies International Forum*, 47, pp. 287–294.

Bacchi, Carol, Eveline, Joan. (2010) *Mainstreaming Politics: Gendering Practices and Feminist Theory*. South Australia: University of Adelaide Press.

Beutel, A.M., Marini, M.M. (1995) 'Gender and values', *American Sociological Review*, 60(3), pp. 436–448.

C40 Cities Climate Leadership Group (2020) *How to Build Back Better with a 15-Minute City* [online]. Available at: https://www.c40knowledgehub.org/s/article/How-to-build-back-better-with-a-15-minute-city?language=en_US [Accessed: 20 November 2020].

de Jong, T., Fyhri, A., Priya Uteng, T. (2018) Gender gap—Perception of safety and cycling use, Workshop 2: Smart Mobilities—Walking and Biking, Nos-Sh workshop series, 21 April. Copenhagen, Denmark: University of Copenhagen.

Eagly, A.H. (1987) *Sex Differences in Social Behavior: A Social-Role Interpretation*. Hillsdale, NJ: Erlbaum.

Eagly, A.H., Wood, W., Diekman, A.B. (2000) *Social Role Theory of Sex Differences and Similarities: A Current Appraisal*. Mahwah, NJ: Lawrence Erlbaum Associates Publishers, pp. 123–174.

Economic, Social Council (ECOSOC) (1997) *Gender Mainstreaming* [online]. Available at: https://www.unwomen.org/en/how-we-work/un-system-coordination/gender-mainstreaming [Accessed: 14 March 2020].

Gilligan, C. (1982) *In A Different Voice: Psychological Theory and Women's Development*. Harvard University Press, Cambridge, Massachusetts.

Heinen, E., Chatterjee, K. (2015) 'The same mode again? An exploration of mode choice variability in Great Britain using the national travel survey', *Transportation Research, Part A: Policy and Practice*, 78, pp. 266–282.

Hjortol, Randi (2008) 'Daily mobility of men and women—a barometer of gender equality?', in T. Priya Uteng and T. Cresswell (eds.) *Gendered Mobilities*. Abingdon and New York: Routledge, pp. 193–210.

Hjorthol, R., Vågane, L. (2014) 'Allocation of tasks, arrangement of working hours and commuting in different Norwegian households', *Journal of Transport Geography*, 35, pp. 75–83.

Hjorthol, R., Engebretesen, Ø., Priya Uteng, T. (2014) *National Travel Survey—Key Results, 2013/14*. TØI-report 1383/2014. Oslo: Norwegian Institute of Transport Economics.

Kim, C., Sang, S., Chun, Y., Lee, W. (2012) 'Exploring urban commuting imbalance by job and gender', *Applied Geography*, 32, pp. 532–545.

Kronsell, A. (2013) 'Gender and transition in climate governance', *Environmental Innovation and Societal Transitions*, 7, pp. 1–15.

Lenz, B. (2020) 'Smart mobility—for all? Gender issues in the context of new mobility concepts', in T. Priya Uteng, H. Rømer Christensen, L. Levin (eds.) *Gendering Smart Mobilities*. Abingdon and New York: Routledge.

Loose, Willi (2010) *Aktueller Stand des Car-Sharing in Europa. Endbericht D 2.4 Arbeitspaket 2; Bundesverband CarSharing* [online]. Available at: http://www.carsharing.info/ima ges/stories/pdf_dateien/wp2_endbericht_deutsch_final_4.pdf [Accessed: 14 March 2020].

Magnusdottir, G., Kronsell, A. (2014) 'The (in)visibility of gender in Scandinavian climate policy-making', *International Feminist Journal of Politics*, 17(2), pp. 308–326.

Matthies, E., Kuhn, S., Klöckner, C.A. (2002) 'Travel mode choice of women. The result of limitation, ecological norm or weak habit?', *Environment and Behavior*, 34(2), pp. 163–177.

Mavisakalyan, A., Tarverdi, Y. (2019) 'Do female parliamentarians make a difference?', *European Journal of Political Economy*, 56, pp. 151–164.

McCright, A.M. (2010) 'The effects of gender on climate change knowledge and concern in the American public', *Population and Environment* 32, pp. 66–87.

McCright, A.M., Dunlap, R.E. (2011) 'Cool dudes: The denial of climate change among conservative white males in the United States', *Global Environmental Change*, 21(4), pp. 1163–1172.

Municipality of Haugesund (2014) *Walking Strategy* [online]. Available at: https://www .haugesund.kommune.no/om-kommunen/beredskap/beredskapsplan/lokaldemokrati/ kommuneplan/1352-gastrategi [Accessed: 14 March 2020].

Municipality of Oslo (2014) *Oslo Bicycle Strategy 2015–2025* [online]. Available at: https ://www.oslo.kommune.no/gate-transport-og-parkering/sykkel/sykkelstrategier-og-do kumenter/#gref [Accessed: 14 March 2020].

Municipality of Oslo (2015) *Municipal Master Plan 2015 Towards 2030* [online]. Available at: https://www.oslo.kommune.no/politikk/kommuneplan/ [Accessed: 14 March 2020].

Municipality of Oslo (2016) *Climate and Energy Strategy for Oslo. Municipality of Oslo* [online]. Available at: https://www.oslo.kommune.no/miljo-og-klima/slik-jobber-vi -med-miljo-og-klima-1/miljo-og-klimapolitikk/klima-og-energistrategi/ [Accessed: 14 March 2020].

Municipality of Oslo (2019) *Municipal Master Plan for Oslo, Societal Element, 2018 Towards 2040* [online]. Available at: https://www.oslo.kommune.no/politikk/kommuneplan/ [Accessed: 14 March 2020].

Plansamarbeidet (2015) *Regional Plan for Land Use and Transport in Oslo and Akershus. County of Viken* [online]. Available at: https://viken.no/tjenester/planlegging/samfun nsplanlegging/regionale-planer/ [Accessed: 14 March 2020].

Polk, M. (2003) 'Are women potentially more accommodating than men to a sustainable transportation system in Sweden?', *Transportation Research Part D*, 8, pp. 75–95.

Priya Uteng, T., Cresswell, T. (eds.) (2008) *Gendered Mobilities*. Ashgate: Aldershot.

Priya Uteng, T., Espegren, H.M., Throndsen, T.S., Bocker, L. (2020) 'The gendered dimension of multi-modality: Exploring the bike sharing scheme of Oslo', in T. Priya Uteng, H. Rømer Christensen, L. Levin (eds.) *Gendering Smart Mobilities*. Abingdon and New York: Routledge.

Ruter. (2016) *M2016. Public transport and mobility strategy, Ruter* [online]. Available at: https://ruter.no/om-ruter/strategier-og-handlingsplaner/M2016/ [Accessed: 14 March 2020].

Saeterøy Maridal, J. (2018) *Projecting Sustainable Mobility Scenarios for Oslo towards 2040: The Potential of Car-Sharing, Ridesharing and Cycling*. Master's thesis in Entrepreneurship, Innovation and Society. Trondheim: NTNU.

Sandow, E. (2008) 'Commuting behaviour in sparsely populated areas: Evidence from northern Sweden', *Journal of Transport Geography*, 16(1), pp. 14–27.

Sandow, E., Westin, K. (2010) 'The persevering commuter—Duration of long-distance commuting', *Transportation Research Part A*, 44, pp. 433–445.

Scheiner, J. (2018) Gender, travel and the life course. Thoughts about recent research, Workshop 3: Smart Mobilities, Planning and policy processes–Gender and diversity mainstreaming 17–18 September. Stockholm, Sweden: VTI.

Scheiner, J., Sicks, K., Holz-Rau, C. (2011) 'Gendered activity spaces: Trends over three decades in Germany', *Erdkunde*, 65(4), pp. 371–387.

Sovacool, B.K., Noel, L., Kester, J., de Rubens, G.Z. (2018) 'Reviewing Nordic transport challenges and climate policy priorities: Expert perceptions of decarbonisation in Denmark, Finland, Iceland, Norway, Sweden', *Energy*, 165, pp. 532–542.

Svanfelt, D. (2018) Gender equality and equity in urban planning, Workshop 3: Smart Mobilities, Planning and Policy Processes—Gender and Diversity Mainstreaming 17–18 September. Stockholm, Sweden: VTI.

Swim, J.K., Vescio, T.K., Dahl, J.L., Zawadzki, S.J. (2018) 'Gendered discourse about climate change policies', *Global Environmental Change*, 48, pp. 216–225.

Winslott Hiselius, L., Smidfelt, L., Kronsell, A., Dymén, C. (2019) 'Investigating the link between transport sustainability and the representation of women in Swedish local committees', *Sustainability*, 11(17), 4728. https://doi.org/10.3390/su11174728

Wittman, Amanda. (2010) 'Looking local, finding global: Paradoxes of gender mainstreaming in the Scottish executive', *Review of International Studies*, 36, pp. 51–76.

12 When gender equality and Earth care meet

Ecological masculinities in practice

Robin Hedenqvist, Paul M. Pulé, Vidar Vetterfalk, and Martin Hultman

Introduction

The global social and environmental problems of today are gendered. They are products of the complex multiple and slow violences of masculinist social injustices and environmental degradation. Accordingly, this chapter considers the flawed aspects of malestream masculinities and research on ecological masculinities, seeking an alternative path forward to a vibrant, living planet that celebrates the intrinsic value of all life. The first part of this chapter provides a conceptual overview of ecological masculinities, which advocates for a transformation from masculine hegemonisation[1] to ecologisation.[2] In their monograph, *Ecological Masculinities: Theoretical Foundations and Practical Guidance*, Hultman and Pulé (2018, pp. 53–54) positioned ecological masculinities as a conceptual shelter for conversations about men, masculinities and the Earth.[3] The arguments presented here are based on the premises conveyed in that text.

As a growing field of study with accompanying pluralised praxes, ecological masculinities prioritise concurrent systemic and personal transformations. Hultman and Pulé (2018, Chapter 8) recognised that further field research testing the practical manifestations of ecological masculinites were needed. This led to the development of an education material[4] and workshop series designed to help accelerate the ecologisation of masculinities throughout civil society, to promote shoulder-to-shoulder work with women and non-binary individuals, and organisations in support of gender equality and ending men's violence against those humans and other-than-humans who are traditionally marginalised. Notably in support of environmental justice, girls and women have been the principle international leaders of grassroots movements. The clear lack of male engagement is problematic since men, being the main perpetrators of violence against the planet and people, carry the lion's share of responsibility to change these fatal and unjust social and ecological realities. With this in mind, the second part of the chapter discusses research and analyses of the education in question piloted at a community garden space in Järna, Sweden, called 'Under Tallarna' (or 'Under the Pine Trees'). An assessment of participant experiences of the education revealed ideological and behavioural changes, indicating possible positive paths forward for men and the mosaic of

DOI: 10.4324/9781003052821-12

masculinities that they harbour. Interviews were conducted with the all-male participants at the end of the workshop series. All interviewees participated voluntarily and were supportive of gender justice and environmental care to varying degrees from the outset of the workshop series. The following personal themes emerged from the interviews as increased outcomes of the education: interconnectivity, accountability, emotional literacy and embodied knowledge. On a systemic level, the efficacy of this education was considered for its capacity to help male participants integrate the realities of our social and ecological crises, and become proactive contributors to creating a gender-equal and ecologically sustainable world for all.

Politics in the (m)Anthropocene

This is a critical moment; a time where the destructive forces of human societies on the Earth's living systems are irrefutable. Anthropogenic global heating has become a disastrous consequence of human industrial activities. What was for decades an intellectual exercise, a concern, a fear, a prediction, an ominous warning, a cautionary tale, is now our cumulative global reality (Lindvall et al., 2020). During the modern industrial era, transnational institutions have categorically avoided moving the planet closer to the necessary solutions (Allen et al., 2018). In fact, awareness and mitigation of these challenges have been met with unapologetic inaction and denial, which accompanies what Hultman and Pulé (2018) referred to as industrial/breadwinner masculinities.[5] Industrial/breadwinner masculinities are defined as 'malestream patriarchal, hegemonic and normative masculinities (which we apply primarily to men, but also to the masculinities adopted by some women and non-binary/genderqueer people as well)' who background the social and environmental implications of industrialisation for the sake of capital growth and its associated accesses to power and privileges (Hultman and Pulé, 2018, pp. 40–41). This definition stands with the tepid actions that attempt to regulate our problems away, which Hultman and Pulé (2018) referred to as ecomodern masculinities. Ecomodern masculinities have their origins in initiatives such as the Brundtland Report (1987; Hultman, 2013). This categorisation is aligned with calls for long-term sustainable development policies and practices that, when combined with measured responses to global emissions[6], have resulted in reforms that pay lip service to the costs of industrialisation on human societies and the ecological integrity of the Earth. Ecomodern masculinities are the gendered identities that dominate systemic pathways that aim to protect and preserve economic growth while also offering nominal care for society and environment (Hultman and Pulé, 2018, pp. 45–46), which, for example, align with the ecological modernisation model for global climate governance. For many individuals (of all gender identities) who benefit from hegemonies that flourish through industrial/breadwinner and ecomodern masculinities, ecologisation can ignite an existential crisis that generates resistance in, at times, extreme forms – for example, the misogyny (and other forms of marginalisation, such as transphobia), pseudo-science and conspiratorial vitriol that commonly accompanies climate denial.

Both aforementioned masculinities categorisations rely heavily on hegemonic masculinist global economics.[7] They also give us insight into the comprehensive barriers that prevent transformational change towards greater care for the planet and people. Hultman and Pulé (2018) noted that the path dependencies that convey power and privileges to industrial/breadwinner are enmeshed with the fossil-fuelled economy (also see Daggett, 2018). They noted the similar inefficacy of ecomodern masculinities, replete with its reformist responses to climate change action through ecological modernisation that pervades global governance, at least in parts of Europe (Chapters 1 and 10, this volume). In combination, both masculinities illuminate the ways that humanity struggles to break free from the gravitational pull that shapes pursuits of consumer-driven self-gratification and, in doing so, places select groups of an advantaged few (particularly Global Northern, straight, white, middle- and older-aged men) ahead of the 'glocal' (global through to local) commons. Seeking solutions beyond these destructive socialisations, Hultman and Pulé (2018) posited a third categorisation called ecological masculinities, which looks beyond the enmeshment between masculinities and the global economic, political and social dependency on industrialised extractivism. They defined this relational categorisation as a material feminist and semiotic 'gathering point for previous conversations on men, masculinities and Earth' which prioritises and dramatises the transformation of masculinities from hegemonisation to ecologisation on political and personal scales (Hultman and Pulé, 2018, p. 53). Ecological masculinities – as conceptualised here and elsewhere (Hultman and Pulé, 2018, 2019; Pulé and Hultman, in press) – seek philosophical alignments with the hard-fought victories of feminists and ecological feminists in controverting men's habitual domination of marginalised groups and individuals along with other-than-human nature.

The view that the world is here for humans, and for a privileged few in particular, to benefit from is an obscene consequence of masculinist hegemonisation (Hultman and Pulé, 2018). Di Chiro (2017) has similarly considered that the (m)Anthropocene with its domination of 'white northern male voices' has created and driven the planet into an epoch of human-induced (male-dominated) global heating along with geological and ecological perturbations, as a nuanced interpretation of what is now broadly referred to as the Anthropocene (also see Raworth, 2014). Ironically, those who stand to lose the most from the knock-on effects of global heating have rushed to pump and mine oil, gas and coal, as if to suck as much wealth from the Earth as quickly as possible (perhaps an indication of their surreptitious acknowledgement of the climate science). In doing so, they have accelerated the severity of the social and ecological consequences needing to be urgently addressed (Anshelm and Hultman, 2014; Dunlap et al., 2016; Pulé and Hultman, in press). Those individuals of the aforementioned, largely Global Northern, straight, white, middle- and older-aged men, are deeply embedded within global industrial and economic systems (Demelle, 2019) and, thereby, are the common embodiments of industrial/breadwinner masculinities and the group most responsible for causing global and intergenerational destructiveness. Notably, some of these individuals have successfully fused ideological climate change denial with right-wing nationalist political agendas around the globe (Anshelm and Hultman, 2014).[8]

The actions of these individuals and the groups they represent, are being contested with increasing levels of fierce grace. Consider Sweden's Greta Thunberg who, since 2018, has gained considerable prominence in the global youth-led climate justice movement. That movement has acutely challenged sceptics and deniers, even employing legal means in some instances (Arnoldy, 2019), making bold demands of international leaders to reverse the social and ecological trends of anthropocentrism (McCall, 2019). Thunberg has applied persistent pressure on international leaders to step up, and in doing so she and her colleagues have brought concerns about the growing climate emergency to the hearts and minds of a broader audience of global citizens – effectively mainstreaming the conversation and inspiring many to action. Youth-generated protests (e.g. Friday's for Future; Extinction Rebellion) have been met by climate change deniers, whose abundance of funds and lobbying efficacy, has flipped into high gear (Crowe, 2019). Commonly associated with right-wing nationalist political agendas and buoyed by denialist pseudo-sciences (Holland, 2014), the climate change denial machine has achieved wide-reaching influence over public debate (Hultman et al., 2019; The Climate Reality Project, 2019). The successes of these vocal minority views are highly funded and organised. They call attention to the need for proactive responses to the growing climate emergency. As an example, consider the increased scholarly attention that exposes these resistances towards effective mitigation politics, such as citizen's democracy, human rights, and the celebration of intra- and trans-species diversity as deeply masculinist (Kaiser and Puschmann, 2017; Forchtner et al., 2018). Some two years into regular youth-led climate strikes around the world, global governments are still in a state of dithering as a result of the bulwark created by these increasingly organised deniers (Thunberg et al., 2020).

Here, consider the global social and environmental terrains to be profoundly gendered precisely because they are steeped in economic inequalities that are correlated with race, class, geopolitics and age. The planet's most privileged groups and individuals are being challenged by the straightforwardness of young people in the climate justice movement who are calling denialism and right-wing nationalist political agendas to account. Their actions have exposed the obfuscating of injustices that accompany industrial/breadwinner masculinities, which have been illuminated by feminist and ecofeminist scholarship and activism for some time (Ruether, 1975, p. 204; Hultman and Pulé, 2018). From these trends, we can conclude that the (m)Anthropocene has a long heritage of gendered Global Northern invention. Our species' energy addiction, which has been fed by fossil fuelled economies for little more than a century, is only one (although a severe) example of the masculinist hegemonies and their associated practices that have weaponised and industrialised masculinist primacy (Hultman and Pulé, 2018). This is not a new phenomenon; the origins of this trend can in fact be traced back to the 17th century as our species transitioned towards the Industrial Revolution (Merchant, 1980). With a seemingly limitless bounty of the earth's resources, boundless Indigenous lands to conquer and machinery to replace human labour, Global Northern imperialists have enacted settler colonial strategies to stretch the reach of their nations, ensuring a one-way flow of wealth from the Earth in its then farthest reaches, through the extracted, colonised, enslaved and dominated, back to Western Europe.

Ecofeminist scholars such as Susan Buckingham (2017), Greta Gaard (2014), Sherilyn MacGregor (2009) and Carolyn Merchant (1980) have illuminated this confluence of gendered characteristics and the dire social and ecological consequences of Western industrialisation over time. Such are the protracted failures of Western modernity. But these failings cannot define our future as a planet or species. Rather than ignoring or freezing in overwhelm in the wake of the social and ecological impacts of the (m)Anthropocene, this chapter argues in favour of masculine ecologisation and considers how this process can be manifested practically. The following sections introduce the case study of select participants from the men's reflective groups that piloted the education material.

The organisation MÄN and its work with gender equality and the environment

MÄN is a feminist non-profit organisation that was founded in Sweden in 1993 to engage men and boys with gender equality and action against men's violence against women. MÄN's vision is a gender-equal world without violence. An important part of its work for change is to challenge destructive masculinities norms as they impact men's and women's relational exchanges, provide a space for self-reflection and discussion, educate in favour of violence prevention methodologies and promote gender-equal parenthood. MÄN is both a national federation with about 1,800 members and 20 local groups and associations around Sweden and a professional organisation with approximately 30 staff members (MÄNa, n.d.). MÄN currently has projects in Sweden and Eastern Europe, and is part of the international network, MenEngage, which involves men and boys in this work globally (MÄNb, n.d.).

In 2014, MÄN began looking at the intersection between gender equality, masculinity and the environment. The organisation acknowledged that ecofeminism has long identified the intersections between patriarchal structures and destructive norms around masculinity, resulting in environmental degradation. This was prompted by the organisation's recognition that men as a group have the largest ecological impact – leading to worsened conditions for women and children in particular – while it is mainly women who are assuming the responsibility to challenge ecological degradation and its social implications. MÄN started to use this insight to bring about change in masculinities norms in pursuit of gender equality and pro-environmental behaviour, concurrently (MÄN, 2019). MÄN proceeded to host seminars on men, gender equality and the environment, joined the Swedish Climate Network, wrote a discussion paper titled *Men, Masculinities and Climate Change* (Kato-Wallace et al., 2015) together with several MenEngage partner organisations, and initiated the Gender and Environment Network for

Inspiration (GeMiNI) – which is a 'meeting place to explore and learn about the interlinked issues of gender, environment, climate and the related theory and practice' that aims to 'contribute to the public debate through regular joint activities, by increasing the awareness about gender and environment as interlinked issues that need to be addressed together' (GeMiNI, n.d.).

By the end of 2017, the #MeToo movement commenced in response to multiple sexual abuse accusations, most acutely (but not exclusively) brought into the light by the sexual violations of Harvey Weinstein. Under the hashtag #MeToo on social media, many women testified to men's sexual harassment and, thus, clarified the magnitude and normalisation of the problem (Sayej, 2017; The Everyday Sexism Project, n.d.). With this, about 1,000 men in Sweden turned to MÄN with questions about what they could do to contribute to change. Therefore, MÄN brought forward the conversation guide #*AfterMeToo* for men's groups, where men got to reflect on their roles in patriarchal structures and the violence towards their surroundings. During four sessions of two hours each, men were encouraged to break with the male culture of emotional restraint and invincibility and feel and reflect on topics of masculinities norms, relations, sex, pornography and objectification of women, with a norm-critical approach. A central element is the specific format that promotes active listening and a fair distribution of sharing/speaking time (MÄN, 2018).[9] The #*AfterMeToo* workshop guide has since been expanded to include an additional four sessions linking men's violence against women and genderqueer people to the structural violence against other-than-human nature. This conversation guide, called *Men in the Climate Crisis*, was piloted at the permaculture garden 'Under Tallarna' ('Under the Pine Trees') in Järna, Sweden.

The men in the climate crisis reflective groups

Data were collected to assess and analyse the effectiveness of the initiative MÄN Under Tallarna; a pro-feminist workshop that was designed and implemented with men at this permaculture venue. In these reflective groups, men were encouraged to embrace the severity of the ongoing global ecological crisis, express feelings about it, reflect on their role in ecological degradation and find diverse ways to contribute to change. The groups explored how to choose reciprocity and care in the making of an equal and sustainable world without violence (MÄN, 2020).[10] *Men in the Climate Crisis* focuses on how to feel strengthened and engaged, rather than afraid and passive, about the crises and challenges of our times (MÄN, 2020).[11]

Between 2018 and 2020, MÄN Under Tallarna has hosted seven reflective groups, leading them through the *Men in the Climate Crisis* workshop – with six to eight participants in each group, spread over eight sessions (one session a week) of two hours each. The participants had actively sought out these groups, which suggests that they were positively attuned to pro-feminist and pro-environmental messaging. Robin Hedenqvist (2020) reached out to all these men to conduct interviews about their experiences. Half of them wanted to participate and, by the end, the research resulted in in-depth interviews with seven participants. Those findings, accounted for in the following section, are based on these interviews and were summarised in his master's thesis (Hedenqvist, 2020). The seven participants were chosen based on their willingness and availability to meet in person for the interviews. The fact that these men were self-selected (interviews were only conducted with participants who were intrigued by the general request) had an impact on the study. For example, it could be that these men were keen to participate in the study just because of their strong opinion about the reflective groups and/or simply because they perceived personal change. If so, there would be a bias against experiences that are less dedicated or transformative.

The men interviewed gave several reasons for participating in the reflective groups: being environmentally and socially concerned; wanting to contribute to gender equality and a healthy environment; being curious about the male role and masculine identity; wanting to associate with other men; to develop as a person; and getting recommended to join. They were quite diverse in terms of age (between 30 and 65) and profession (e.g. in healthcare, professional communication, construction and engineering). Most, however, could be described as Western, white, middle-class, middle-aged, cisgender men. This is a privileged group within prevailing power structures that tend to have a socialised prepossession (or internalised sense) of feeling superior and entitled. In general, this group benefits from the socio-cultural role of industrial/breadwinner masculinities, reinforces it, and has a large socio-ecological footprint (Hultman and Pulé, 2018). Therefore, it is particularly vital to promote personal and structural change amongst this group of men. To study their experiences in the reflective groups is a potential pathway towards ecologisation.

Perceptions of global social and ecological crises, and men's roles in them

There was a view amongst several participants that patterns of personal and internalised superiorisation of male domination, and the patriarchal structures that manifest them, are responsible for the abusive culture of industrial/breadwinner masculinities; a culture replete with its justifications of violence against marginalised people (especially women and genderqueer people) and other-than-human nature. In this culture of abuse, other-than-human nature is rendered a mere resource, deprived of intrinsic value, and is accompanied by disproportionately large levels of travel, meat and energy consumption. Caring for the glocal commons tends to be associated with traditionally feminine and womanly traits, thus offering a legitimate critique of hegemonic masculinities and manhood.

The emotionally restricting norms of toughness and stoicism promoted by hegemonic masculinist socialisations shun vulnerability internally and externally through social constraints that condition boys to suppress feelings and lose touch with their inner emotional life, and other people, as they grow up (Hultman and Pulé, 2018). Upon reaching adulthood, this lack of emotional comfort and ability to be earnest is in many men's lives expressed through domination and control, effectively oppressing their surrounding environments in a similar manner to the suppression of personal feelings. To adopt a bully mentality and be violent against oneself, other men in dominant groups, marginalised people, and other-than-human nature can become a strategy to compensate for such emotional and personal insecurity. The double-ended suppression noted here is indicative of the emotional emaciation that commonly accompanies masculinist socialisations which are, of course, most acutely internalised by cis-males but can pervade the lived experiences of women and genderqueer people as well. These masculinist norms restrain feelings and intuition by reifying logical problem-solving, trust in economic rationalism, seeking technological fixes to social and ecological problems, and amplifying competitive and individualistic approaches to life that infer a materialistic narcissism (Mellström, 1995). These values have been imperatives for the institutionalisation of industrial capitalism and its associated ecological destruction, in which people (their labour) and other-than-human nature are considered sole resources for profit maximisation (Merchant, 1980).

The patriarchal/capitalist principle of domination, combined with the masculinist norms championed by industrial/breadwinner masculinities, block many men's abilities to acknowledge and deal with the pain and suffering that comes from abusive behaviour, effectively obstructing personal and structural change. In order to counteract these patterns of command and control that aim to 'tame' and subordinate the surroundings in similar ways to those who are 'otherised', it is vital to focus on and reconcile the lack of emotional and relational proximity amongst men as an important starting point in the transition from hegemonisation to ecologisation (Hultman and Pulé, 2018). In other words, men must feel the despair and loss associated with our global social and ecological problems (Pease, 2019).

Vulnerability and emotional literacy

Interestingly, the participants in the workshops viewed the ability to care for the Earth, others and ourselves as deeply linked to the ability to be emotional and feel empathy and connectivity with others. One participant noted that:

> How we are taking care of our Earth is linked to how we are taking care of our inner emotional life and how we are taking care of our relationships. It has a very strong connection. (Interviewee 6, our translation).
>
> (Hedenqvist, 2020, p. 35)

Through the workshop series at Under Tallarna, it became evident that a vital first step towards ecologisation is to expose the costs of social inequalities and ecological destruction on the men themselves and the relationships that they hold most dear. The case study demonstrated that, from this starting point, facilitating broader, deeper, and wider care was not only made possible but became a natural series of next steps to be taken as the participating men were able to touch their inner despair about the state of the world. This may, on some level, seem like more of the same; of making it all about the individualised man *in situ* and attempting to redefine masculinities norms, which runs the risk of enabling the reinvention of masculinist hegemonies dressed up in new and more caring forms (Pease, 2019). However, our research demonstrated that to go too far, too fast by breaking with masculinity altogether runs a great risk of cauterising many men's curiosity and willingness to find ways to be integral to solving our conundrums, beyond being blamed for the planetary problems. The male reflective groups welcomed the creation of a safe space in order to let one's guard down and be vulnerable for a change. In effect, they created a forum for the kind of 'learn[ing] how to express their emotions and concerns about the looming ecological crisis facing the planet' that Pease (2019, p. 121) called for. The participants were asked to reflect on the sense of entitlement, encouraged by masculine socialisations, that stands at the core of men's violence (against the Earth, others and the self) and made explicit through MÄN's #*AfterMeToo* foundational training. This resulted in expressions of grief concerning social and ecological exploitation and a will to scrutinise personal behaviour more closely.

> Because … I think as a man … and I myself have previously in my life … have taken a lot of criticism like it is me who is wrong. I've done wrong, I am wrong. And when you end up in that place, there is no room for development. And for further … to open up and dare to be vulnerable. (Interviewee 4, our translation)
>
> (Hedenqvist, 2020, p. 40)

Some participants discussed the ways that, before participating in the reflective groups, in-grained defence mechanisms laid into their psyches from a young age prohibited attempts to be open-hearted and get truly close to another person. A similar defensiveness was felt whenever criticism of specific (and relatable) behaviours was raised – invoking emotionally charged responses such as guilt, shame and anger, which were commonly suppressed, but through the workshop were encouraged and brought forth. These expressions of autonomy and toughness, aligning with traditionally masculine characteristics ('manliness' being defined by being 'unwomanly' [sic]), make it difficult to acknowledge the personal impacts of one's choices and/or behaviours on others. The research demonstrated that, for some men, this was a key source of resistance to change. Criticism became a threat to self-identity control and accumulated power, prompting some to aggressively reject or resist the ecologisation process. Notably, violent and oppressive behaviour was exposed for the way some men can use these traits to conceal the negative feelings evoked by criticism and accountability to planetary problems. These gendered responses preserved the status quo and the patriarchal social order.

The workshop series helped to illuminate the banal normality of industrial/breadwinner masculinities as a globally dominant unmarked categorisation and the role of these socialisations in securing power and control through the institutionalisation of masculinist and industrialised capitalism (Anshelm and Hultman,

2014). These revelations helped participants register the degree to which climate change denial was directly correlated in destructive ecological, familial and personal actions that disregarded the levels of care needed in a truly sustainable world. In this sense, the participants struggled to defend identities that are deeply tied to the self-gratifying and presumed right to dominate.

Relationality, care, and embodiment

Several participants adopted a wider relational point of view through the reflective groups. The practices of active listening, problematising norms of masculinity, discussing how they have affected oneself and others while embracing the facts and experiences of oppressed 'others', and sharing intimate stories and feelings on these topics seem to have contributed to a greater sense of interconnectivity (illuminating the ways that everyone/everything is connected and constitutive of nature) and emotional literacy. It was made evident to participants in the workshop series that those behaviours and actions that they considered normal and harmless, might be experienced entirely different by others. In other words, the adage that 'no man is an island' became increasingly self-evident. When they became aware of the structural problems and the limitations they imposed, their willingness to challenge prevailing power structures increased.

While the workshops had the effect of exposing blind spots associated with white male privilege, shifting attention away from blaming and shaming the men made it easier to move towards more attentive and ecologically attuned sentiments and actions. It facilitated an openness where criticism could be constructively considered, not as a threat, but as an opportunity for self-development towards better social and ecological impact individually and societally. This mental shift was transformational and a vital ingredient in fostering accountability for personal actions, awakening the courage to question those who sustain an abusive culture and minimising the risks of future harmful social and environmental actions.

> But then it also helped me … the first male reflective group I joined last fall. It helped me a lot … the step from the relationship with people, with take-off points in #Metoo, to the relationship with nature. That it wasn't just intellectual, but rather … I just felt that I wanted to act in other ways. I stopped flying because it felt right, more than it was an intellectual decision. … It helps me a lot that it doesn't feel like a sacrifice not to fly or not to watch porn for example, but it just feels … right. (Interviewee 1, our translation)
>
> (Hedenqvist, 2020, p. 38)

Most of the participants interviewed perceived themselves as conscious about issues of gender and the environment. A common assertion after completing the workshop series suggested that accumulating more *factual* knowledge about the climate crisis was not a priority, but rather emphasises the importance of developing emotional literacy and relational connectivity with the other participants. This increased the 'contact awareness' (as referred to by one interviewee) and enhanced participants' abilities to *practise* their knowledge in everyday life, particularly with their loved ones. This represented a starting point from which the participants could then develop deeper emotional bonds with the Earth, others and themselves, motivating the participants to move away from destructive practices as a virtuous act rather than as a sacrifice. For example, awareness about heavy pollution associated with flying has become wider knowledge, but it was not until this knowledge became grounded in emotional connections to the environment that the pledge to never fly again could be made by some of the men who participated. Similarly, awareness existed about the domestic and societal violence and abuse associated with pornography but consumption and support for that industry had become a habit for some and this only subsided once emotional connectivity was forged in the self and with the group. Thus, building proximity with others (rather than dwelling in isolation) made it easier to break with destructive, privileged patterns and truly act according to values considered soul-serving rather than soul-destroying. Hence, to go from an isolated and intellectualised approach to building mutual and emotional ties to the Earth, others and the self, accelerated the development of 'embodied knowledge' and from that the implementation of caring behaviours in everyday life. To feel relational (as opposed to autonomous) became synonymous with feeling included, connected and part of something bigger than the self, making it harder to 'otherise' and dominate and easier to care.

(Re)connecting with nature

Seemingly having improved the relational and emotional comprehension of participants, the workshop series served as an effective counterpoint to the myopic care and masculinist sentiments of industrial/breadwinner masculinities. Most participants expressed increased enthusiasm and action towards (re)connections with human and other-than-human others. This included reduced support for the physical and emotional distancing and alienation promoted by globalised markets and global ways of being (e.g. worldwide commodity chains, globetrotting, etc.) accompanied by a reclaimed and localised sense of place. One participant described how he discovered this underlying urge to be in the forest – a missing piece from his childhood that he had lost contact with – which has inspired him to go into the woods as often as possible:

> the sessions have probably made me realise just how much nature means to me … I grew up in the woods basically. But I came to the insight that I live so far away nowadays. I live so far away from nature. I'm never in the forest anymore. And it became very powerful for me that I feel a longing for those parts. So it's had the effect that I now on a principally daily basis look for the opportunities to go outdoors. … And that, in turn, brings about more … that you might want to live even more consciously than I did maybe. Yeah.

> You think twice. In the small things. I don't take the car short distances. I
> try to go by public transport. A little more in that direction. And then also
> like I began to do with gender equality, to raise the issue in my everyday life
> and question those in my near surroundings. (Interviewee 7, our translation)
>
> (Hedenqvist, 2020, p. 46)

Apart from bringing him more energy, he proclaimed that the strengthened physical relationship with the outdoors gave him an increased sense of engagement
with pro-environmental behaviour; to adopt continuous personal and interpersonal engagements that reduced ecological footprints and influenced others to
do the same. He also stated that he used to think all problems were other men's
fault, which stopped him from seeing his role in patriarchal structures, awakening
personal accountability and responsibility, which he had not experienced before.
Realising that he, too, must actively work against these structures in order to
change things – not just point at others – became a pivotal motivator to become
part of a core team at his male-dominated workplace to promote gender equality.

Another participant asserted how he previously sought personal and professional affirmations by being conscious and enacting what he believed to be progressive masculine traits. Through participation in the workshop series, he noted
that personal appearances were previously overriding his deeper understanding
and support for feminist and environmental progress on the structural level.
This kind of superficial activism was registered as similar to ecomodern business
strategies where more resources are put on 'pink-' and 'green-washing' than on
contributing to authentic structural change. Think, for example, of how international apparel companies brand their products as sustainable and pay feminist
celebrities to promote them, while having women and girls in South East Asian
sweatshops making them for low wages under dreadful conditions that also pol-
lute the surrounding environment with chemical discharge (ITGLWF, 2011). He
noted that being open, vulnerable and self-reflexive developed his awareness of
the differences between shallow and deep activism, and with this insight, enabled
him to focus on actual change – such as practising a gender-equal relationship
and parenthood with his partner.

> It has definitely been a start to … a work, a process and maybe above all, a
> spiritual part of the environment. Like it also was a lot about … like con-
> necting, getting in touch, feeling a birch or something. Finding something
> … that we also belong together in some way. More gratitude about every-
> thing I get … that I eat less meat than I did before since it just feels right. But
> if I do, not always, but quite often, I can think that this has been a cow or a
> deer and send a small thought of thankfulness that I get to eat a little piece.
> (Interviewee 1, our translation)
>
> (Hedenqvist, 2020, p. 45)

Other participants inferred an elevated concern for (other-than-human) animal
welfare. This awareness brought with it a greater inclination to acknowledge and

bless the other-than-human animal on the plate, reducing meat and dairy consumption, as well as a desire to become self-sufficient where meat was still produced by developing skills in animal husbandry in order to raise the animals at home, under less cruel conditions than in industrial farming. These expressions indicate a further relational viewpoint where the connection to – and care for – other-than-human animals increased and represented a first step towards recognising their intrinsic value. However, a reticence towards veganism and plant-based diets was noted, suggesting that patterns of masculinist entitlement to the 'otherised' bodies of other-than-human animals persisted in this instance (cf. Aavik, in press). A less anthropocentric perspective was held by one participant who promoted the rights of other-than-human animals as well as plants, after having claimed that the lack of biodiversity visible in monocultures implicates a kind of suffering for them. The small-scale (g)local community was, nonetheless, viewed as a place to challenge alienation and social and ecological exploitation through elevated engagement in meaningful relations with the environment. The closer relationality and emotional bond to nature (animals included) were believed to facilitate caring behaviours as the impacts of one's actions become increasingly visible and relatable. It was highlighted that such integrated (g)localised living would be simpler but also healthier and more fun than the idealised but de-animated (Global Northern) lifestyles and conditions that are revered today.

Ultimately, the way in which these men narrated their experiences at Under Tallarna, the global ecological crisis and their role in it, positions social and environmental justice as ideal. Their narratives indicate that pro-feminist and pro-environmental reflective groups can be used to strengthen the relational and emotional bonds to nature; bonds that facilitate a sense of accountability characterised by a heightened sensibility towards the surroundings that guides a better ecological and social impact. We conclude that the education considered here took participants some way to what Latour (2014) inferred is a purposeful life in the epoch of the (m)Anthropocene, where:

> all agents share the same shape-changing destiny. A destiny that cannot be followed, documented, told, and represented by using any of the older traits associated with subjectivity or objectivity. Far from trying to 'reconcile' or 'combine' nature and society, the task, the crucial political task, is on the contrary to distribute agency as far and in as differentiated a way as possible.
> (Latour, 2014, p. 17)

At least until transformational experiences, such as these workshops, enable us to transcend the object/subject relationship. In this particular case study, the participants shared a desire to support the rise of such Earth-inclusive approaches to lives of action that help reconcile the root causes of social injustices and ecological degradation towards truly sustainable (ecologised) thinking. This goes beyond (ecomodern) end-of-pipe solutions that temporarily alleviate guilt through deflection and keep organised and structural change in the space of reform. Their (perceived) personal changes counteract the alienation from, and abuse of, marginalised (eco)systems, groups and beings – that so commonly is encouraged by industrial/breadwinner masculinities – and enable the institutional transition towards norms and practices in favour of broader, deeper and wider care for our glocal commons.

Conclusion

The devastating social and ecological crises of the (m)Anthropocene have exposed industrial/breadwinner masculinities as heavily influencing the perpetuation of a world order based on domination over our glocal commons. Western masculine socialisations are deeply tied to values and identities aligning with industrial modernisation and the extractivist global political economy. There is, thus, a vital and urgent need to reconsider the ways that men and masculinities are constructed away from the reification of hegemonic patriarchal domination towards broader, deeper and wider care. Building on ecofeminist insights, the conceptualisations of ecological masculinities have emerged as one way to facilitate such a transition. One of the pluralised ways in which this concept seeks practical change is through an educational material and workshop series for male reflective groups that has been piloted in Järna, Sweden, which we have considered in this chapter.

Interviews with participants revealed that they experienced personal changes towards more caring values and behaviours as a result of the workshop series. Some common themes crystallised as increased outcomes of the education: interconnectivity, accountability, emotional literacy and embodied knowledge. The interviewees asserted that they were conscious about issues of gender and environment even before joining the groups, and that it was one of the reasons for participating, but that it often felt difficult to act in line with their awareness. To access suppressed feelings about personal, familial, social and ecological crises and listen to other people's stories in a safe space seem to have had a positive effect on their emotional and relational capabilities. Several participants implied achieving a heightened sense of emotional literacy and interconnectivity (how everyone/everything is connected, and constitutive of, nature), which made it feel good and less of a sacrifice to engage in pro-feminist and pro-environmental behaviours and act according to their awareness. Hence, emotional and relational literacies were necessary additions to the knowledge/factual comprehension in promoting empowerment and personal change in the first instance and to, in turn, foster challenge to the deep structures of patriarchy, including a shift from entitlement to care and collective responsibility. To transcend the rationality of the mind and the emotionality of the body can be described as developing 'embodied knowledge' that facilitates accountable actions towards more caring ecological and social impacts. While the following outcomes are not assured, the participants did indicate these personal changes from the reflective group workshop series: increased engagements for gender equality at home and work; more attentiveness to the lives of human and other-than-human others; developed closer bonds with nature, both emotionally and physically; made commitments to stop flying; reduced meat and dairy consumption; larger willingness to organise.

Through the reflective groups the process of masculine ecologisation, as constructed by these men, reinforced the personalised and structural transitions towards norms and practices aligning with social and environmental care. In this sense, what the participants shared in the interviews after they had been part of the reflective groups may offer a glimpse of hope that men's care can be extended to glocal environmental issues, they can change, and the gendered aspect of our social and ecological problems is transformable. The study did have notable limitations. Specifically, the education was tested with progressive men who, although clearly in need of overcoming passive awareness, were already receptive to visions of a gender-equal and environmentally caring world. In this sense, they represented a favourably biased test group. Consequently, future research could focus on how other, less progressive groups of men experience the *Men in the Climate Crisis* workshop series, along with mixed groups, women only and non-binary only groups as well. Developments in alignment with these recommendations are underway. Version two of the education material is currently in development. This development of the education material in the direction of broader gender inclusivity will provide fresh opportunities to study the ways that women, genderqueer people and mixed groups experience this education material and workshop series. This level of diversification offers rich ground for further theoretical and methodological testing of the education materials and its efficacy in awakening broader, deeper and wider care, not only in men, but throughout the masculinities that dwell in us all. In support of this development, future studies would benefit from pre- and post-participant interviews, in order to reach beyond the self-perceived changes of participants, to include codable metrics, and with those, quantitative analyses.

Acknowledgement

We would like to express our deep gratitude to Annica Kronsell, Gunnhildur Lily Magnusdottir and Karen Morrow for offering us treasured feedback on the draft version of this text. Catharina Landström and the departmental faculty and students at Chalmers University's Science, Technology and Society also graciously offered detailed feedback on a seminar version of this paper, which we are deeply thankful for. We also wish to praise the anonymous participants who were interviewed in this study. We additionally note that this work was generously supported by Chalmers Gender Initiative for Excellence (Genie).

Notes

1 Masculine hegemonisation is used here to indicate the active process of reifying select masculinities norms, such as insertion into spaces, being competitive, righteous, violent and dominant as ways of playing win–lose games with the planet and people.
2 Ecologisation is used to refer to the process of prioritising a relational approach to the Earth, others and the self, that honours the intrinsic value of all life.

3 Earth is intentionally capitalised as an acknowledgement of the sentience of the planet as a single, self-regulating organism, aka. Gaia (Lovelock, 1979).
4 'Education material' refers to an eight-part workshop process, developed by Hultman and Pulé at Chalmers University, with Vetterfalk from the feminist non-profit organisation MÄN: Men for Gender Equality, founded in Sweden in 1993. The version of the material discussed here focuses on information and embodiment practices customised for men's behaviour and values changes away from violence and towards greater care for the Earth, others and the self.
5 We use 'industrial/breadwinner masculinities' close to the concept of breadwinner masculinity often used in feminist studies (Stacey, 1990), to supersede the more familiar, but older, term 'hyper-masculinity' that is defined as the amplification of traditional expressions of masculinities through physical strength, aggression, sexual prowess, military might, domination, being a 'winner', and not being a 'girl' (Kivel, 1999; hooks, 2004; Katz, 2006, 2012, 2016; Hultman and Pulé, 2018). We do so to emphasise the entangled relationship between male domination and industrialisation, and the social and environmental consequences which have proved to be severe. This categorisation represents the most obtuse of 'unmarked' (i.e. normalised) masculinities, the social and environmental impacts of which are obscured by the systems that created and continue to maintain them (Barthes, 1967, p. 77; Reeser, 2010, pp. 8–9; MacGregor and Seymour, 2017).
6 Global emissions are taken here to be the principle anthropocentric effluent of modernity.
7 Consider that a decade ago, the extractive industry response to a global financial crisis that resulted in a big drop in the use of coal, gas and oil, and the lowering greenhouse emissions was to usher in renewed and accelerated research, development and exploitation of natural resources. In the wake of recent economic crises brought on by COVID-19, the reflex of global regulators has been to prop up international financial mechanisms as the primary line of defence against severe economic collapse once again (Quéré et al., 2020).
8 Consider the political intonations along with social and ecological implications of Donald Trump in the US, Jair Bolsonaro in Brazil, Narendra Modi in India, Viktor Orbán in Hungary, Benjamin Netanyahu in Israel, Vladimir Putin in Russia, Recep Tayyip Erdoğan in Turkey and more, while these individuals were, or remain, in power (Kyle and Gultchin, 2020).
9 For detailed instructions and methods on the #AfterMeToo conversation guide, see: MÄN (2018).
10 For detailed instructions and methods on the Men in the Climate Crisis conversation guide, see: MÄN (2020).
11 Important inspiration and input has been received from Hultman and Pulé (2018), Macy and Johnstone (2012), Swedish ecofeminist Karin Styrén, gendered sustainability researcher Pernilla Hagbert, Transition Network and Earth Rights activist Pella Thiel, Transitions journalist Abigail Sykes and climate justice educators Gustaf Sörnmo and Frida Ekerlund from the organisation Vardagens Civilkurage (Everyday Moral Courage).

References

Aavik, K. (in press) 'Vegan Men: Towards Greater Care for (Non)human Others, Earth, and Self', in P.M. Pulé and M. Hultman (eds.) *Men, Masculinities and Earth: Contending with the (m)Anthropocene.* Camden: Palgrave Macmillan.
Allen, M.R., Dube, O.P., Solecki, W., Aragón-Durand, F., Cramer, W., Humphreys, S., Kainuma, M., Kala, J., Mahowald, N., Mulugetta, Y., Perez, R., Wairiu, M. and

Zickfeld, K. (2018) 'Framing and context', in V. Masson-Delmotte, P. Zhai, H.-O. Pörtner et al. (eds.) *Global Warming of 1.5°C: An IPCC Special Report on the Impacts of Global Warming of 1.5°C Above Pre-Industrial Levels and Related Global Greenhouse Gas Emission Pathways, in the Context of Strengthening the Global Response to the Threat of Climate Change, Sustainable Development, and Efforts to Eradicate Poverty.* Geneva: IPCC, pp. 47–91.

Anshelm, J. and Hultman, M. (2014) 'A green fatwā? climate change as a threat to the masculinity of industrial modernity', *NORMA: International Journal for Masculinity Studies*, 9(2), pp. 84–96.

Arnoldy, B. (2019, September 23) 'Greta and 15 kids just claimed their climate rights at the UN', [Article], *Earthjustice* [online]. Available at: https://earthjustice.org/blog/2019-september/greta-thunberg-young-people-petition-UN-human-rights-climate-change (Accessed: 13 October 2020).

Barthes, R. (1967) *Elements of Semiology*. New York: Hill and Wang.

Brundtland, G. (1987) *Our Common Future: Report of the World Commission on Environment and Development*. Oxford: Oxford University.

Buckingham, S. and Le Masson, V. (eds.) (2017) *Understanding Climate Change Through Gender Relations: Routledge Studies in Hazards, Disasters and Climate Change*. Oxon: Routledge.

Crowe, K. (2019, September 27) 'How 'organized climate change denial' shapes public opinion on global warming', [Article], *CBC News* [online]. Available at: https://www.cbc.ca/news/technology/climate-change-denial-fossil-fuel-think-tank-sceptic-misinformation-1.5297236 (Accessed: 12 October 2020).

Daggett, C. (2018) 'Petro-masculinity: fossil fuels and authoritarian desire', *Millennium: Journal of International Studies*, 47(1), pp. 24–44.

Demelle, B. (2019) Top 10 climate deniers [Home Page]. *Before the Flood* [online]. Available at: https://www.beforetheflood.com/explore/the-deniers/top-10-climate-deniers/ (Accessed: 13 October 2020).

Di Chiro, G. (2017) 'Welcome to the white (m)anthropocene? A feminist-environmentalist critique', in S. MacGregor (ed.) *Routledge Handbook of Gender and Environment*. Oxon: Routledge, pp. 487–507.

Dunlap, R., McCright, A.M. and Yarosh, J.H. (2016) 'The political divide on climate change: partisan polarization widens in the US', *Environment: Science and Policy for Sustainable Development*, 58(5), pp. 4–23.

Forchtner, B., Kroneder, A. and Wetzel, D. (2018) 'Being skeptical? exploring far-right climate change communication in Germany', *Environmental Communication*, 12(5), pp. 589–604.

Gaard, G. (2014) 'Towards new ecomasculinities, ecogenders, and ecosexualities', in C. Adams and L. Gruen (eds.) *Ecofeminism: Feminist Intersections with Other Animals and the Earth*. New York: Bloomsbury, pp. 225–240.

GeMiNI (no date) *About GeMiNI.* Available at: https://www.facebook.com/pg/Gender-and-Environment-Network-Sweden-Gemini-106487417379196/about/?ref=page_internal (Accessed: 2 March 2020).

Hedenqvist, R. (2020) *Exploring Ecological Masculinities Praxes: A Qualitative Study of Global Northern Men Who Have Participated in Pro-Feminist and Pro-Environmental Reflective Groups.* Master's Thesis. Stockholm University. Available at: hhttps://drive.google.com/file/d/1SYk4akHlZsLj5Qx9WepOl2G658vEvuQY/view?usp=sharing

Holland, J. (2014, May 16) 'Eight Pseudoscientific Climate Claims Debunked by Real Scientists', [Article], *Moyers on Democracy*. Available at: https://billmoyers.com/2014

/05/16/eight-pseudo-scientific-climate-claims-debunked-by-real-scientists/ (Accessed: 13 October 2020).

hooks, b. (2004) *We Real Cool: Black Men and Masculinity*. New York: Routledge.

Hultman, M. (2013) 'The making of an environmental hero: a history of ecomodern masculinity, fuel cells and Arnold Schwarzenegger', *Environmental Humanities*, 2(1), pp. 79–99.

Hultman, M. and Pulé, P.M. (2018) *Ecological Masculinities*. London: Routledge.

Hultman, M. and Pulé, P.M. (2019) 'Ecological masculinities: a response to the manthropocene question?', in L. Gottzén, U. Mellström and T. Shefer (eds.) *Routledge International Handbook of Masculinity Studies*. Oxon: Routledge, pp. 477–487.

Hultman, M., Björk, A. and Viinikka, T. (2019) 'The far right and climate change denial: denouncing environmental challenges via anti-establishment rhetoric, marketing of doubts, industrial/breadwinner masculinities enactments and ethno-nationalism', in B. Forchtner (ed.) *The Far Right and the Environment*. London: Routledge, pp. 121–135.

ITGLWF (The International Textile, Garment and Leather Workers' Federation) (2011) *An Overview of Working Conditions in Sportswear Factories in Indonesia, Sri Lanka and the Philippines* [Report]. ITGLWF Report.

Kaiser, J. and Puschmann, C. (2017) 'Alliance of antagonism: counterpublics and polarization in online climate change communication', *Communication and the Public*, 2(4), pp. 371–387.

Kato-Wallace, J., van der Gaag, N., Barker, G., Santos, S., Doyle, K., Vetterfalk, V., Pisklakova-Parker, M., van de Sand, J. and Belbase, L. (2015) Men, masculinities and climate change: a discussion paper [Discussion Paper]. *MenEngage Alliance*. Available at: http://menengage.org/wp-content/uploads/2016/04/Men-Masculinities-and-Climate-Change-FINAL.pdf (Accessed: 12 October 2020).

Katz, J. (2006) *The Macho Paradox: Why Some Men Hurt Women and How All Men Can Help*. Naperville, IL: Sourcebooks.

Katz, J. (2012) 'Violence against women—it's a men's issue', [Video file], *TedxFiDiWomen*, November. Available at: http://www.ted.com/speakers/jackson_katz (Accessed: 12 October 2020).

Katz, J. (2016) *Man Enough?: Donald Trump, Hillary Clinton, and the Politics of Presidential Masculinity*. Northampton: Interlink.

Kivel, P. (1999) *Boys Will Be Men: Raising Our Sons for Courage, Caring and Community*. Gabriola Island: New Society Publishers.

Kyle, J. and Gultchin, L. (2020) 'Populists in Power Around the World', [Report], *Tony Blair Institute For Global Change*. Available at: https://institute.global/policy/populists-power-around-world (Accessed: 11 October 2020).

Latour, B. (2014) 'Agency at the time of the anthropocene', *New Literary History*, 45(1), pp. 1–18.

Lindvall, D., Vowles, K. and Hultman, M. (2020) *Upphettning: Demokratin i klimatkrisens tid*. Stockholm: Fri Tanke förlag.

Lovelock, J. (1979) *Gaia: A New Look at Life on Earth*. Oxford: Oxford University Press.

MacGregor, S. (2009) 'A stranger silence still: the need for feminist social research on climate change', *Sociological Review*, 57(2), pp. 124–140.

MacGregor, S. and Seymour, N. (eds.) (2017) *Men and Nature: Hegemonic Masculinities and Environmental Change*. Munich: RCC Perspectives.

Macy, J. and Johnstone, C. (2012) *Active Hope: How to face the mess we're in without going crazy*. Novato, CA: New World Library.

McCall, R. (2019, September 27) 'Why climate deniers hate activists so much: guilt', [Article], *Newsweek*. Available at: https://www.newsweek.com/climate-change-denia lists-hate-activists-vulnerability-1461543 (Accessed: 12 October 2020).

Mellström, U. (1995) *Engineering Lives: Technology, Time and Space in a Male-Centred World*. PhD dissertation. Linköping University.

Merchant, C. (1980) *The Death of Nature: Women, Ecology and the Scientific Revolution*. New York: HarperCollins.

MÄN (2018) *#AfterMeToo—Reflective Groups for Men*. Available at: https://mfj.se/assets/ documents/english/reflective-groups-guide-man.pdf (Accessed: 2 March 2020).

MÄN (2019) *Attachment to Motion 6*. Available at: https://mfj.se/assets/uploads/2019/04/ motion-6.pdf (Accessed: 2 March 2020).

MÄN (2020) *Conversation Guide: MÄN in the Climate Crisis*. Available at: https://mfj.se/ assets/documents/english/Men-in-the-climate-crisis-(prototype).pdf (Accessed: 2 June 2020).

MÄNa (no date) *About MÄN*. Available at: https://mfj.se/en/home (Accessed: 2 March 2020).

MÄNb (no date) *Our Partnerships*. Available at: https://mfj.se/en/our-partnerships (Accessed: 2 March 2020).

Pease, B. (2019). Recreating Men's Relationship with Nature: Toward a profeminist environmentalism. *Men and Masculinities*, 22(1), 113–123.

Pulé, P.M. and Hultman, M. (eds.) (in press) *Men, Masculinities and Earth: Contending with the (m)Anthropocene*. Camden: Palgrave Macmillan.

Quéré, C.L., Jackson, R.B., Jones, M.W., Smith, A.J.P., Abernethy, S., Andrew, R.W., De-Gol, A.J., Willis, D.R., Shan, Y., Canadell, J.G., Friedlingstein, P., Creutzig, F. and Peters, G.P. (2020) 'Temporary reduction in daily global CO2 emissions during the COVID-19 forced confinement', *Nature*, 10, pp. 647–653.

Raworth, K. (2014, October 20) 'Must the Anthropocene be a Manthropocene?', [Article], *The Guardian* [online]. Available at: https://www.theguardian.com/commentisfree/201 4/oct/20/anthropocene-working-group-science-gender-bias (Accessed: 12 October 2020).

Reeser, T. (2010) *Masculinities in Theory*. Malden, MA: Wiley-Blackwell.

Ruether, R.R. (1975) *New Women New Earth: Sexist Ideologies and Human Liberation*. Minneapolis, MN: Winston Press.

Sayej, N. (2017, December 1) 'Alyssa Milano on the #MeToo movement: 'We're not going to stand for it any more'', [Article], *The Guardian* [online]. Available at: https:// www.theguardian.com/culture/2017/dec/01/alyssa-milano-mee-too-sexual-harassment -abuse (Accessed: 12 October 2020).

Stacey, J. (1990) *Brave New Families: Stories of Domestic Upheaval in Late-Twentieth-Century America*. Berkeley: University of California Press.

The Climate Reality Project (2019, September 5) 'The climate denial machine: how the fossil fuel industry blocks climate action', [Article], *Climate Reality Leadership Corps* [online]. Available at: https://climaterealityproject.org/blog/climate-denial-machine -how-fossil-fuel-industry-blocks-climate-action (Accessed: 13 October 2020).

The Everyday Sexism Project. (no date) *The Everyday Sexism Project*. Available at: https:// everydaysexism.com (Accessed: 7 October 2020).

Thunberg, G., Neubauer, L., De Wever, A. and Charlier, A. (2020, August 19) 'After two years of school strikes, the world is still in a state of climate crisis denial', [Article], *The Guardian* [online]. Available at: https://www.theguardian.com/commentisfree/202 0/aug/19/climate-crisis-leaders-greta-thunberg (Accessed: 11 October 2020).

13 Pathways for inclusive wildfire response and adaptation in northern Saskatchewan

Heidi Walker, Maureen G. Reed, and Amber J. Fletcher

Introduction

Extreme climate-related events are often stimuli for the development of disaster risk reduction and climate change adaptation strategies (IPCC, 2012; Wise et al., 2014). Developing these strategies is complex, involving multiple actors and institutions across scales and jurisdictions that may have differing priorities, objectives and decision-making authorities. While scientific-technical approaches that reduce the proximate causes of risk often dominate these processes, scholars and practitioners are increasingly recognising the need for transformational change to address the structural and systemic causes of risk and vulnerability (Nalau and Handmer, 2015; O'Brien et al., 2015; Pelling et al., 2015; Heikkinen et al., 2018).

Responding to this need, 'adaptation pathways' approaches examine

> whether systemic change is needed and the role of incremental adaptation in achieving this; and raising awareness and understanding of the interplay between knowledge, values, power and agency to inform responses to change, particularly in dynamic, complex and contested contexts.
>
> (Wise et al., 2014, p. 327)

Pathways approaches recognise that past responses to hazards shape current and future trajectories. Such responses are also imbued with power relations, which affect who and what is represented in adaptation processes (Fazey et al., 2016), and also *how* they are represented (Cox et al., 2008). Feminist scholarship on climate change impacts and adaptation, for example, recognises that institutional policies and practices are shaped by power relationships associated with gender, class and race (Kaijser and Kronsell, 2014; Ravera et al., 2016). These relationships enable and constrain the availability of certain adaptation pathways and result in different outcomes and experiences across social groups.

In this chapter, we apply insights from literature on adaptation pathways and feminist literature on climate change to examine how past responses to a major wildfire event in a jurisdictionally complex region of northern Saskatchewan, Canada, inform current and future adaptation processes and trajectories. Drawing

DOI: 10.4324/9781003052821-13

from interviews with community residents and government representatives, we examine how emergency and wildfire management institutions facilitated or constrained pathways for adaptation and how these were shaped by dominant knowledges and values. Ultimately, we consider the implications of these predominant pathways for diverse social groups and whether a transformation is required for building inclusive adaptation processes as communities continue to live with fire in the future.

Adaptation pathways and the role of institutions

Institutions are an important site for understanding how the development of various adaptation pathways are constrained or enabled (Pelling et al., 2015). In this chapter, we define institutions as the '"rules of the game" – the rules, norms and practices – that structure political, social and economic life' (Chappell and Waylen, 2013, p. 599). Formal institutions include policies, programmes, regulations and authorities which are influenced by informal institutions such as sociocultural norms, values and knowledge systems (Reed et al., 2014). Institutions for hazard response and adaptation are shaped by historical processes and power relations, resulting in institutional inertia that may constrain policy and practice (Burch et al., 2014; Clarke et al., 2016). Furthermore, it is important to critically examine the informal relations that underpin formal institutional practice in order to recognise whether and what fundamental shifts are needed to achieve desirable futures (O'Brien, 2012).

Gender is an institutionalised system of power that, in turn, shapes other institutions (Chappell and Waylen, 2013). Informal institutions are gendered, and these gendered norms and expectations become embedded within formal institutions for climate change mitigation and adaptation (Ravera et al., 2016). For example, the dominant framing of climate change as a technical, scientific and security problem is underpinned by masculine norms, which has meant that solutions have often been relegated to traditionally male-dominated sectors, such as public safety and infrastructure (Macgregor, 2010). Accordingly, responses and adaptation to climate hazards have centred on strategies to reduce physical exposure and structural damage, which are led by male-dominated professions and activities (e.g., engineered solutions, firefighting) (Enarson, 2016).

Feminist intersectionality theory recognises that gender interacts with other axes of identity and power, such as race, ethnicity and socio-economic status, to influence climate change institutions (Kaijser and Kronsell, 2014). Pease (2016) noted that the masculinised nature of climate change response intersects with Western rationalist and instrumentalist ideologies, resulting in hierarchical, rule-based procedures based on external 'expertise' (e.g., professional firefighting authorities) over local knowledges and priorities (Paveglio et al., 2015). As a consequence, the contributions, values and knowledge of people who fall outside the norms of 'external experts', such as women and Indigenous people, are often excluded.

Gendered and culturally hegemonic wildfire institutions have very real, but often difficult-to-see, influences on adaptation trajectories. For example, Tyler and Fairbrother (2013, p. 113) pointed out that wildfire response, policy and practice in Australia are typically 'assumed to have emanated from objective and empirical, if not scientific, bases but this does not take into account the fact that emergency management, bushfire response and fire-fighting, remain overwhelmingly male-dominated areas'. They found that where national policy promoted the household decision between early evacuation and 'stay-and-defend', men were far more likely than women to stay-and-defend–largely due to pressure to conform to dominant masculine norms and expectations. Early evacuation, by contrast, was viewed as a weaker, feminised response. Interestingly, Tyler (2019) observed that despite greater fatality rates among men during past wildfire events in Australia, the 'extremely gendered' nature of wildfire management institutions has meant a continued emphasis on masculinised 'fight' approaches to wildfire preparedness and planning, with less attention on preparedness for early evacuation. The study noted community engagement programmes geared towards empowering women to stay-and-defend. However, such programmes are based on the assumption that women either lack the knowledge of wildfire preparedness or are more amenable than men to behaviour change (Tyler, 2019), highlighting how masculine norms underpin formal institutions, creating an institutional 'stickiness' that reinforces certain development pathways (Mackay et al., 2010). Programmes promoting stay-and-defend presume an ideal solution, which is based on masculine practices. Importantly, such dynamics are context-specific. For example, in countries like Canada where mandatory evacuation is the preferred policy response to major wildfire events, gendered institutional dynamics likely operate in different ways.

The adaptation pathways approach identifies key decisions and interventions over time, along with the informal institutions that underpin these processes, to understand how various adaptation pathways are enabled or constrained (Pelling et al., 2015; Fazey et al., 2016; Heikkinen et al., 2018)[1] (Table 13.1). Adaptation pathways are analysed by the extent of change including resistance, incremental adjustment, or transformation. Resistance is characterised by 'business-as-usual' approaches to adaptation, often involving additional investment in existing infrastructure, social institutions and/or economic flows. Incremental change refers to marginal shifts in these same areas, which enhances flexibility without causing major systemic disruption. While there is still significant debate in climate change scholarship about what is meant by transformation, it can be broadly characterised as a fundamental change to societal structures and power relations (O'Brien, 2012; Heikkinen et al., 2018). Resistance and incremental adaptation are far more common than transformational strategies that address underlying social causes of vulnerability (Wise et al., 2014; Heikkinen et al., 2018).

Dominant adaptation pathways may be identified through the priorities for addressing risk (as well as the framing of risk), the types of knowledge that are centred, and the extent to which existing power relations are maintained or challenged (Table 13.1). Resistance is characterised by a reliance on existing practices, centralised knowledge and the maintenance of existing power structures. Incremental change focuses on the proximate causes of risk, professional

Table 13.1 Characteristics of adaptation pathways (based on Nalau and Handmer, 2015; Pelling et al., 2015; Heikkinen et al., 2018)

	Resistance	Incremental change	Transformation
Description	Follows existing development pathways; enhances investment in existing infrastructure, social institutions and/or economic flows	Marginal shifts to infrastructure, social institutions and/or economic flows	Fundamental changes to existing societal structures
Priorities for addressing risk	Risk viewed as routine, relies on existing methods and practice; resources devoted to maintaining status quo and system stability	Risk viewed as non-routine, acknowledges need for improvement to existing practice; addresses proximate causes of risk and vulnerability (e.g. physical infrastructure)	Risk viewed as complex and unbounded; addresses structural and root causes of risk and vulnerability (e.g. power hierarchies, social and economic relationships)
Knowledge	Conventional; centralized control over information	Professional and specialized expertise; may include some participatory elements	Trans-disciplinary, whole-of-society response; highly participatory
Power relations	Existing structures maintained	Existing structures maintained; new issues may be introduced	Fundamental changes to existing structures
Advantages	Provides stability and certainty	Enhances system diversity and flexibility and can incrementally open new policy options without major systemic disruption	Opens up new policy options; re-orients development pathways towards sustainability and social justice
Disadvantages	Reduces flexibility over time; adopts narrow options and worldviews	Prioritizes functional persistence; does not address underlying institutions that contribute to systemic vulnerability	May introduce unexpected outcomes and instability as systems reach new equilibria; requires constant monitoring, reflection and renewal

expertise and maintenance of existing power structures, though it may include some new issues and participatory elements. Transformation is characterised by an emphasis on the root causes of risk and vulnerability, highly participatory approaches and major changes to existing power structures. The priorities for addressing risk, the types of knowledges recognised, and power relations shape, and are shaped by, formal and informal institutions for climate change adaptation. From the transformational perspective, pathways that reinforce existing forms of inequality and marginalisation are increasingly seen as undesirable and an indicator of the need for major structural change (e.g. Tschakert et al., 2013; O'Brien et al., 2015). Therefore, it is necessary to 'shed light on and problematise norms and underlying assumptions that are naturalised and regarded as common sense, but build on and reinforce social categorisations and structures of power, not least through institutional practices' (Kaijser and Kronsell, 2014, p. 428).

2015's wildfires in northern Saskatchewan

In 2015, prolonged dry conditions resulted in a rash of over 700 wildfires across the Canadian province's northern boreal forest region and the largest evacuation the province had ever experienced. The La Ronge region of northern Saskatchewan was one of the areas most impacted by the 2015 fires; it is also a place particularly notable for its jurisdictional complexity. Its population centre is a tri-community that includes a town (La Ronge), northern village (Air Ronge) and First Nation (Lac La Ronge Indian Band [LLRIB]), with a total population of approximately 6,000. While the three communities within the population centre are politically distinct, they are to some extent socially integrated and share some key services. In addition to its main population in the tri-community, the LLRIB also comprises five smaller satellite communities with a total population of over 3,000.

The community leaders jointly issued a mandatory wildfire evacuation order in early July of 2015, spurring residents to evacuate on their own to stay with family and friends elsewhere or by bus to urban evacuation centres. Four of the five LLRIB satellite communities were also either partially or fully evacuated, as were nearby unincorporated subdivisions and lakeside cabins. Once the evacuation was called, in addition to local governments and organisations, provincial and federal government departments and agencies, the national police service, the Canadian Army, municipal firefighting agencies and non-government organisations (e.g. the Red Cross) became involved with the evacuation and emergency response. Residents spent anywhere between two weeks and one month outside of their community. Ultimately, about 100 cabins were destroyed in surrounding areas, but the tri-community itself experienced no major physical damage and no serious injuries were incurred by either residents or emergency responders.

Method

Our research was conducted three years after the fires. The lead author lived in the region for 16 months and conducted 44 semi-structured interviews with community members and local government representatives who had experienced, or been involved with, the wildfires. The interviewees were identified through established key community contacts, archived media reports and other relevant documents (e.g. local emergency plan). Ten interviews were with key local and provincial government and agency representatives, including from local governments (municipal, village), LLRIB government, local and First Nation agencies (Fire service, Indigenous social service agency) and provincial government agencies (e.g. wildfire management). These actors represent a unique viewpoint, as they were both local residents of the tri-community and key actors in response efforts during or after the wildfire event. Interviews focused on the roles, responsibilities and responses of the representatives' respective organisations during and after the wildfire event. The remaining 34 interviews took place with local residents from the tri-community and surrounding areas and focused on their experiences of the fires. Participants were selected based on stratification by location (jurisdiction of residence at the time of the fires). We also ensured the participation of women, men, Indigenous and non-Indigenous residents. Additional demographic characteristics, such as age, income and length of residency were collected through a participant checklist following each interview. Interviews were transcribed verbatim and coded using a primarily inductive approach aided by NVivo 12 qualitative data analysis software. Direct quotes that illustrate the findings indicate the speakers' gender and community of residence, but names and identifying information have been removed to maintain participants' confidentiality.

Looking to past wildfire responses to learn for future adaptation trajectories

At the local level, adaptation pathways in the La Ronge area appear to be predominately characterised by resistance and incremental change to infrastructure and institutions. During and after the wildfire event, the emphasis was on the proximate causes of risk and professional and specialised expertise was most heavily relied upon. Though some participatory elements and new issues, such as impacts to the local business community, were evident, existing institutional power relationships have largely remained intact and have resulted in the continued exclusion of certain voices, knowledges and values.

Priorities for addressing risk

The risk priorities for the La Ronge wildfires were characterised by elements of resistance and incremental change. The emphasis remained on the proximate causes of risk and vulnerability during and after the event, as evidenced by the framing of wildfire as an immediate threat and the prioritisation of physical infrastructure to mitigate future exposure to fire. During the fires, a local Emergency Operations Centre committee was convened, which was comprised of local government leaders and administrators with advising and supporting roles from other regional (e.g. the tri-community fire service), provincial (e.g. Wildfire Management, Emergency Management & Fire Safety, Transportation, Health, Social Services) and federal (e.g. police) jurisdictions. The committee appears to have adopted hierarchical structures and a focus on the immediate physical risks associated with the fires. In the words of one participant who was involved in the emergency operations:

> [Emergency Operation Centre meetings] are wonderful things to watch and listen to. People are quickly identified and everybody gives a brief report, but if somebody asks a question and says 'Can you give me help?', 'Okay, you two handle that on a sidebar [conversation] when we are done, but let's get through the common briefing'. So, it is well managed and I didn't feel there were obvious gaps. Now that is in terms of the emergency management.
>
> (male emergency operations representative)

As the participant continued, it became clearer that emergency management in this case was framed primarily as an issue of ensuring the health, safety and welfare of citizens as related to the *immediate* threat (i.e. the proximate cause of risk) of fire within the community:

> There were necessary sidebar conversations about social services support for the evacuees, which was a concern of the Lac La Ronge Indian Band that could and should be addressed in separate conversations. I think the Chief wanted some of that addressed at the big table, but we didn't need to all be there for that. … She has a valid point, but at the briefing I think raise the matter, identify who do I get help from and then afterwards get that help. My approach to it is much more of a command-and-control – we are here to share information and then go do my work.

From this perspective, social services and community supports are considered peripheral to 'big table' conversations.

Since 2015, additional measures have been taken to enhance the preparedness for, and mitigation of, wildfire risk within the tri-community area. When asked about adaptation measures that had occurred in and around their communities, participants most commonly referred to a significant uptick in fuel management projects. Fuel management is a core component of FireSmart, a nation-wide programme aiming to balance the 'natural role [of fire] with the protection of human life, property and economic values' (FireSmart Canada, 2020). It involves the thinning or removal of vegetation to mitigate the severity, intensity and spread rates of fire around communities. Many local residents and government representatives expressed enthusiasm for the additional fuel management projects that had been implemented, which is likely, to some extent, due to its successful application in the nearby unincorporated subdivision of Wadin Bay. Relatively few homes and cabins were lost in that community in 2015, which was largely attributed to the three fuel management treatments completed as part of its FireSmart planning efforts in the previous year (Ault et al., 2017).

> But the FireSmart program is great and I think it should do more.
>
> (woman, Air Ronge)

> they call it FireSmart, you know [LLRIB] actually started cleaning up in the trees a bit between the communities, which I think is good.
>
> (man, LLRIB)

As discussed later in this chapter, however, FireSmart was not uniformly supported by community members, some of whom questioned its effects on the landscape and associated values.

The prioritisation of the proximate causes of risk during the fires appeared to set the stage for the continued emphasis of these issues in the recovery and longer-term adaptation phases following the event, indicating an adaptation pathway characterised by resistance. The risk was narrowly defined through a professional emergency management discourse and subsequent adaptation efforts have been technical in nature. Through a feminist lens, such approaches value masculinised realms of expertise while community concerns are problematically sidelined (Pease, 2016). However, the increasing scale to which these wildfire mitigation initiatives are applied and somewhat expanding opportunities for public participation – as discussed in the following section – suggest elements of incremental change.

Knowledge

Professional and specialised knowledge was prioritised after the wildfire event, though as time progressed, there were some additional opportunities for public engagement in debriefing and wildfire mitigation efforts. For example, two local public engagement sessions occurred shortly after the event, which provided opportunities for debriefing between local residents and key decision-makers. These sessions focused primarily on the frontline fire operations, economic impacts of the fires and the role of the local business community. The town of La Ronge held the first session, which focused on providing information and answering questions about how fires had been handled within the community. Experts from sectors that played key roles in frontline firefighting efforts were present to provide additional information.

> We did a presentation on behalf of [the] mayor and council on what our response was and what we did and why we did certain things. There were also representatives from the fire department, wildfire management, conservation officers, not in an official capacity but there in case there was a question that we needed background information.
>
> (male municipal government representative)

The La Ronge and District Chamber of Commerce – an organisation that advocates for businesses in La Ronge, LLRIB, and Air Ronge – provided a second opportunity for resident engagement. The meeting addressed feelings that the local business community had been excluded from decision-making throughout the fire event, as the following resident indicated:

> One of my beefs is what I would consider businesses that had a role to play were shut down and kicked out of town. I think that is a huge mistake. I mean the first thing you do in a panic situation is try to surround yourself with people who can contribute and help you out. Kicking out the guy who owns the hardware store and has the parts, the rental store that rents water pumps – kicking those people out of town I would say actually hinders your efforts and doesn't have to.
> (male resident, La Ronge)

The meeting provided an opportunity for some community residents – in this case, local business owners – to express their concerns and provide suggestions for enhanced inclusion going forward. To some extent, recognition of these concerns and the need to maintain economic stability throughout a major event gained attention within government agencies.

> We have had some in the business community offer their insights.
> (male municipal government representative).

> One of the things we would like to do is involve our local businesses … so that they are not handcuffed when an evacuation does happen and they can still make a bit of a profit or keep their doors open when they come back to the community after the evacuation.
> (male provincial government representative).

As mentioned in the previous section, when the immediate urgency of the event faded and the communities began a return to normal life, action for local wildfire preparedness and adaptation predominately involved fuel treatment activities within the tri-community region. Two local municipal leaders pointed out that relatively little action had occurred across local jurisdictions since 2015, indicating that the region primarily relied on the knowledge and expertise of the provincial wildfire management agency for wildfire planning and development of adaptation strategies.

> I don't think we've sat down with the three communities and said this is what we're going to do next time this happens. So, I think that definitely needs to be done.
> (male municipal government representative)

> The main concern is that we are going to wait too long. It is kind of going to be like the Shakespearean tragedy where we are going to be sitting there

and we are going to be thinking about what we should do rather than actually taking some action items seriously and by the time the next fire rolls through, we don't have a solid plan. So that is definitely the number one concern that I have.

(male municipal government representative)

The strong existing institutional arrangements for FireSmart practice have enabled some opportunities for community participation. The FireSmart Communities Program, for instance, supports local residents in the development of a local FireSmart Board, a community protection plan and community FireSmart events (FireSmart Canada, 2020). Before the 2015 fires, the resort subdivision of Wadin Bay was the only community in the La Ronge region to hold FireSmart Community Recognition status. Spurred by the success of the programme in Wadin Bay, two other unincorporated subdivisions in the La Ronge region have also joined the FireSmart Communities Program since 2015, providing some opportunity for local participation in wildfire preparedness.

Power relations

While some participatory elements were evident, existing power dynamics were more or less maintained, resulting in the exclusion of some social groups, experiences and values during and after the event. Differentiated experiences at the intersection of location, culture and gender were apparent, as participants reported prioritisation of central communities over satellites, marginalisation of Indigenous and local knowledge and values and failure to acknowledge women's contributions.

During the fires, LLRIB was not only concerned about its main population centre within the tri-community, but also the four of its five outlying satellite communities that were partially or fully evacuated. Historical inequalities associated with colonisation in Canada have resulted in significant economic discrepancies between Euro-Canadian and Indigenous populations, especially those living on reserves (Wilson and McDonald, 2010), which meant that compared to La Ronge and Air Ronge, LLRIB played a larger role in organising transportation to evacuation centres and ensuring the welfare of its residents throughout the wildfire event. Frustration with the lack of attention to wellbeing – for example, 'where 3,000 people went and whether 3,000 people were okay and whether this mom who was nine months pregnant was sleeping on the floor' (female LLRIB government representative) – at local emergency operations committee meetings was clear. This participant elaborated to say:

The focus very much became La Ronge. We [LLRIB staff] were having a meeting in the morning with regards to our outlying communities, then we would go to [La Ronge] council chambers and have a meeting there with regards to an update on the tri-community, and then we would go to [the evacuation registration hub] and continue to do registrations and coordination and answer calls. Then there would be another meeting at about 6:00 pm with regards to what was happening with the outlying communities. So it got to be a lot of meetings, but what happened was that [La Ronge and Air Ronge] couldn't have cared less about what was happening in Grandmother's Bay, Hall Lake, or Stanley Mission – they just wanted to talk about what was happening in the tri-community. That is all that mattered to them. … It was very interesting to see our tri-community really separate and have different opinions on what was important and what wasn't important.

The dominant approach to wildfire management may, therefore, have prioritised the immediate threat of fire in the areas surrounding the main population centres, rather than the secondary impacts residents experienced after evacuation, which may have been particularly pronounced for many LLRIB residents from the satellite communities who relied on evacuation by bus and accommodation at large urban evacuation centres. While the prioritisation of response to the proximate causes of risk was successful insofar as no serious physical injuries were sustained, such models may deny local agency, knowledge and culturally relevant mechanisms for dealing with risk (Scharbach and Waldram, 2016). These models may also lead to the exclusion of certain communities – in this case, smaller Indigenous communities located further away from centres of decision-making.

The public engagement sessions after the event, to some extent, opened up new policy directions (e.g. consideration of impacts to the business community). However, the retrospective discussions about how the fires were handled within the community again revealed an institutional inertia that emphasised response to the immediate risk of fire and centring of the male-dominated sectors who responded to this risk (e.g. municipal and wildland firefighting agencies). Contributions such as cooking, cleaning, delivering meals, caring roles, rebuilding social fabric and responding to mental and emotional health impacts were most often performed by women, but often less recognised than the male-dominated sectors' 'frontline' responses. For some women, these contributions also occurred at the intersection of gendered roles and Cree cultural norms, such as for the following woman who volunteered at the LLRIB evacuation hub to provide meals for residents and emergency workers:

> [helping in this way] is something I feel is a cultural thing. That would be something my grandmother would do – at her cabin and you would see a boat coming, [she would] turn the coffee on and get cookies out … Anybody that came over had to have coffee or food and for our family that is such a big thing, so to me it was just absolutely normal – when you go anywhere and see somebody helping you, you do what can what you can to give back to them.
> (woman, LLRIB member)

As she continued, her frustration at the lack of acknowledgement such contributions received was clear: 'not once were [LLRIB and the volunteers] recognised in the media or recognised anywhere'. Likewise, two women who were interviewed together had engaged in caring work during the fires and noted the lack of attention that their contributions received in formal engagement and decision-making processes.

Participant 1: I mean in all fairness, for sure protecting the property and people is number one, but we just wanted them to be inclusive of everything. (woman, La Ronge)

Participant 2: but as far as afterwards, the leadership – they all had their own experience with the wildfire situation too. For sure it was stressful for everybody and they had lots on their plates, but if you have a group of people that have stepped up you would think there would be some acknowledgement of that and just learning from the experience because it hadn't been done before. (woman, La Ronge)

As an alternative to the more top-down engagement sessions, some community members expressed a desire to see more collaborative engagement inclusive of the diverse experiences and contributions of community residents.

I don't feel we are going to be any better prepared for the next time because they are not asking the right questions to the right people ... I would have liked to see a public consultation from the groups that were here. What worked for you? What didn't work for you? Like a think tank or brainstorming session ... rather being from a finger-pointing point of view, more of a constructive look at it, but there hasn't been anything like that.

(woman, LLRIB member)

We need to identify both strengths during the last event and build on them and areas to improve in the future. We need to include dialogue with trappers and traditional people.

(man, La Ronge)

Longer-term adaptation initiatives fell within FireSmart's mandate to protect life, property and economic values, indicating a continued emphasis on the proximate causes of risk and pathways characterised by resistance and incremental change. Many of the fuel management projects completed in and around the tri-community have not fallen under the FireSmart Communities Program, which means that project development and completion were led by the provincial wildfire management agency with input and approval by community leadership. Shortly after a fuel treatment was implemented near her neighbourhood, one resident shared the grief she felt through the loss of place where she frequently went to collect medicinal plants, lead land-based education programmes for children and spend time with family and friends. She especially lamented the lack of engagement with local residents prior to completion of the fuel management project.

We have these spaces within our community that are like stepping off grid and into an untouched piece of paradise. I spend much of my time exploring these places and sharing this love of the boreal with my family and with families in my community. The threat of forest fires has always been a part of my life here in La Ronge and four years ago the threat caused us to evacuate our home and leave most of what we had behind. The important things were with me though, the irreplaceable family members. Since the fires, I have seen our tree population around town dwindle as we become a 'fire smart' community. This has involved the thinning of trees in places that surround residential areas and will supposedly slow the spread of fire as it will have less fuel. This past winter, an area by my house that I frequent daily fell victim to this practice. ... The heart-breaking part about it is that there was no consultation or warning before the trees were taken. The land was stripped of diversity, left defenceless against the winds, and all in the name of saving 'things'.

(woman, Air Ronge)

Another resident expressed concern that a singular reliance on fuel treatments has eclipsed other approaches for building adaptive capacity, such as proactive emergency preparedness and planning.

I am worried about people taking fire protection to the extreme now. I think we need an emergency plan and an emergency response and an understanding that this is climate change and this is what we have to come to face, but not to battle it [by] clear cutting the whole area.

(woman, unincorporated subdivision).

These concerns challenge an underlying assumption that the protection of physical structures and economic assets are the principal – or only – values residents wish to maintain. It also indicates that the prioritisation of these issues may exclude the diverse social groups and the non-material values they hold for areas immediately adjacent to their communities. One participant explicitly cited the need to bridge diverse knowledges and experiences as wildfire preparedness and adaptation progresses:

It is not to go back 50 years to live like we did, but there are things that were done for a reason and that wisdom I think can be brought, but we have sort of let too much of that go. ... computers and hard evidence-based scientific knowledge are not everything. It is wonderful – I am not saying that we shouldn't have those kinds of statistics and that kind of work, we need that also – but we need more common sense and what worked in the past and let's carry that forward.

(woman, unincorporated subdivision)

This sentiment, along with other noted examples, indicates that there is an apparent appetite for more transformative change to include a broader range of voices and issues.

Discussion

That adaptation pathways were characterised by resistance and incremental change is not overly surprising, as major wildfires are commonly framed as technical problems and approached with technical solutions and sectoral expertise (Bosomworth, 2015; Nalau and Handmer, 2015). A feminist institutional lens indicates that this framing itself is gendered and, although historically male-dominated sectors may have formal rules that include diverse groups, masculinised norms can shape the development of institutions and sectors to the disadvantage of women and other non-conforming groups (Kronsell, 2016). The resulting norms, then, give rise to institutional inertia that is 'reinforced over time and often come to be deeply embedded in organisations and dominant modes of political action and understanding' (Pierson, 2004, p. 11).

The emphasis on the proximate causes of risk proved highly effective for preventing fatalities during the 2015 fire season, but institutional inertia resulted in a continued singular emphasis on proximate risks over the longer term. Although some opportunities for public participation emerged, focus remained 'on creating solutions within domains dominated by men' (Cox et al., 2008, p. 477). In this case, these solutions arose from and supported wildfire management agencies and local business. This also meant that the active contributions made by groups outside of 'frontline' work, such as caring roles often performed by women due to gendered expectations and socialisation, were largely excluded from public discussion and debriefing following the fires.

Our data also revealed the intersectional nature of power embedded within institutional practice for wildfire response and adaptation. The primary emphasis on reducing physical harm through the rapid removal of residents from their communities was reminiscent of Scharbach and Waldram's (2016, p. 66) observation that wildfire risk and response in Saskatchewan is based on a Western model of risk, 'conceptualised synchronically in response to acute challenges (i.e., an impending, immediate threat); once out of harm's way, risk is no longer the operative factor for dealing with evacuees'. How histories of colonisation contributed to different responsibilities across local jurisdictions, and how impacts and losses were experienced in diverse ways by women, men, youth and Elders after removal from their communities, were either absent or considered peripheral to local emergency management discussions. Women's contributions to wildfire response often went unrecognised while longer-term adaptations have neglected broader systems of meaning and knowledge associated with the landscape. Intersectionality, therefore, shaped not only the material impacts of the wildfire, but also the dominant discourse shaping wildfire response and adaptation (Moosa and Tuana, 2014).

Many of the impacts and losses experienced due to environmental change and hazard events are nonmaterial and nonquantifiable, such as those to mental and emotional health, self-determination and influence and sense of place (Turner et al., 2008; Tschakert et al., 2019). In a previous analysis, we found that these nonmaterial impacts and losses in La Ronge were experienced differently across intersections of location, gender, ethnicity, age, and access to financial resources and influenced by broader structures such as gendered norms and expectations and histories of colonisation (Walker et al., 2020). However, diverse experiences remained largely invisible in formal debriefing and decision-making processes, such as in the public engagement sessions following the fires.

Power dynamics have consequences for which and whose experiences, knowledges and values were included or excluded as communities prepare for fire in the future. Longer-term adaptation efforts (e.g., FireSmart) integrate some level of community participation. Globally, such programmes have played a role in enhancing fire preparedness and, in some circumstances, have also contributed to social cohesion, self-efficacy and resilience (Haynes et al., 2020). However, fire management is conceptualised by this programme primarily through a physical-infrastructural lens, which may not necessarily account for local ecological knowledge and values attached to the natural environment (Bosomworth, 2018). Fuel management projects are located in close proximity to La Ronge tri-community neighbourhoods – often the most accessible natural spaces for residents, which may be used and valued in diverse ways by different social groups within the community. Reliance on FireSmart as the sole means of engagement limits the kinds of considerations that are accounted for in future planning for wildfires or long-term adaptation measures that communities might adopt. If adaptation initiatives only respond to risks to physical and economic assets and are implemented primarily by external agencies, the nonmaterial values that diverse social groups hold for the natural areas immediately adjacent to their communities, which contribute to individual and community health, wellbeing, and resilience, may also continue to be impacted. Where local knowledges and values are not recognised, mistrust toward external agencies may also impede future response efforts (Paveglio et al., 2015).

While no evidence of transformative change was evident in the La Ronge study, residents' cited need for more inclusive planning, including the bridging of local, traditional and external knowledge indicated a desire for transformation. Findings that demonstrated the exclusion of certain values and experiences also indicate a need for transformation – 'fundamental shifts in power and representation of interests and values' (Pelling, 2011, p. 84). Implementing both incremental and transformative forms of adaption is one way forward, though attention is needed to how these may constrain one another (Wise et al., 2014; Hadarits et al., 2017). It may also be necessary to examine equity in power relations by 'questioning and challenging fixed beliefs, values, stereotypes, identities and assumptions especially about who is able to participate in adaptation planning, whose knowledge counts and who suffers the consequences of specific choices' (Ajibade and Adams, 2019, p. 856). As described in the results, this includes explicit attention to how informal institutions, including gendered and western norms and power relations, intersect to shape institutional policy and practice for emergency and wildlife management (Waylen, 2014).

Transformation requires more than just greater participation in existing institutions (Haynes, Bird and Whittaker, 2020). Rather, it necessitates changes to participation itself, including highly participatory visioning and decision-making that a) centres local knowledges, values and experiences, especially those that exist at the intersections of multiple forms of oppression, and b) takes a learning-oriented approach. First, centring local knowledges and experiences can make institutionalised forms of inequality visible, and can even transform them (Jacobs, 2019). Historically marginalised groups, such as Indigenous women, are also actively responding to the effects of climate change and hazards in their communities, as was evident in the La Ronge study. Prioritising these forms of leadership in environmental decision-making can also help to rebalance power relations that are steeped in the legacies of colonisation (Whyte, 2014; Dhillon, 2020).

Second, chasms may exist between the 'expert' knowledge of disaster risk reduction and the local knowledge and embodied experiences of altered local spaces, raising the need for strategies that foster learning and knowledge co-production (Tschakert et al., 2016). Both individual and social learning are necessary for effective adaptation processes and can be facilitated through collaborative planning that fosters multi-directional knowledge transfer, critical reflection and collective dialogue (Armitage et al., 2011; Tschakert et al., 2013). Even highly participatory processes, however, do not necessarily result in transformative outcomes as they are not immune to intra- and inter-group power dynamics or the tendency to focus on technical/infrastructural solutions, which may reproduce existing forms of inequality and vulnerability (Godfrey-Wood and Naess, 2016; Tschakert et al., 2016). Explicitly integrating consideration of structural inequalities into dialogue and reflection processes and employing methodological innovations that draw on embodied and affective experiences may help address these challenges (Tschakert et al., 2016).

Conclusion

Past responses to wildfire in the La Ronge region have focused on the proximate causes of risk and the actors and specialised knowledges that respond to those risks. Some limited opportunities for public engagement expanded the range of issues considered (e.g. to include economic impacts on the local business community). However, the authoritative emphasis on proximate causes of risk, physical and economic impacts and professional knowledge indicated that adaptation trajectories continue to be characterised by resistance and incremental change. While highly effective for basic harm reduction, this emphasis has constrained transformation during recovery and longer-term planning, thus missing opportunities to address root causes of risk and vulnerability. The pathways characterised by resistance and incremental change have been constructed with Western and masculine norms, resulting in the exclusion of certain voices, knowledge and experiences. These include the contributions of women – especially Indigenous women – to the emergency response efforts and the nonmaterial impacts of wildfire and evacuation that are experienced in diverse ways across intersections of gender, ethnicity, socio-economic class and age.

Lack of transformation at the local level to date does not necessarily mean that it is not occurring at other societal scales or over longer time periods (Heikkinen et al., 2018). Indeed, a short-term study undertaken three years after an event may be insufficient to track transformative change. However, current approaches that continue to exclude certain voices, knowledges, and experiences, and which reinforce existing forms of inequality, indicate the need for more immediate transformative approaches to inform inclusive adaptation processes at the local scale. Such approaches include the development of highly participatory decision-making processes that are learning-oriented and centre local knowledges, values and experiences. The establishment of participatory processes in the quiescent times between major events can both contribute to more inclusive and transformative adaptation and inform improved immediate response to future wildfire events.

Acknowledgement

This chapter draws on research supported by the Social Sciences and Humanities Research Council of Canada.

Note

1 Heikkinen et al (2018) delineate categories of incremental, reformistic and transformative change, which are comparable to Pelling et al's (2015) categorisations of resistance, incremental change and transformation, respectively.

References

Ajibade, I., and Adams, E.A. (2019) 'Planning principles and assessment of transformational adaptation: Towards a refined ethical approach', *Climate and Development*, 11(10), pp. 850–862.

Armitage, D., Berkes, F., Dale, A., Kocho-Schellenberg, A. and Patton, E. (2011) 'Co-management and the co-production of knowledge: Learning to adapt in Canada's Arctic', *Global Environmental Change*, 21, pp. 995–1004.

Ault, R., Baxter, G. and Hsieh, R. (2017) *Wildfire Tested Fuel Treatments 2015: Weyakwin and Wadin Bay, Saskatchewan*. Point-Claire, Canada: FP Innovations.

Bosomworth, K. (2018) 'A discursive–institutional perspective on transformative governance: A case from a fire management policy sector', *Environmental Policy and Governance*, 28, pp. 415–425.

Bosomworth, K. (2015) 'Climate change adaptation in public policy: Frames, fire management, and frame reflection', *Environment and Planning. part C*, 33, pp. 1450–1466.

Burch, S., Shaw, A., Daleb, A. and Robinson, J. (2014) 'Triggering transformative change: A development path approach to climate change response in communities', *Climate Policy*, 14(4), pp. 467–487.

Chappell, L. and Waylen, G. (2013) 'Gender and the hidden life of institutions', *Public Administration*, 91(3), pp. 599–615.

Clarke, D., Murphy, C. and Lorenzoni, I. (2016) 'Barriers to transformative adaptation: Responses to flood risk in Ireland', *Journal of Extreme Events*, 3(2), pp. 1650010.

Cox, R.S., Long, B.C., Jones, M.I. and Handler, R.J. (2008) 'Sequestering of suffering: Critical discourse analysis of natural disaster media coverage', *Journal of Health Psychology*, 13, pp. 469–480.

Dhillon, C.M. (2020) 'Indigenous feminisms: Disturbing colonialism in environmental science partnerships', *Sociology of Race and Ethnicity*, 6(4), pp. 483–500.

Enarson, E. (2016) 'Men, masculinities, and disaster: An action research agenda', in Enarson, E. and Pease, B. (eds.) *Men, Masculinities, and Disaster*, pp. 219–233. Abingdon, Oxon and New York, NY: Routledge.

Fazey, I.R.A., Wise, R.M, Lyon, C., Campeanu, C., Moug, P. and Davies, T.E. (2016) 'Past and future adaptation pathways', *Climate and Development*, 8(1), pp. 26–44.

FireSmart Canada. (2020) *What is FireSmart?* [online]. Available at: https://firesmartcan ada.ca/what-is-firesmart/ (Accessed: 29 October 2020).

Godfrey-Wood, R. and Naess, L.O. (2016) 'Adapting to climate change: Transforming development?', *IDS Bulletin*, 47(2), pp. 49–62.

Hadarits, M., Pittman, J., Corkal, D., Hill, H., Bruce, K. and Howard, A. (2017) 'The interplay between incremental, transitional, and transformational adaptation: A case study of Canadian agriculture', *Regional Environmental Change*, 17, pp. 1515–1525.

Haynes, K., Bird, D.K. and Whittaker, J. (2020) 'Working outside 'the rules': Opportunities and challenges of community participation in risk reduction', *International Journal of Disaster Risk Reduction*, 44, p. 101396.

Heikkinen, M., Ylä-Anttila, T. and Juhola, S. (2018) 'Incremental, reformistic or transformational: What kind of change do C40 cities advocate to deal with climate change?', *Journal of Environmental Policy & Planning*, 21(1), pp. 90–103.

IPCC. (2012) *Managing the Risks of Extreme Events and Disasters to Advance Climate Change Adaptation*. Cambridge and New York: Cambridge University Press.

Jacobs, F. (2019) 'Black feminism and radical planning: New directions for disaster planning research', *Planning Theory*, 18(1), pp. 24–39.

Kaijser, A. and Kronsell, A. (2014) 'Climate change through the lens of intersectionality', *Environmental Politics*, 23(3), pp. 417–433.

Kronsell, A. (2016) 'Sexed bodies and military masculinities: Gender path dependence in EU's common security and defense policy', *Men and Masculinities*, 19(3), pp. 311–336.

Macgregor, S. (2010) 'A stranger silence still: The need for feminist social research on climate change', *Sociological Review*, 57, pp. 124–140.

Mackay, F., Kenny, M. and Chappell, L. (2010) 'New institutionalism through a gender lens. Towards a feminist institutionalism?', *International Political Science Review*, 31(5), pp. 573–588.

Moosa, C.S. and Tuana, N. (2014) 'Mapping a research agenda concerning gender and climate change: A review of the literature', *Hypatia*, 29(3), pp. 677–694.

Nalau, J. and Handmer, J. (2015) 'When is transformation a viable policy alternative?', *Environmental Science & Policy*, 54, pp. 349–356.

O'Brien, K. (2012) 'Global environmental change II: From adaptation to deliberate transformation', *Progress in Human Geography*, 36(5), pp. 667–676.

O'Brien, K., Eriksen, S., Inderberg, T.H. and Sygna, L. (2015) 'Climate change and development: Adaptation through transformation', in *Climate Change Adaptation and Development: Transforming Policies and Practices*. pp. 273–289. Abingdon, Oxon and New York, NY: Routledge.

Paveglio, T.B., Carroll, M.S., Hall, T.E. and Brenkert-Smith, H. (2015) 'Put the wet stuff on the hot stuff': The legacy and drivers of conflict surrounding wildfire suppression', *Journal of Rural Studies*, 41, pp. 72–81.

Pease, B. (2016) 'Masculinism, climate change and 'man-made' disasters', in Enarson, E. and Pease, B. (eds.) *Men, Masculinities, and Disaster*. Abingdon and New York: Routledge, pp. 21–33.

Pelling, M. (2011) *Adaptation to Climate Change: From Resilience to Transformation*. New York: Routledge.

Pelling, M., O'Brien, K. and Matyas, D. (2015) 'Adaptation and transformation', *Climatic Change*, 133(1), pp. 113–127.

Pierson, P. (2004) *Politics in Time. History, Institutions and Social Analysis*. Princeton: Princeton University Press.

Ravera, F., Iniesta-Arandia, I., Martín-López, B., Pascual, U. and Bose, P. (2016) 'Gender perspectives in resilience, vulnerability and adaptation to global environmental change', *Ambio*, 45(S3), pp. 235–247.

Reed, M.G., Scott, A., Natcher, D. and Johnston, M. (2014) 'Linking gender, climate change, adaptive capacity, and forest-based communities in Canada', *Canadian Journal of Forest Research*, 44(9), pp. 995–1004.

Scharbach, J. and Waldram, J.B. (2016) 'Asking for a disaster: Being 'at risk' in the emergency evacuation of a Northern Canadian Aboriginal Community', *Human Organization*, 75(1), pp. 59–70.

Tschakert, P., Ellis, N.R., Anderson, C., Kelly, A. and Obeng, J. (2019) 'One thousand ways to experience loss: A systematic analysis of climate-related intangible harm from around the world', *Global Environmental Change*, 55, pp. 58–72.

Tschakert, P., Das, P.J., Shrestha Pradhan, N., Machado, M., Lamadrid, A., Buragohain, M. and Hazarika, M.A. (2016) 'Micropolitics in collective learning spaces for adaptive decision making', *Global Environmental Change*, 40, pp. 182–194.

Tschakert, P., van Oort, B., St. Clair, A.L. and LaMadrid, A. (2013) 'Inequality and transformation analyses: A complementary lens for addressing vulnerability to climate change', *Climate and Development*, 5(4), pp. 340–350.

Turner, N.J., Gregory, R., Brooks, C., Failing, L. and Satterfield, T. (2008) 'From invisibility to transparency: Identifying the implications', *Ecology and Society*, 13(2), p. 7.

Tyler, M. and Fairbrother, P. (2013) 'Bushfires are 'men's business': The importance of gender and rural hegemonic masculinity', *Journal of Rural Studies*, 30, pp. 110–119.

Tyler, M., Carson, L. and Reynolds, B. (2019) 'Are fire services 'extremely gendered' organizations? Examining the Country Fire Authority (CFA) in Australia', *Gender, Work & Organization*, 26, pp. 1304–1323.

Walker, H.M., Reed M.G., and Fletcher, A.J. (2020) 'Applying intersectionality to climate hazards: A theoretically informed study of wildfire in northern Saskatchewan', *Climate Policy*,21(2), 171–185. .

Waylen, G. (2014) 'Informal institutions, institutional change, and gender equality', *Political Research Quarterly*, 67(1), pp. 212–223.

Whyte, K.P. (2014) 'Indigenous women, climate change impacts, and collective action', *Hypatia*, 29(3), pp. 599–616.

Wilson, D. and MacDonald, D. (2010) *The Income Gap Between Aboriginal Peoples and the Rest of Canada*. Ottawa: Canadian Centre for Policy Alternatives.

Wise, R.M., Fazey I., Stafford Smith, M., Park, S.E., Eakin, H.C., Archer Van Garderen, E.R.M., Campbell, B. (2014) 'Reconceptualising adaptation to climate change as part of pathways of change and response', *Global Environmental Change*, 28, pp. 325–336.

14 Moving forward: Making equality, equity, and social justice central

Gunnhildur Lily Magnusdottir and Annica Kronsell

Fortunately, the climate crisis is increasingly recognised across the world. Associated with this recognition is a crisis discourse. While crises may encourage action, the crisis discourse also conveys a message that all individuals are implicated in the crises on equal terms and in equal ways. This is not the case. Paraphrasing Cynthia Enloe[1] (see also Chapter 3, this volume):

> we are not all in this together. We're on the same rough seas, but we're in very different boats. And some of those boats are very leaky. And some of those boats were never given oars. And some of those boats have high-powered motors on them. We are not all in the same boat.

Enloe calls attention to the danger of a world where patriarchal and anthropocentric norms dominate and social differences are overlooked. This book highlighted those issues, and the relation between gender and social justice issues and climate issues, in particular, by concentrating on how institutions and policy-makers who are compelled to deal with climate issues can connect this activity with a sensitivity to social differences. Climate institutions are significant – we have argued – as they are responsible for steering toward climate objectives through strategies, incentives, guidelines, policies, roadmaps, and other actions that aim to mitigate as well as adapt to climate change. Furthermore, climate institutions enact power and affect power relations by promoting certain norms and values through policies and by distributing resources and, in turn, engaging specific actors.

This book explored if and how climate institutions in industrialised states recognise and understand the relevance of gender and other social differences in their climate policy-making. In democracies, they have an obligation to do so. We find that they do, but mostly from a very narrow understanding of gender. We have argued that equality, equity, and social justice are so central to climate mitigation and adaptation, that they should be key considerations in climate policies and actions, not only for democratic and justice reasons but also for substantive reasons. Not least on the occasions when gender, social justice, and climate objectives can be mutually supportive, such as if women's transport and waste management behaviour becomes normative it could also lead to reduced emissions (Chapters 6, 7, 8, 9, and 11, this volume), but is more tenuous at other times

DOI: 10.4324/9781003052821-14

such as in hegemonic masculine areas – for example, in the construction sector (Chapters 9 and 10, this volume), responses to natural disasters (Chapter 13, this volume), and in shaping opinions on climate change, masculinity, and care for Earth (Chapter 12, this volume).

Path dependence, 'sticky' institutions and eco-technological framing

We departed from a feminist institutional framework to locate the main institutional challenges and barriers to further inclusion of climate-relevant social factors and, in particular, gender in climate policy-making, and applied the concepts of path dependence and 'stickiness', which lays out the course of action of climate institutions and limits the scope of policy-makers and other actors for innovation and new paths in climate policy-making.

We identified path dependence in most chapters. Ecological modernisation emerged as a path dependence that is structuring climate institutions in industrialised states and can be said to have substantial power of the climate agenda in general. Ecological modernisation centres around the idea that climate issues can be resolved in tandem and harmoniously with continued economic growth, and increase wealth and prosperity, as the market adjusts resource and energy use through prize mechanisms and from a continuum of innovations. This is not explicitly or discussed at length in any of the chapters, but rather forms a type of normal condition. This powerful normality informs and structures climate policy-making so that it will privilege efficiency arguments and technical knowledge over other types of knowledge that could help address social issues.

Second, patterns of path dependence can be traced in climate policy-making at all levels, from intergovernmental actors such as the UN and EU to local authorities and actors. At the UN level, efforts toward social inclusivity are hampered by path dependencies related to the characteristic of the international legal system (Chapter 2, this volume). In particular, state sovereignty means there is limited enforcement capacity when it is based on state consent, thus, even when international institutions such as the UNFCCC have established rules on gender they are politically constrained in getting states to follow through, even on the commitments that they have agreed to. At the EU level, climate change path dependence concerned the topic of climate change and its framing as a developmental issue which later became tinged by foreign and security concern leading to a prioritisation overriding gender concerns. On the other hand, the fact that in the EU gender equality is a fundamental value, formal commitments to gender equality and gender mainstreaming which has mainly been related to the labour market, can possibly come into play in cases where climate change is framed as a problem in internal policies (EU green deal). So far, 'efforts to address gender inequality and efforts to address climate change continue to run in parallel rather than being integrated into each other' due to institutional stickiness and path dependence (Chapter 3, this volume).

Second, climate policy is a very broad field of action, and climate institutions are many. State and local authorities, such as the case of Germany (Chapter 4, this volume) and Sweden (Chapters 5 and 6) may very well develop ways to integrate equality and equity issues, but whether they will be realised, or not, is constrained by priorities and previous choices in other ministries and other sectors, like construction, transport, and energy (Chapters 7, 9, 10, and 11). The focus tends to be on economic interests and technical solutions in line with the dominant norms of ecological modernisation (Chapter 12), when what is likely needed is more focus on public interest and welfare. This means that research on climate change, in general and within sectors, has not advanced the studies of social effects and implications. As argued, there is a huge gap in knowledge among policy-makers.

Third, a barrier found in many of the institutions and sectors, as highlighted in the book, is a hegemonic masculine environment which results in the imbalanced representation of men and women – and sometimes also limited recognition of marginalised social groups. This barrier is not only found in formal climate institutions such as in Swedish, Danish, and German ministries (Chapters 4, 5, and 7), but also in different sectors, such as the transport sector, European waste management (Chapter 8), and in European and North-American construction (Chapters 9 and 10). There are a number of tools available for climate policy-makers, such as gender mainstreaming, but the low representation of women and marginalised groups makes gender mainstreaming and intersectionality appear less appropriate and might make policy-makers insensitive to climate-relevant differences in behaviour as, for example, transport patterns (Chapter 11) and responses to and effects of climate change, such as severe wildfires in Canada, on different social groups (Chapter 13). Imbalanced representation and limited recognition of intersectional social factors, relevant to climate change, do not only risk making climate strategies less effective, when targeting the general population but can also be seen as a signal of democratic deficit and limited environmental and even economic justice in climate politics.

Fourth, civil servants who work in formal climate institutions and often find themselves in path dependent institutional environments, characterised by eco-modernisation and limited intersectional understanding are faced with democratic challenges. Those civil servants are limited by the legal framework of the work within (Chapter 2, this volume), the administrative context and logic and the mandate they have from their governments. They are often aware that in order to maintain validity and credibility in policy-making, their actions and ideas have to fit within a given framework that is deemed legitimate by both the public and politicians (Chapters 4, 5, 6, and 7).

Finally, climate change action is set in a frame of time urgency. There is an urgent need to reduce carbon emissions even by the set dates of 2030 and 2050, which puts stress on the policy-makers to find ways to enact this. At the same time, social transformation takes time and as Alber et al. (Chapter 4, this volume) state; 'climate protection specialists in the ministry are chronically overloaded'. Hence, even if they are willing and able to work with gender it cannot take too much time, and a genuine intersectional approach would make it even more time-consuming. Such approaches require that gender and intersectional research are taken into account and that developed policies are monitored and evaluated. A very important issue, then, is how is this doable within the stressful timeframe of climate policy-making today? Is it possible to move this quickly? Is it even desirable, or are stressed and overworked policy-makers more likely to follow a path-dependent logic of appropriateness and stick to familiar strategies in their policy-making, which might favour certain social groups, and knowledge.

Potentials and suggestions for advancing equality and social justice

Moving on to the opportunities to advance equality and social justice issues in climate policy-making, we revisit the following question:

What and where are the main opportunities to advance gender equality, equity, and social justice within climate institutions?

Issues relating to social inclusion and climate change do surface in climate institutions in industrial states, but they tend to run in parallel with limited coordination between them. These issues must be addressed in cross-cutting ways at many levels, in terms of formal policy-making in climate institutions at different levels, cooperating on a global level with the UN at the forefront, to local initiatives at the municipal level. Furthermore, a just and socially sustainable climate policy-making requires further coordination between different policies and sectors vital to climate mitigation and adaptation, and a common understanding of the intersectional nature of climate change and its effects.

Apart from cross-cutting cooperation, coordination, and a common understanding, gender experts, femocrats, and other civil servants – with a profound understanding and knowledge of social justice issues – who might perceive themselves as inside-activists, are important. Their leeway for innovation is, however, limited due to the aforementioned institutional barriers, but existing formal legal frameworks on gender and social inclusion are crucial building blocks for pursuing equality and equity in institutions.

The strategy of gender mainstreaming is, for example, of importance when discussing formal remedies, but it has to date often merely been a limited addition to existing policies. Gender mainstreaming still has the potential to advance socially just climate policies since it is a dynamic tool which can be broadened to include both new gender perspectives, as well as knowledge of other intersecting social factors, in everyday policy-making. However, none of these potentials and suggested approaches are sufficient on their own and what is being called for is a new thinking and a transformative and sustainable way of integrating gender and intersectionality with all climate policy-making initiatives, which challenges the current path dependencies of the institutions in which climate action is enabled.

Note

1 The Quaker European Office (Brussels) April 13, 2020; https://www.youtube.com/w atch?v=Gaif6mTwFw8

Index

Page numbers in "italic" indicate a figure and page numbers in "bold" indicate an illustration.

15-minutes post-COVID-19 cities 185
21st century 1
2030 Agenda for Sustainable
    Development (SDGs), UN 28

abuse culture 216
Action Alliance for Climate Protection 56
Action on Climate and the SDGs 29
adaptation 36, 37
Adaptation Committee 22
adaptation pathways 227–230, **229**,
    238–241
adequate information and awareness, lack
    of, women and 146–147
administrative practices 83–84
African Gender Institute 76
age 1; age distribution 10
Agenda 2030, and gender 28–29
Alber, Gotelind 7, 247
Alberta Council of Turnaround
    Industry Maintenance Stakeholders
    (ACTIMS) 155
Allwood, Gill 7, 47
Alston, M. 186, 187
Andersson, Carolyn Hannan 76
androcentrism 63
(m)Anthropocene 208–211; social and
    ecological crises of the 220
anthropogenic global heating 208
anti-gender movements 78
appropriateness: gendered logic of 6;
    logic of 6
Arora-Jonsson 90
Association for Women's Rights in
    Development (AWID) 69
ATLAS 8 qualitative analysis software 92

backlash, global 31
Baruah, Bipasha 9
Bee, Beth 130
behaviour, masculine and feminine
    forms of 6
Beijing Declaration and Platform for
    Action (BDPA) 18; strategic objectives
    for governments and 18
Beijing Declaration and Platform for
    Equality, Development and Peace 127
Beijing Platform of Action 169
Bhopal, Kalwant 130
bicycle strategy 192
bike-sharing scheme, in Oslo
    (Norway) 202
Biskupski-Mujanovic, Sandra 9
Black and minority ethnic women 130
black feminist theory 5
blind spots and oversights, and
    employment equity in 151–153
BMU (German Ministry for the
    Environment, Nature Protection and
    Nuclear Safety 64
Braid, K. 155
Breengaard, Michala Hvidt 9
'brown' or traditional (fossil fuel–based)
    sectors 157
Brundtland Report 208
Buckingham, Susan 9, 211
BUILD Inc. 156
building production, social aspects o 165
Building Trades Union (CBTU),
    Canada's 156
Building Up 156
'Build Together' 157
Build Up Skills initiative (BUS) 169

BUS EU initiative 172
business-as-usual (BAU) scenario 193;
   long trips 194–195; medium trips 195;
   projections 194–197; short trips 196
Butler, Judith 107

Canada 77; green transition in 151–153;
   non-profit sector in 156; policies and
   programmes 153–154; as a 'post-gender'
   and 'post-racial' society 152
Canadian Apprenticeship Forum (CAF-
   FCA) 152–153; labour market report of
   153–154
Canadian Building Trades 155
Canadian green economy: findings
   on research on women in 145–157;
   methodology used in studying women in
   144–145; women in 143
Canadian provincial programme, and
   women employment in transportation
   154–155
carbon emissions 2, 6, 247
car culture 119; gendered dimensions
   of 106
car ridership vis-à-vis other modes 196
car traffic growth 197
cases and methods, research methods for
   book 90–92
CEDAW 23, 28
Centre for Skills Development and
   Training 154
Chant, Sylvia 76
Chappell, L. 45
Charter of Fundamental Rights, Article
   23 of 170
Christensen, Hilda Rømer 9
Christian Democratic Union (CDU) 53
circular economy, European Union
   and 124
Circular Economy Action Plan, European
   Union and 139
city council cooperation, in Oslo
   (Norway) 188–189
civil servants 97; in Swedish climate
   institutions 86–87
civil society 8
Clarke, Linda 10, 150
class 1
Climate Action Directorate, EU
   Commission's 1
climate action plans 183
Climate Action Programme, German
   Parliament and 58
Climate and Energy Package 37

climate and energy strategy, Oslo
   (Norway) 191
climate authorities, gender representation
   and 3
climate awareness 80
climate budget, Oslo (Norway) 191
climate change: diversity and
   intersectionality into 48–49; EU's policy
   agenda and 36; externally 38; floods and
   heatwaves and 37; gender intersectional
   approach and 2; masculinist dominance
   in 19; security challenges 1
climate change action 247
climate change adaptation strategies 226
climate change deniers 86
climate change events, gendered
   vulnerabilities of 2
climate change governance, gender
   engagement and 20
climate change impacts and adaptation,
   feminist scholarship on 226
climate change mainstreaming,
   organisational structures and 46–47
climate change policy, waste management
   and 126–127; climate change process
   63; policy-making, reach of 124
climate change work, social justice and 92
Climate Council 106, 120n2
climate crisis 245; men in reflective groups
   212–213
climate diplomacy 36–37
climate emergency, construction industry
   and 164
climate emission targets: and equity and
   justice 3; and gender equality 3
climate events, violence of 2
climate framework document 43
climate-friendly motorised transport 119;
   transport practices/strategies 9, 108–112
climate institutions 2, 245; barriers
   and 247; and bringing gender into
   climate policy 55; female policy-
   makers in 5; gender 3; Germany 8;
   intergovernmental and governmental
   7–8; power relations and 2–3; sectoral
   8–10; and social differences in climate
   policy-making 2
climate laws 36, 37
climate mainstreaming 39
climate mitigation challenges 4
climate neutrality 37; continent 36;
   economy 42
climate objectives: proposed strategies for
   1; social differences and 2

climate policies: gender and 55; gender-equitable 54; gender mainstreaming and 64; in Germany 53–55; and institutions, gender and intersectional perspectives on 1; rich industrialised countries and 2; social differences and 1

climate policy-making/makers: equality and 2; gender and 185–187; Germany 8; justice issues and 2; key actors in 1; plans for 183; representation in 3–5; social differences and 2; social equity and 2

climate protection plan, citizen participation and 57; Climate Protection Plan 2050 56–58

climate transition 10; social inclusion and 1

CO_2 emissions 84; CO_2 reductions, transport and in Denmark 105–106

Cohen, M. 155

Collins, Patricia Hill 5

Commission Communication (2003) 37

Commission Directorate-General for Climate Action (DG CLIMA) 5, 38

Committee on Capacity Building 22

Common Market 48

communicative events 107–108

Conestoga College's School of Engineering and Technology 154

Conference of Parties (CoP) 19, 23, 27; in Decision 3/CP.23 26; to the Framework Convention on Climate Change (COP1) 53

conservative politics, and gender equity 78–80

constituted bodies, UNFCCC's 24

construction 164; lack of gender diversity 176–177; women in 165

construction and transportation: challenges and opportunities for women in 145–157; employment in 143–144; wages in 150; women's underrepresentation in 145; *see also* construction

content analysis, qualitative method of 187–188

'Contribution of gender justice to successful climate policy, The" 52

Convention on Biodiversity (CBD) 20

Convention on the Elimination of All Forms of Discrimination Against Women (CEDAW) 18

COP (Conference of the Parties) 79

Cornwall, Andrea 45, 76

Council Conclusions on Climate Diplomacy 38, 42, 47

COVID-19 pandemic 91; and biking in Denmark 119; gendered impact of 44; gender inequality and 130; and job losses in feminised sectors 152

critical acts: networks and 75–76; understanding alongside politics 76–78

critical mass 4

cross-border nature, of climate change 37

daily mobilities, gendered patterns of 184–185

Dainty, A. 167–168

Danish Cyclist Association 119

Danish Infomedia's online archive 108

Debusscher, P. 44

Decision 3/CP.23 23, 24; Conference of Parties (CoP) and 26; Decision 3/CP.25 25; Decision 18/CP20, on Lima Work Programme on Gender (LWPG) 23; Decision 21/CP.22 23; Decision 23/CP.18, at UNFCCC CoP 20

democracy, ill-functioning 3

democratic deficit 3

democratic legitimacy 98

DEMOTRIPS 184, 188, 203; analyses of 193–194

Denmark 105–106; non-motorised transport in 119; public transit system 119

Department for Human Rights 2012–2017, MFA 75

descriptive representation 3, 4, 22

desk-based synthesis report 24

Development Assistance Committee (OECD-DAC) 76

DG Development of Gender and Development 41, 47–48

Directorate-General (DG) for Development 41

disability, inclusive language on 155

disaster risk reduction 226

Diversity and Inclusion Lens and Toolkit 154–155

Dubrovnik 139

ecofeminist scholars 211

ecological destruction, men and 11

ecological footprints 6

ecological masculinities 11, 207

Ecological Masculinities (Hultman & Pulé) 207

ecological modernisation 246

ecologisation 214, 221n2
Eco-Management and Audit Scheme (EMAS) 132
ecomodern masculinities 10, 208
ECOSOC 186
eco-technological framing 246–248
EIGE (European Institute for Gender Equality) 40, 43; gender mainstreaming in the EU ad 128; methodology 132
elderly 10
electrification, in Norway 183–184
Elson, Diane 76
ELWPG (Enhanced Lima Work Programme on Gender), gender action plan and 25
emergency, institutions for 11
emotional literacy, and vulnerability 214–216
employment equity, blind spots and oversights about 151–153
Employment Equity Coordinator 155
employment equity measures 9
ENAREDD+ 130
energy-efficiency policies 53; gender and 168–171
energy sector 159n1; women employed in 166
engineering, manufacturing and construction, graduates in **165**
Engström, Stefan 77
Enhanced Lima Work Programme on Gender (ELWPG) 25
Enloe, Cynthia 44, 245
Environment, Nature and Climate Protection 56
Environmental Action Programme 37
environmental carbon emissions, European Union's (EU) strategy to reduce 10
environmental degradation 207
equal pay for equal work 40
ethnicity, inclusive language on 155
EU climate policy, components of 36–39
EU Commission, Climate Action Directorate 1
EU external climate policy 7; feminist institutionalist analysis of 47; gender and 39–43
EU policy documents, gender mainstreaming and 48
Euro-Canadian and Indigenous populations 235
Europe, as a climate-neutral continent 36
European climate action 95

European Commission 38–40, 44; gender and development within 37
European Consensus on Development (2017) 39, 41
European Council Strategic Agenda 2019–2024 42
European External Action Service 41, 42, 47; creation of 48; gender composition at the management level of 45; shifting of 45
European External Action Service Gender and Equal Opportunities Strategy 45
European External Action Service Principal Advisor 46
European Green Deal 36, 37, 127; (2019) 38, 42
European Institute for Gender Equality (EIGE) 40, 43, 170
European Parliament 40, 43; and the Council 44; draft report on Gender Equality 46; gender and development within 37–38; mainstreaming of issues and 46–47
European Parliament Development Committee 44
European Performance of Buildings Directive (EPBD) 165
European Union (EU) 41; Circular Economy Action Plan 139; circular economy and 124; Foreign and Security Policy 2019/2167 (INI) 46; gender mainstreaming and 127–131; green transition policy 164
European Union Delegations (EUDs) 46
executive and managerial positions, women's underrepresentation 146
Executive Board of the Clean Development Mechanism 22
external climate policy 44; EU 36, 39–43; feminist institutionalist analysis of 43–47; gender and 39–43

Fairclough, Norman 107
Federal Ministry for the Environment (BMU) 53, 54–55
female civil servants 70
female participants 24
female policy-makers, in climate institutions 5
female representation, on UNFCCC constituted bodies 20, *21, 22*
feminist discursive institutionalism 43–44
feminist environmental literature, intersectionality and 2

Feminist Foreign Policy (FFP) 77
feminist institutional framework 246
feminist institutionalism 5, 6, 36, 43,
  87, 90; analysis 70; analysis of gender
  43–47; EU external climate policy 47;
  external climate policy and 43–47;
  performativity and 107
feminist institutionalist literature 98
feminist intersectional research 2
feminist movements 4
feminist scholars 17, 40; and climate
  institutions 5
feminist theory 2; intersectionality 6–7
feminist transport research 106
FEMM Committee, European
  Parliament's 40
femocrat 4, 8, 90
FireSmart 232–233; FireSmart
  Communities Program 235, 237;
  FireSmart Community Recognition
  status 235
first-generation (civil and political)
  rights 31
Fletcher, Amber J. 11
Foreign and Security Policy 2019/2167
  (INI), EU's 46
Forum Syd 73
Fourth United Nations World Conference
  on Women (Beijing) 71, 127
frames, limiting 81–83
Fraser, Nancy 129
fuel management 232–233

Gaard, Greta 211
GAP, national 64
GAP II 42, 45
gender 1; approaches to 30; climate policy-
  making and 185–187; energy-efficiency
  policies and 168–171; feminist
  institutionalist analysis of 43–47; in
  Germany 53; inclusive language on
  155; Paris Agreement 27–28; UNFCCC
  documents on 26; World Health
  Organisation (WHO) definition of 30
Gender, Power and Climate Change
  conference 69
Gender+ 65n8
Gender Action Plan 23–25; (2010) 41, 48;
  (2017) 43; GAPII 41; UNFCCC 54
gender activist 90
gender advisors 84
gender analysis 41; from feminist theory 2
'Gender and Climate' research project 55,
  58, 63, 64

gender and human rights, global backlash
  and 31
gender and intersectional perspectives 1
gender-awareness, sustainable
  environmental practices 135–138
gender balance 24
gender bias and gender stereotyping,
  women and 147–148
gender binary 4, 32
gender blindness 42, 171; construction
  industry and 165
GenderCC-Women for Climate Justice
  52, 56
Gender Constituency (2011) 7
gender coverage 26
Gender Day event 27
gender diversity, lack of in construction
  176–177
gendered climate policies 3
gendered daily mobilities 10
gendered logic of appropriateness 6
gender effects, limiting frames 81–83
gender engagement, in global climate
  change governance 20
gender-equal approach, institutional
  challenges and 8
gender equality 9, 10, 32, 40–43; in the
  Canadian green economy 143–144;
  conservative politics and 79; embedding
  in policy 173–176; environmental
  performance and **136**; European
  Union and 40, 48; Gender Equality,
  draft report 46; in Germany 55;
  importance of political buy-in 30;
  men–women binary and 40; perspective
  9; and sensitivity case studies **136**;
  and UNFCCC action 29–30; United
  Nations ad 17–18
'Gender Equality as a Prerequisite for
  Sustainable Development' (Johnson-
  Latham) 73, 82
Gender Equality Strategy 42, 128, 170;
  EU and 39–40
gender-equitable climate policy 54
gender equity 80; and Canada's green
  economy 153; conservative politics and
  78–80; guidelines 4
gender expert, critical acts of a 70–72
gender expression, inclusive language
  on 155
gender-focused civil servants 70
gender greenwashing 64
gender identity, inclusive language on 155
'Gender Impact Assessment' (GIA) 52

Gender Impact Assessment (GIA) tool 8, 55, 58–59, 64–65; applying intersectionality to 59–62
gender inclusion, United Nations Framework Convention on Climate Change (UNFCCC) 19–20
Gender in environment and climate change 170
gender inequality, in UNFCCC constituted bodies 22
gender issues: UNFCCC engagement with 30–31; United Nations and 18
gender-just climate policy 36, 39
gender justice 52
gender landmarks, UNFCCC regime and 20–22
genderless sustainable paradigm, policy and planning and 119
gender machinery 63
gender mainstreaming 4, 36, 39, 40, 43, 44, 46, 55, 84, 248; in climate policy 64; critiqued of 127–131; European Union and 127–132; in institutions and processes 44; requirements of 109; as a transformative process 138–139; transport policy and 106; Urban Waste project and 132–134; waste management and 9; and waste management in the EU 131–132
gender-proofing methods, and waste reduction measures 134–135
gender representation, monitoring and reporting on 20–22
gender-responsive climate policies 27, 55; and the UNFCCC 53
gender-sensitive change, institutional resistance to 45
Gender Strategy 64
gender transformation 45
gender visibility, in the UNFCCC regime 26–27
gender wage gap 150
geography 1
Gerd see Johnsson-Latham, Gerd
German Bundestag 56
German climate policy 52, 53
German climate protection programmes 55
German Federal Environment Agency (UBA), 'The contribution of gender justice to successful climate policy." 52
German Federal Government 62, 91
German Ministry for the Environment, Nature Protection and Nuclear Safety (BMU) 52

German Parliament, Climate Action Programme and 58
German Women's Council 57, 63
Germany 246; climate policy in 53; climate policy response 52–53; Gender Impact Assessment (GIA) tool and climate policy 58–59; gender in 53; mainstreaming gender into climate policy 53–55
Gilligan, Carol 130
Global Climate Change Alliance Plus initiative (GCCA+), EU's 38
global climate change governance 7; gender and 31; gender engagement and 20
global climate regime 7; progress on gender issues in 31–32
global financial crisis, impact of on women 130
Global North/Northern 1, 40, 52, 209–211
global social: and ecological crises 213–214; and environmental problems 207
Global South 37, 40
Global Strategy 38–39, 42
global warming 80
Google Scholar citations 74
government officials, challenges of promoting gender equality 70
Great Belt Bridge 115
green activists 5
green construction policies 171
green economy, climate institutions and 2
green growth and development strategies 143
green growth sectors, women's underrepresentation in 153
greenhouse gas emissions 1, 37, 124; curbing 9
Green Party 188
Greenpeace 57
Green Radicals 86
green transition, in Canada 151–153
green transition policy, European Union (EU)'s 164
Griffin, Penny 129, 130
Guerrina, Roberta 45

Halck, Jørgen 108
Hankivsky, Olena 64
Hanson, Susan 106
Hatch, M.J. 92
Haugesund 193
Health and Safety of Women in Construction (HASWIC) Workgroup 149

Hedenqvist, Robin 10, 212
hegemonisation 214; masculine 221n1
High Commissioner for Refugees 71
High heels: building opportunities for
   women in the construction sector 167
High Representative, creation of 48
Horizon 2020 policy development
   team 129
Hultman, Martin 11, 208; *Ecological
   Masculinities* 207
human rights, equal 31
human rights agenda 17
Hummel, Diana 7

ICT use 135
identities, role expectations and 89–90
ill-functioning democracy 3
impacts, vulnerability to 1
Indigenous peoples, inclusive language
   on 155
individual policy-makers 6
industrial/breadwinner masculinities 208,
   222n5
industrialised states, carbon emissions
   reduction and 2
Industrial Revolution 210
inequities and/or fomenting public
   resistance (iSTA06) 96
institutional analysis, intersectionality as a
   tool in 6–7
institutional change: functionalist views of
   17; implications for in SEPA and STA
   96–99; need for 86–87
institutional culture 22–27; shift 41, 45
institutional environments 247
institutional inertia, Johnsson-Latham,
   Gerd 84
institutional path dependencies 80–81, 97,
   99–100
institutional resistances 62
institutions: gendered structural
   characteristics of 17; international law
   17; 'stickiness' of 87–89
integrated land use 183
intergovernmental and governmental,
   climate institutions 7–8
international climate governance, shaping
   of 17
international environmental law 29
international human rights law 29; efficacy
   of 31
international law, gender of 17
international law institutions,
   representation of men in 17

intersectional analytical lens 40
intersectionality 8, 40; climate change and
   2; Gender Impact Assessment (GIA)
   tool and 59–62; as a tool 2; as a tool in
   institutional analysis 6–7
Intersectionality Based Analysis (IBA) 64
interview study 90
ISOE-Institute for Social-ecological
   Research 52

Johnsson-Latham, Gerd 8; critical acts of
   70–72; expertise on gender and 72–75;
   feminist institutional theory and 69–70;
   'Gender Equality as a Prerequisite for
   Sustainable Development' 73; goal-
   setting at a political level 75–76

Kabeer, Naila 76
Katalytik, M. 150
Kelleher, C. 45
Kenny, Meryl 17; feminist analysis 30–31
Klimatriksdagen (The Climate
   Parliament) 70
Knapskog, Marianne 10
Kronsell, Annica 8, 40, 69, 70, 130, 186;
   Commission Directorate-General for
   Climate Action 5
Kt CO$_2$e 127
Kvinna till Kvinna 69
Kyoto Protocol 20

labour and VET 171–173
Labour Party 188
Landström, Catharina 119
land use: integrated 183; plans 183;
   and transport, regional plan for Oslo
   189–190
Latour, B. 219
LCIPPFWG (Local Communities and
   Indigenous Peoples Platform Facilitative
   Working Group 25
Least Developed Countries (LDCs) 38
LEC programmes 177
Lévy, Caren 106
Leyen, Ursula von der 36, 38
LGBT interventions 80
LGBTQ+: inclusive language on 155;
   persons, disadvantages of 31; rights,
   similarity to women's rights 31, 79
Lima Work Programme on Gender,
   UNFCCC 43; Decision 23/CP.18
   on 23
limiting frames, gender effects 81–83
Lisbon Treaty 39, 41, 44, 47, 48; (2009) 38

Local Communities and Indigenous Peoples Platform Facilitative Working Group (LCIPPFWG) 25
logic of appropriateness 6
Lombardo, E. 139
long trips: business-as-usual (BAU) scenario 194–195; growth rates and aggregate mode shares 199; older persons and 197; per capita, business-as-usual (BAU) scenario 195; select modes by age group and gender 199; zero growth objective (ZGO) and 197
low-carbon economies 146
low energy construction (LEC) 165
LWPG (Lima Work Programme on Gender) 23

MacGregor, Sherilyn 49, 211
Magnusdottir, Gunnhildur Lily 8, 130; Commission Directorate-General for Climate Action 5
mainstreaming gender, Johnsson-Latham and 71
male-biased work culture and working conditions, women and 148–149
male dominance 4; industry, construction as a 164
malestream masculinities 207
Mama, Amima 76
MÄN 211–212; MÄN Under Tallarna 212
Manners, I. 44
March, J.G. 6, 88
Maridal, Jomar Saterøy 10
masculine ecologisation 211, 221
masculine hegemonisation 221n1
masculine sectors 10
masculinist social injustices 207
masculinities 207–208; ecological 11, 207; ecomodern 208; industrial/breadwinner 208; malestream 207
masculinities research 11
Mason, Karen 71
McFarland 158
medium trips: age groups and 198; business-as-usual (BAU) scenario 195–196; per capita, business-as-usual (BAU) scenario 195; zero growth objective (ZGO) 197–198
Men in the Climate Crisis 212, 221
Merchant, Carolyn 211
Mergaert, L. 129, 139
Merkel, Angela 53

METRAC (Metropolitan Toronto Action Committee on Violence against Women and Children) 154
Metropolitan Toronto Action Committee on Violence against Women and Children (METRAC) 154
MFA Department for Human Rights 2012–2017 75
migration-security-climate nexuses, gender equality 47
Mikkelsen, Sonja 106, 108, 112–114; number of articles about 109; traffic policy and 115–119
Ministry of Economic Affairs 53
Minto, R. 129
mitigation 36–37
mobilities: daily gendered patterns of 184–185, 196; gender and 106; per capita 196–197; plans 183
mobility-as-a-service (MaaS) 193
modernisation, ecological 246
mode substitution factors 204n2
Morrow, Karen 7
municipal master plan, for Oslo 190–191

National Energy Efficiency Action Plans (NEEAPs) 169
Nationally Determined Contributions 38
national zero-growth objective (ZGO) 184
Nation Transport Plan (NTP) 197
nature, connecting with 217–220
neo-classical growth-oriented paradigm, policy and planning and 119
networks, and critical acts 75–76
NGOs, Swedish 70
non-motorised transport, in Denmark 119
non-profit sector, in Canada 156
non-state actors 19
non-traditional occupation (NTO), definition of 146
non-unionised programmes 157
normalisation 5
Norway: city council cooperation n Oslo 188–189; electrification in 183–184; zero growth objective (ZGO) in 10
Norwegian planning and climate policy agenda 183
NVivo 12 qualitative data analysis software 230–231

Office to Advance Women Apprentices 156
Olsen, J.P. 6, 88

One Million Climate Jobs Campaign 151, 153
Organisation for Economic Co-operation and Development (OECD) countries, employment in 143
Oslo (Norway): bicycle strategy 192; climate and energy strategy and 191; land use and transport regional plan 189–190; municipal master plan 190–191; population by gender and age groups *194*; public transport strategy 191–192; strategies for modifying travel behaviour 201–203
Oslo City Council 192
own bodies and sexuality, rights to 78

Paris Accord 79
Paris Agreement 54; (2015) 37, 38; and gender 27–28
path dependencies 5, 6, 70, 88, 246–248; feminist lens to 17; institutional 45, 80–81, 97
pathways approaches 226
'Patriarchal Violence as a Threat to Human Security' 73
'pattern-bound' effects 5
Perez, Caroline Criado 32
performativity, feminist institutionalism and 107
Peters, J. 150
Phillip, Anne 3, 9; Anne, *Politics of Presence, The* 108
plans 192–193
Poland 37
Policy Coherence for Development 38
policy documents, content analysis of 189–193
policy-making process 4
political buy-in, and gender equality 30
political institutions, balance between male and female 3–4
political leaders, and traditional gender roles 79
political participation 1
Political Studies Review, on gender and external action 44
politics, importance of 76–78
politics of presence 3
Politics of Presence, The (Phillips) 108
Pomeranzi, Bianca 76
population: elderly in 10; gender and age groups (Oslo) *194*; growth, Oslo's 10
post-COVID-19 cities 185

Power and Privileges (Johnson-Latham) 73
pre-apprenticeship programmes 154
presence, politics of 3
privileges: disruption of 114–115; "invisibility" of 63
professional identities 88–89
public transit system, in Denmark 119
public transport strategy, Oslo (Norway) 191–192
Pulé, Paul M. 10, 208; *Ecological Masculinities* 207

qualitative method of content analysis 187–188
quantitative prognosis model 187–188

race 1; inclusive language on 155
Rao, A. 45
REDD+ 130
red–green coalition 188
Red Seal Interprovincial Standards Program 152–153; Red Seal trades 154
Reed, Maureen G. 11
relational point of view 216–217
religious and ultra-conservative groupings 31
representation: categories of 3; descriptive 4, 22; substantive 4, 22–23; of women and social group 4
Republican administrations (USA) 78
resident engagement: power dynamics and 235–238; and the wildfires in norther Saskatchewan 234–235
rights-based coverage 29; the United Nations and 18
Rio+20 Conference 20; Rio Conventions 7, 20
Rivas, A. 45
Roggeband, Connie 60
Röhr, Ulrike 7
role expectations, identities and 89–90
Russian Orthodox Church 78

Sahin-Dikmen, Melahat 10, 150
Sahlin, Mona 77
Saskatchewan, wildfires in 230
Saudi Arabia 78, 81
Scharbach, J. 239
Schori, Pierre 76
Schulze, Svenja 54
second-generation (social and economic) rights 31

security challenges, climate change as 1
security-focused foreign policy 7
self-initiated analysis 94
SEPA: climate change action of 92–96; implications for institutional change in 96–99
sewage, employment by gender in **131**
sexual harassment and violence against women 149–150
sexuality–related diverse perspectives 32
sexual orientation, inclusive language on 155
short trips: business-as-usual (BAU) scenario 196; growth rates and aggregate mode shares 201; number of trips on select modes by age group and gender *202*; per capita *198*; zero growth objective (ZGO) and 198
Sijapati, B.B. 90
Singleton, Ben 8
Skype for Business software 91
Small Island Developing States (SIDS) 38
Social-Democratic Party 53
Social Development Goals (SDGs) 43
social differences 1
social inclusion, climate transition and 1
social inequalities, men and 11
Socialist Left Party (Oslo) 188
social justice 248; and Canada's green economy 153; climate change work and 92
state behaviour, shame as a driver of 29
Steering Committee of the Green Economy Network 153
'sticky' institutions 246–248
Stieß, Immanuel 7
Stout, Margaret 99
Structural Adjustment Lending (SAL) 71
structural inequality 17
structuration, process of 88
Subsidiary Body for Implementation (SBI) 19
substantive representation 3, 4, 22–23
sustainability, human society and change 86
Sustainable Development Goals (SDGs), UN 1: and gender 28–29
sustainable environmental practices, and gender-awareness 135–138
Sweden 40, 247; Swedish Church community 73; Swedish civil servants 8; Swedish climate institutions 90, 98; Swedish Environment Advisory Council 73

Swedish Environmental Protection Agency (SEPA) 90, 100; civil servants in 86–87; implications for institutional change in 96–99
Swedish Government 69; impact of top-down directions from 99–100
Swedish Ministry for Foreign Affairs (MFA) 69, 70
Swedish Transport Administration (STA) 90, 95, 100; civil servants in 86–87; climate change action of 92–96; implications for institutional change in 96–99

Task Force for Equality in the Commission 44
TFEU, Article of 39
third-generation (solidarity) rights 31
Thunberg, Greta 210
TiNNGO 109
top-down directions, of Swedish Government 99–100
Toronto Police Department 154
Toronto Transit Commission (TTC), Diversity and Inclusion Lens and Toolkit of 154
trade, priority of 81
trade unions and industry associations, in Canada 156
traditional (fossil fuel–based) sectors 158
'traditional' gender roles, misogyny and 79
traffic policy, Mikkelsen and 115–119
transport: and climate change policy and CO_2 reductions 105–106; in Denmark 119; environmentally friendly 9; gendered dimensions of 106; social identities of users of 106; women's employment 154
transportation and construction, women employed in 157
transport development 183
transport patterns, gender/intersectionality and 10
transport policy 105; gendered representations in 108–112; gender mainstreaming and 106; institutional gendered dimensions of 106; women and 108–112
transport system: men and women usage of 202–203; safety audits of 154
travel behaviour 183, 193; gendered 183; in Oslo (Norway) 201–203; social differences in 10

travel patterns, for the city of Malmö in Sweden 184
Treaty of Rome 40
Trøjborg, Jan 118
Turnbull, Peter 147, 148

UBA 55
UN 69
UN BDPA 23
UN Commission for Human Rights 75, 79
UN Commission on Sustainable Development 77, 83
UN Conference on Environment and Development 127
'Under Tallarna' (or 'Under the Pine Trees') 207–208, 214
UN Division for the Advancement of Women 76
UNFCCC 37, 246; decisions on gender 54; engagement with gender issues 30–31; Gender Action Plan 8; gender implications and 25; gender-responsive climate policies and 53; Lima Work Programme on Gender 43; provisions on gender 52
UNFCCC constituted bodies 24; Adaptation Committee 22; average percentage female and male membership of 21; female representation on 20, 21, 22; gender inequality and 22; politically constraints and 32
UNFCCC CoP, decision 23/CP.18 20
UNFCCC documents, on gender 26, 30
UNFCCC Gender Action Plan 8
UNFCCC regime 19; gender landmarks in 20–22; institutional activity of 29–30; visibility of gender in 26–27
UN human rights machinery, LGBTQI+ rights and 31
Union for Foreign Affairs and Security Policy 42
United Arab Emirates 78, 81
United Nations 1, 71; gender issues and 18; Social Development Goals (SDGs) 43; *see also* United Nations
United Nations Charter 18
United Nations Framework Convention on Climate Change (UNFCCC) 3, 7, 17, 37; engagement with gender issues 30–31; gender composition reports from 20; gender equality issues and 29–30; gender inclusion in 19–20; inclusivity and equality and 19

United Nations Framework Convention to Combat Desertification (UNCCD) 20
United Nations institutions, gender disparity within 17
United Nations Sustainable Development Goals 1; and gender 28–29
Universal Declaration of Human Rights 18
UN's 2030 Agenda 41–42
UN World Conference on Women, Fourth 78
urban mobility plan 192
Urban Waste project 124–126, 131, 136; gender considerations and 125; gender mainstreaming and 132–134
Uteng, André 10
Uteng, Tanu Priya 10

Vancouver Island Highway Project (VIHP) 155
Vatican 78
Verloo, Mieke 60, 128–129
Vetterfalk, Vidar 10
vulnerability, and emotional literacy 214–216

wages, in construction and transportation industries 150
Waldram, J.B. 239
Walker, Heidi 11
Wängnerud, Lena 22
wasteapp, gendered use of 135
waste behaviour 133
waste management 9, 124; climate change policy and 126–127; employment by gender in 131; gendered attitudes and behaviour towards 132; gender mainstreaming and 9; gender mainstreaming in the EU and 131–132
waste prevention actions 124
waste reduction measures, gender-proofing of 134–135
water, employment by gender in 131
West, Bjørn 118
Westh, M., number of articles about 109
white male privilege 216
white supremacy 79
wildfire management 11, 227
wildfires: in northern Saskatchewan 230; past responses to 231; priorities for addressing risk 231–232, 241–242; professional and specialised knowledge used after 233–235, 241–242; research methods used for (northern Saskatchewan) 230–231

women: and access to new technologies
143; Black and minority ethnic women
130; in the Canadian green economy
143–144; construction industry
and 165; discriminated against 78;
employed in the energy sector 166;
employment in construction and
transportation industries 143–144, 157;
employment opportunities and 150–151;
empowerment of 41, 42; experiences,
mainstreaming of 4; human rights 18;
issues and similarity to LGBTQI+
rights 31; lack of information about
employment opportunities 146–147;
and low participation in construction
165–168; male-biased work culture
and working conditions 148–149;
marginalised 9; participation rights 18;
policies and programmes for 153–154;
in senior, managerial levels 83; sexual
harassment and violence against
149–150; sexual harassment and violence
against women 159; so-called traditional
role for women 79; transport and waste
management behaviour 245–246; *see also*
transport; waste management
Women and Gender Constituency
(WGC) 20
Women in Development' (WID) 71
'Women in Non-traditional Occupations:
Stories to Inspire' 155

Women in Skilled Trades (WIST)
programme 154
Women in the Building Trades 156
Women in Trades Awards/Bursaries
Program 155
Women's Rights and Gender Equality
and the Committee on Development,
FEMM Committee on 38
workforce certification, in Canada 154
working conditions 166
World Bank 69, 71
World Health Organisation (WHO),
definition of gender and 30
World War Two 31
Wuppertal Institute Environment,
Climate, Energy 52

Yemen 78
youth-generated protests 210

Zacchi, Ole 108
zero energy building (NZEB) 164,
165, 171
zero-growth goal (Ruter) **192**
zero growth objective (ZGO) 193, 196,
204n1; achieving 197; and climate
policy-making in 10; long trips and 197;
medium trips and 197–198; short trips
and 198; and sustainable modes of travel
198–201
Zoom for Business software 91

Printed in the United States
by Baker & Taylor Publisher Services